The Clayton Type 1 Bo-Bo
Diesel-Electric Locomotives – British Railways Class 17

Front Cover:
D8587 and D8586, 64A St Margarets (Edinburgh), 9 July 1966. (RCTS Archive)

Back Cover:
D8587, Millerhill Yard, 18 August 1972. (Anthony Sayer)

Below: **D8588, 51L Thornaby, 23 May 1964.** (Colour-Rail)

The Clayton Type 1 Bo-Bo
Diesel-Electric Locomotives – British Railways Class 17

Development, Design and Demise

ANTHONY P. SAYER

AN IMPRINT OF PEN & SWORD BOOKS LTD.
YORKSHIRE – PHILADELPHIA

First published in Great Britain in 2021 by
Pen & Sword Transport
An imprint of Pen & Sword Books Ltd
Yorkshire - Philadelphia

Copyright © Anthony P. Sayer, 2021

ISBN 978 1 52676 200 9

The right of Anthony P. Sayer to be identified as Author of this work has been asserted by him in accordance with the Copyright, Designs and Patents Act 1988.

A CIP catalogue record for this book is available from the British Library.

All rights reserved. No part of this book may be reproduced or transmitted in any form or by any means, electronic or mechanical including photocopying, recording or by any information storage and retrieval system, without permission from the Publisher in writing.

Typeset in Palatino by SJmagic DESIGN SERVICES, India.
Printed and bound by Printworks Global Ltd, London/Hong Kong.

Pen & Sword Books Ltd incorporates the Imprints of Archaeology, Atlas, Aviation, Battleground, Discovery, Family History, History, Maritime, Military, Naval, Politics, Railways, Select, Transport, True Crime, Fiction, Frontline Books, Leo Cooper, Praetorian Press, Seaforth Publishing, Wharncliffe and White Owl.

For a complete list of Pen & Sword titles please contact:

PEN & SWORD BOOKS LTD
47 Church Street, Barnsley, South Yorkshire, S70 2AS, England
E-mail: enquiries@pen-and-sword.co.uk
Website: www.pen-and-sword.co.uk

Or

PEN AND SWORD BOOKS
1950 Lawrence Rd, Havertown, PA 19083, USA
E-mail: Uspen-and-sword@casematepublishers.com
Website: www.penandswordbooks.com

CONTENTS

Preface 7
Acknowledgements 9
Abbreviations 10

Chapter 1	BR First-Generation Type 1s 11
Chapter 2	The Route to Standardisation 17
Chapter 3	The 'Standard' Diesel-Electric Type 1: Order Placement 21
Chapter 4	The Next 100 Type 1's: D8617-D8716 or Back to Plan A? 32
Chapter 5	Some Technical Aspects 37
Chapter 6	Appearance Design and Styling 54
Chapter 7	Delivery and Acceptance Testing 58
Chapter 8	Allocations 62
Chapter 9	Overhaul and Repairs 74
Chapter 10	Locomotive Histories 76
Chapter 11	Performance and Service Problems: Engines and Associated Equipment 210
Chapter 12	Other Performance and Service Issues 238
Chapter 13	Accident and Fire Damage 244
Chapter 14	Operations: High-Level Summary 251
Chapter 15	Detail Differences 266
Chapter 16	Liveries 270
Chapter 17	Storage and Withdrawal 282

Chapter 18	Storage Locations: Depots and Yards	291
Chapter 19	Disposal: The Private Yards (1968-71)	311
Chapter 20	Disposal: St Rollox Works	319
Chapter 21	Disposal: The Private Yards Return (1975-76)	351
Chapter 22	Post-Withdrawal Use	361
Chapter 23	Industrial Service	368
Chapter 24	Preservation	370
Chapter 25	Concluding Remarks	372
	References and Sources	374

PREFACE

Unlike my previous books in the modern traction *Locomotive Portfolios* series, this is not the first book published which has been devoted exclusively to the Claytons. In 2016, Book Law Publications published a book entitled *An Illustrated Historical Review of the Clayton Type 1 Bo-Bo Diesel-Electric Locomotives - Class 17* by John Hooper. This excellent book was, as the title implies, a largely pictorial book, albeit with extended captions, together with chapter introductions including information covering 'Tenders, Specifications, Options' (for D8588-D8616), 'Allocations', 'Works Visits', 'Scrapping', etc. Photographs of the Claytons operating in the North-East of England were particularly welcome.

This book attempts to complement and develop elements of Hooper's book and introduce some new components to further develop our knowledge of the Clayton fleet. This has been possible by the use of new material derived from:

(a) the Diesel Locomotive Record Cards for 101 of the 117 locomotives, particularly Works visit information.
(b) extensive personal sightings from numerous sources to broaden the information available on the class, particularly with respect to Works visits for those locomotives where no DLRCs have been found.
(c) archive material regarding the background to the ordering of the two batches of Claytons, together with details surrounding the potential ordering of a further 100 Type 1 locomotives and why the decision was ultimately made to return to the English Electric product.
(d) previously unpublished archive material regarding the storage of the Claytons in 1968/69, and,
(e) minutes of meetings held between British Railways and Paxman between 1967 and 1971 enabling a significantly deeper understanding of the problems experienced by the Class and the remedial actions pursued.

All this information has enabled detailed listings of the history of each of the 117 Claytons to be produced; however, due to severe space constraints, these histories have had to be substantially abridged. I know from some of the comments made about my first book covering the NBL Type 2s (D6100-57) that this will be appreciated by some and disliked by others. If you remember your mathematics classes at school, you will remember the phrase 'Show your workings!' and this is what I did with both the NBLs and the Metrovicks, as it was the detailed histories which helped to uncover many of the previous, and frequently repeated, errors. Unfortunately, with 117 locomotives, something had to give, and so abridged histories will have to suffice; however, all available information regarding fire/accident damage, works visits and most sightings of locomotives whilst in store or after withdrawal are included. Availability of the DLRC and personal sighting information regarding works visits has enabled the locomotive livery histories to be substantially developed.

The generic and frequently used 'Clayton' name can be confusing. The Clayton Equipment Company Ltd. was the main contractor for the first order of 88 locomotives; the second batch of 29 was constructed by Beyer Peacock (under license) but, despite this, all 117 locomotives have been, and indeed still are, referred to as Claytons. To complicate matters further, many Scottish engineman referred to the locomotives as 'Paxmans' which is accurate in 115 cases, but misses the fact that D8586 and D8587 were fitted with Rolls-Royce engines.

More accurate is 'Class 17' and the sub-classes (Class 17/1 – Clayton/Paxman/GEC (D8500-85); Class 17/2 – Clayton/Rolls-Royce/GEC (D8586/7); Class 17/3 – Beyer Peacock/Paxman/Crompton Parkinson (D8588-D8616), although these are less memorable descriptions in practice. Every enthusiast knows what a Clayton is, but a Class 17/3....? In the following text, the generic term 'Clayton' is used extensively with additional clarification where necessary.

With the Claytons having two engines in one locomotive, it is necessary to very pedantic about the use of the word 'engine'. Thus 'engine' refers to the Paxman (or Rolls-Royce) prime-mover *not* the complete locomotive; 'engine' in the context of the Claytons *cannot* be used interchangeably with 'locomotive'.

References to individual locomotives in the text and photograph captions use the D-prefix throughout even though a considerable number of locomotives lost their prefix after the end of steam in August 1968. To accurately specify locomotive identities during the period August 1968 up to withdrawal is frankly impossible, hence the decision to use the prefix throughout. There is one exception to this, however; Section 12.5 lists the final situation of locomotives at withdrawal (i.e. whether D or no-D).

To avoid confusion, the title 'St Rollox Works' is used throughout despite the fact that, with the closure of Cowlairs Works in 1968, St Rollox became 'Glasgow Works'. The correct spelling of the name 'J. MacWilliam', the owner of the Shettleston scrapyard, has been used (except within quotes), rather than perpetuating the incorrectly spelt 'J. McWilliam' as used in the railway press since time immemorial.

Anthony Sayer

D8530, Niddrie North Junction, 13 October 1970. "Target E18" consists of ten LGW (Leith General Warehousing) vans loaded with grain from Leith Docks to Gorgie (North British Distillery). Note the wooden-bodied, un-braked 'cottage' wagons with their distinctive apex roofs. (Bill Jamieson)

ACKNOWLEDGEMENTS

Archive sources have provided a major input in developing the class and individual locomotive histories and I would like to personally thank the teams of people at the National Archives and the National Railway Museum for their kind assistance.

My thanks are also due to Richard Carr of the Paxman Archive Trust for providing brochures and numerous official photographs of the 6ZHXL engine used in all bar two of the Clayton fleet. Richard's 'Paxman History Pages' are a positive cornucopia of information.

A considerable number of quotes have been taken from the *Railway Observer*, the magazine of the Railway Correspondence and Travel Society (RCTS); these have been reproduced with the very kind permission of W. Gordon Davies of the RCTS.

Inputs from Dave Coddington, Laurie Mulrine and Bill Hamilton have provided a considerable amount of local information and context. Dave has at last explained why so many withdrawn Claytons exhibited '1S83' in their headcode boxes after withdrawal!

My continued use of the 'back to basics' approach has, by necessity, relied heavily on a massive amount of personal observation information in various forms. I have made every attempt to credit everyone concerned in the 'References and Sources' section. My thanks go to you all.

This book has used over 280 images to support the text. My thanks go to the following organisations and individuals for the photographs used herein:-

Paul Appleby (Rail Photo Archive), Richard Carr (Paxman Archive Trust), John Chalcraft (Rail-Photoprints), Paul Chancellor (Colour-Rail), Geoff Corner (TOPtical Digital Memories), W. Gordon Davies and John Broughton (RCTS Archive), Robin Fell (Transport Treasury), Martin High (Transport Topics), Mick Mercer (Rail Image Collections), Steve Montgomery (Exe-Rail), Rail-Online and Old-Maps.co.uk.

Also Edward Bather, Malcolm Best, Stewart Blencowe, David Dippie, Stephen Fisher, Bill Hamilton, Dave Harlott, Bill Jamieson, Doug Kirk, Brian Lee, Dave Lennon, Keith Long, Jonathan Martin, Laurie Mulrine, Michael Philips, Kevin Redwood, Hugh Searle, Steve Thorpe, Allan Trotter (Eastbank Model Railway Club), John Grey Turner, Alan Walker, Grahame Wareham, and griffith_p (flickr).

There are images from a very small number of photographers where it has proved impossible to obtain the appropriate permission to use their photographs despite every endeavour being made to track down the individuals concerned. In addition, there are one or two images where the identity of the photographer is totally unknown; lack of accreditation on slides or prints has prevented any possibility of determining their provenance. In both cases, anyone who feels that they have not been adequately credited should please contact me via the publisher to ensure that the situation is corrected in future editions.

ABBREVIATIONS

A	Amps
AEI	Associated Electrical Industries.
BLS	British Locomotive Society.
BR	British Railways / British Rail.
BRB	British Railways Board.
BRB SC	BRB Supply Committee.
BRB W&EC	BRB Works & Equipment Committee.
BRCW	Birmingham Railway Carriage & Wagon Co. Ltd.
BTC	British Transport Commission.
BTH	British Thomson-Houston.
CM&EE	Chief Mechanical & Electrical Engineer.
CP	Crompton Parkinson.
DLRC	Diesel Locomotive Record Card.
DRC	Derby Research Centre.
ECS or ecs	Empty coaching stock.
EE	English Electric.
EHC	Engine History Card.
ER	Eastern Region.
FDTL	*'Fires on Diesel Train Locomotives'* Reports.
G&SW	Glasgow & South Western.
GN	Great Northern.
GEC	General Electric Company.
hp	Horsepower.
LAMA	Locomotive and Associated Manufacturers' Association
LCGB	Locomotive Club of Great Britain.
LMR	London Midland Region.
LMS	London Midland & Scottish.
MV	Metropolitan-Vickers.
NBL	North British Locomotive Co. Ltd.
NER	North Eastern Region.
NTP	National Traction Plan.
RCTS	Railway Correspondence and Travel Society.
rpm	Revolutions per minute.
RSL	Rolling Stock Library.
ScR	Scottish Region.
SLS	Stephenson Locomotive Society.
s.p. or sp	Stabling point.
S(s)	Stored serviceable.
S(u)	Stored unserviceable.
TOPS	Total Operations Processing System.
T&RS	Traction & Rolling Stock.
V	Volts.
w/e or we	Week ending.
WCML	West Coast Main Line

Chapter 1
BR FIRST-GENERATION TYPE 1s

1.1 Background: Early BR Type 1 Locomotives and Initial Attempts at Standardisation.

The 'Pilot Scheme' provided fourteen types of diesel locomotives amounting to 174 locomotives including three Type A (later Type 1) classes. These Type A 'Pilot-Scheme' Orders (all ostensibly constructed under the 1957 Building Programme) were authorised by the Works & Equipment Committee (W&EC) and the British Transport Commission (BTC) as follows:

W&EC Meeting 12/10/55 (Min.571/16) / BTC Meeting 15/12/55 (Min.8/590)				
20	English Electric	D8000-D8019	Ordered 16/11/55	Delivered 06/57-03/58
10	BTH	D8200-D8209	Ordered 16/11/55	Delivered 11/57-10/58
10	NBL	D8400-D8409	Ordered 16/11/55	Delivered 05/58-09/58

The 'Pilot Scheme' designs were to be tested over a three-year period with the deliberate objective of determining a limited number of types for volume production commencing 1961/62 based on the operational experience gained. However, facing deteriorating financial results, the BTC abandoned the three year trial period, and accelerated the introduction of diesel locomotives as fast as British production capacity would allow, believing that dieselisation and the consequent elimination of steam would dramatically improve the position.

Minute (Min.) 9/384 of the BTC Meeting of 26 July 1956 recorded that:

'The Commission . . . discussed the purchase of additional main line locomotives and agreed that they would be prepared to consider requests for a number of these without further trials, provided that:-

1. There is sufficient technical evidence to show that the type of locomotive desired is fully and without doubt able to meet requirements which are comparable to those in the service for which it is intended.
2. The substitution of the diesel locomotives for steam locomotives is economically justified by the manner in which they will be operated.'

The dangers of giving up the trial period were made clear by R.C. Bond (Chief Mechanical Engineer) but Chairman Sir Brian Robertson insisted that the Board's decision was adhered to, with the specific condition that reliable locomotives were introduced. The Commission also insisted that the number of different designs of locomotives should be reduced to an absolute minimum. This, as already mentioned, was one of the key objectives of the 'Pilot Scheme' process but it had now become necessary to recommend the smallest possible number of types without any operating experience having been obtained with the locomotives then on order. The only way of now achieving this was to base recommendations on engineering judgement, knowledge of various firms' products, and the operating experience of other railways. As far as the 1958 Building Programme was concerned, no Type A locomotives were included, building being restricted to fifty-six Type B (Type 2) locomotives.

A Memorandum to the BTC dated 16 May 1957 acted as a cover note for a Report entitled *Modernisation of British Railways: Report on Diesel and Electric Traction and the Passenger Services of the Future*. This report highlighted 'the very much more rapid introduction of diesel locomotive traction than had first been intended', stating that the Plan for diesel locomotive manufacture up to 1962 'sets a production task of magnitude' envisaging the ordering of a further 1,889 main-line diesel locomotives during the period 1957-62.

The BTC at their Meeting on 23 May 1957 (Min.10/212) approved the general concept of the extension and acceleration of introduction of diesel traction as contained in the report, without commitment to the exact pace of extension, stating that:

'The Commission would be prepared to go further than they have already gone in regard to ordering diesel main-line locomotives, in spite of the risk of unsatisfactory performance in the early stages, if the Regions presented them with a limited number of firm plans for their use in specific areas, containing as clear a justification as possible.'

The Commission again asked that the question of limiting the number of main-line diesel locomotive types be specifically addressed, and to advise on what was practicable. The Report *Main Line Diesel Locomotives: Limitation of Variety* (R.C. Bond & S. Warder, 26 July 1957) was produced in response to this request. The principles used in the *Limitation of Variety* report governing the selection of locomotive types for ordering beyond those already ordered were (i) reliability and (ii) as much standardisation as British production capacity would allow. On the basis of these key considerations, the recommendation was that any diesel-electric locomotive orders placed in the 1959 and subsequent Building Programmes should be limited to the following types (plus the BRCW Type 3s for the Southern Region which were seen to be a special case), subject to the phasing of the Regions' specific requirements and manufacturing capacity:

English Electric and Sulzer engines featured strongly. It was considered that English Electric had the largest experience and productive capacity of any British manufacturer and had the resources to ensure the delivery of a reliable product, whilst the Sulzer engine was the best known and widely used product outside of the USA, and was recognised for its excellent design and workmanship. Paxman engines were proposed as a reserve type.

Subsequent discussions with industry soon showed, however, that it was not possible to adhere strictly to the recommendations and, to meet BR's demands for diesel-electric locomotives and to capitalise on available production capacity, non-recommended contractors were subsequently awarded orders for more of their equipment (see later).

At the BTC Meeting on 19 September 1957 (Min.10/400), the Commission approved, in principle, a 1959 Building Programme of the order of 750 to 800 locomotives composed of the types recommended in the *Limitation of Variety* report. It was recognised that this was a large requirement and that available production capacities may dictate some deferral into 1960. However, a subsequent capital investment restriction in late 1957 severely limited the purchase of locomotives for delivery in 1959 and it was, therefore, impossible to take full advantage of building capacity for main-line diesel locomotives in that year. Thus the revised 1959 Building Programme included eighty-four Type 2 diesel-electric locomotives (plus shunting and electric locomotives); all of these Type 2s were to be built

Type/hp	Engine	Transmission	Mechanical Parts
Type 1 (1000hp)	English Electric	English Electric	English Electric
Type 2 (1160hp)	Sulzer	BTH	BR
Type 3 (1750hp)	English Electric	English Electric	English Electric
Type 4 (2500hp)	Sulzer	Crompton	BR

in BR Workshops employing Sulzer engines (subsequently D5030-D5113). The W&EC approved the building of these locomotives at their meeting on 22 January 1958 (Min.1110/20) and authorised by the BTC on 13 February 1958 (Min.11/53). The remaining diesel locomotive requirements were deferred into 1960 and 1961. Contractors were, however, invited to tender in anticipation of these later Building Programmes.

The financial situation eased during 1958 and as a consequence there were two Supplements to the 1959 Building Programme, with locomotive types selected on the basis of the tenders already received. The First Supplement (ninety locomotives), including no Type 1s; however the Second Supplement (for forty-nine locomotives), including ten Type 1s.

Contractors had previously offered available capacity for the construction of additional locomotives during 1959; in the Type 1 range, English Electric had offered thirty locomotives and BTH ten. A Memorandum to the W&EC dated 3 November 1958 proposed acceptance of these offers, but the W&EC at their meeting on 12 November decided to recommend only the ten BTH for construction in the 1959 Building Programme, agreeing 'that no action should be taken to place orders for the 30 Type 1 locomotives from the English Electric Co. at the present time'. Reasons given for this recommendation to the BTC were that the potential deployment of Type 1s in Scotland was still under investigation and that axle-loading restrictions in the London area of the Eastern Region favoured the BTH design. The W&EC also added the caveat that their proposal was 'subject to any decision reached by the Chairman's Conference on Modernisation to be held on 13/11/58'. The BTC meeting of 20 November (Min.296) accepted the W&EC proposal for the ordering of ten BTH Type 1s having satisfied themselves of the economic and local operating advantages of continuing to order Type 1 locomotives.

W&EC Meeting 12/11/58 (Min.1300) / BTC Meeting 20/11/58 (Min.11/496(d))
10 BTH   D8210-D8219   Ordered 17/12/58   Delivered 11/59-02/60

None of the locomotives in the First and Second Supplements featured in the *Limitation of Variety* recommendations but were repeat orders for locomotives of types already in service.

In late 1958, a memorandum was produced for discussion at the above-mentioned Chairman's Conference on Modernisation on 13 November 1958; this was produced by R.C. Bond (by now BTC Technical Advisor) and entitled *Main Line Diesel Locomotives: The Approach to Standardisation* (dated 10 November 1958). The recommendations concerning the preferred diesel-electric manufacturers for each of the Types 1-4, as detailed in the *Limitation of Variety* report, were reiterated. However, this memorandum went a stage further by explicitly listing the designs which it was proposed should be specifically *excluded* from any future orders; this included the NBL/Paxman Type 1 (D8400-9). Notes of the Chairman's Conference record that 'The Chairman . . . fully agreed with the proposals set out in the Technical Advisor's paper for the elimination of a number of locomotive types.' Bond's memorandum also supported repeat orders for types of locomotives not included in the *Limitation of Variety* recommendations, including the BTH Type 1, to enable financial capital allocations to be used to the full during 1959/60.

A further report covering limitation of variety was produced in 1959 entitled *Standardisation of Main Line Diesel Locomotives* (BR General Staff, 8 June 1959). By this date, approximately 150 diesel-electric main line locomotives were in service; however, there was still insufficient experience with the new traction to statistically challenge the logic of the *Limitation of Variety* report recommendations. This report, once again, explicitly listed the 'excluded' types of diesel-electric locomotives. Following on from this report and in a Memorandum to the BTC from the General Staff dated 22 June 1959, the Commission were recommended to:

(a) endorse the principles embodied in the Report on *Limitation of Variety* of main line diesel locomotives (nearly two years after it was published in 26 July, 1957!), and,
(b) approve the proposal to specifically exclude the four specified types from the forthcoming 1960 and 1961 programmes.

The BTC Minutes of the Meeting on 25 June 1959 (Min.12/253) recorded that the Commission accepted these recommendations with only a few minor caveats.

The backdrop of expanded and accelerated production pressures, combined with individual manufacturing company constraints across the UK, allowed the BTH Type 1 diesel-electric fleet to be expanded from the ten 'Pilot Scheme' locomotives to forty-four.

Type 1 orders included as part of the 1959-61 Building Programmes were authorised by the W&EC/BTC as follows:

W&EC Meeting 06/01/59 (Min.1342) / BTC Meeting 08/01/59 (Min.12/4)

30 English Electric   D8020-D8049   Ordered 14/01/59   Delivered 10/59-03/60
Previously deferred

17 BTH   D8220-D8236   Ordered 05/02/59   Delivered 03/60-10/60
Dictated by London-area 'operating factors & in the interests of (local) standardisation'

Both orders to be constructed as part of the 1959/61 Building Programme.

W&EC Meeting 24/11/59 (Min.1601/6) / BTC Meeting 10/12/59 (Min.12/497)

7 BTH/AEI   D8237-D8243   Ordered 18/12/59   Delivered 11/60-02/61
Taking advantage of declared BTH/AEI production capacity available in 1960

To be constructed as part of the 1960 Building Programme.

W&EC Meeting 03/12/59 (Min. 1611/7) / BTC Meeting 17/12/59 (Min.12/510(b))

78 Unspecified Type 1   Subject Supply Committee (SC) allocation of construction contractor

SC Meeting 17/12/59 (Min.721) / BTC Meeting 31/12/59 (Min.12/528)

78 English Electric   D8050-D8127   Ordered 05/02/60   Delivered 03/61-07/62

To be constructed as part of the 1961 Building Programme.

Through their 3 December 1959 meeting, the W&EC requested BTC authorisation for a further 88 Type 1s as part of the 1962 Building Programme. Approval was given at the BTC of 17 December, subject to allocation of construction contractor by the Supply Committee. The Supply Committee deferred this allocation decision 'until more reliable information can be obtained regarding the availability of new designs' (see Section 1.2).

On the face of it, the BTC quest for standardisation could have had an adverse longer-term impact on the speed of diesel locomotive acquisition by effectively restricting production capacity to their selected manufacturers only. However this issue was addressed by the BTC relatively early on. In the BTC Meeting of 23 May 1957 (Min.10/212) it was stated that:

'On the question of the capacity of the locomotive manufacturing industry to meet the Commission's future requirements for locomotives . . . it was agreed that after decisions had been reached on the (*limited*) number of types required there should be consultations with the industry at which the broad intentions of the Commission could be indicated . . .

'Consideration could then be given to spreading orders among a fairly wide range of firms, and construction under licence, with a view to ensuring that future production capacity would be sufficient to meet the ultimate requirements of the Commission.'

This approach was taken a stage further in BTC Min.10/400 (Meeting dated 19 September 1957) whereby 'The General Staff were authorised to see that early negotiations were opened with the Locomotive and Associated Manufacturers' Association (LAMA) with the aim of making the maximum use of the potential manufacturing capacity of the locomotive industry, and also to enlist the interest of industry in general in the manufacture by some firms of others' products under licence. It was (*also*) agreed that the possibility should be explored of making use of firms not at present associated with locomotive building.'

A Memorandum to the W&EC dated 17 March 1959 entitled *Supplementary Locomotive Building Programme 1960 - Main Line Diesel Locomotives* commented, 'Ultimately it is hoped to establish a standard BR design of locomotive in each of the power ranges when enquiries for future requirements from, say, 1962 onwards will be issued to <u>all</u> manufacturers . . . for quotations on the basis of the standard designs.'

1.2 The Need for an Alternative Type 1.

By the close of 1959, the BTC was faced with a dilemma. All of the Type 1s introduced to date were intended for trip freights and shunting duties, typified by low mileage and 'non-continuous' working, and were single-cab designs with high bonnets with consequent forward-visibility issues. The safety consequences of this arrangement necessitated the employment of a second-man for effective signal sighting.

A Memorandum to the Technical Committee dated 8 January 1960 entitled *Standardisation of Main-Line Diesel-Electric Locomotives* described the problem quite succinctly:

'These (*Type 1*) locomotives have hitherto been designed with a single cab at one end, giving unrestricted vision to the footplate staff in one direction, but with vision by the driver somewhat hampered in the other direction on right-hand curves by the presence of the bonnet covering the diesel engine. The double cab arrangement of locomotives of Type 2 and upwards is inconvenient where a considerable proportion of the locomotive's duties consists of shunting. The present arrangement of a single cab tends to prevent full implementation of the Manning Agreement, and, in order to permit one man operation in all cases permitted by the agreement, the Chief Traffic Officer has requested that future Type 1 locomotives should have a high cab and low bonnet so as to permit vision in all directions from the driving position. Such an arrangement is impossible within the British loading gauge with the medium speed vertical engines hitherto employed, and a redesign using high speed and/or flat engines is necessary to meet this requirement, a solution widely in use on the Continent to meet similar circumstances.'

Thus, although the English Electric Type 1 had proved to be British Railway's most successful 'Pilot-Scheme' investment in terms of reliability, its operating disadvantages were not insignificant, to the extent that alternative designs were now proposed to improve operating safety and enable the cost-saving Single-Manning arrangements to be fully deployed.

The lead time for the introduction the 78 Type 1s required in the 1961 Building Programme necessitated the continued construction of the English Electric design (D8050-D8127). However, the focus for the 1962 Building Programme shifted to a completely new and ultimately untried design.

D8080, 66B St Rollox, 6 August 1962. The English Electric Type 1 design as first introduced in 1957. It was a highly successful class in availability and reliability terms. However, the full-height bonnet severely restricted the driver's view when travelling No.1 end first, the end nearest here. As a consequence, these locomotives fell from grace, albeit temporarily! (Rail-Online)

Chapter 2
THE ROUTE TO STANDARDISATION

2.1 Standardisation of Types.

A Memorandum to the Technical Committee dated 8 January 1960 entitled *Standardisation of Main-Line Diesel-Electric Locomotives* provided specific recommendations on standardization and limitation of variety with respect to future orders for Types 1 to 4 inclusive (restricting each diesel-electric Type to a single design of locomotive), in particular dealing with the 'New requirements as to the form to be taken in future by Type 4 and Type 1 locomotives'.

With respect to the Type 1 design, it was hardly surprising that the new Type 1 design was proposed with the following characteristics:

- Bo-Bo wheel arrangement with 17-ton axle load,
- the provision of flat engines, or high-speed vertical engines, to permit lowest bonnet height,
- single cab with all round vision, and,
- maximum speed of 60mph.

The authors of the memorandum, S. Warder and J.F. Harrison, recommended that all British firms should be able to quote for and produce all of the Standard Types, and that the contractor, whose design was accepted, must allow British Railways or any other contractor to build locomotives to the accepted design, thereby ensuring sufficient overall manufacturing capacity.

The Technical Committee of Meeting 15 January 1960 (Min.788) fully accepted the proposals in the Warder/Harrison memorandum and the BTC General Staff submitted it to the BTC for ratification on 26 January. The BTC also accepted the memorandum at their Meeting on 11 February (Min. 13/51) with only caveats regarding the Type 2 design.

2.2 The New 'Standard' Type 1(s).

On 15 September 1960 a memorandum was forwarded to the Technical Committee entitled *Standardisation of Main Line Diesel-Electric Locomotives - Types 1 and 4*. The memorandum preamble clearly described the events surrounding the development of the new Type 1 design proposal in the period between January and September 1960; these included:

- tenders invited and bids received for the supply of Type 1 diesel-electric locomotives as per the specification in Section 2.1 above,
- proposals made by the Western Region Area Board for an alternative Type 1 locomotive of 650hp,
- discussions with the Eastern Region in connection with their General Manager's proposal for a number of 500hp Type 1 diesel-electric locomotives, and,
- the outstanding proposal to obtain ten locomotives of around 800hp with a modified gear ratio specially for hump-shunting on the Eastern Region, as approved by the W&EC (Min.1110, 22 January 1958) but never implemented.

Full details of the Type 1 (Bo-Bo, single-cab, flat-bed engine) bids, together with details surrounding the selection of the preferred design, are covered in Section 3.1.

The Western Region made a separate case for up to 400 Type 1 0-6-0 650hp diesel-hydraulic locomotives, without train heating, as a cheaper alternative to the Bo-Bo option (by circa £15,000).

D8519, 66A Polmadie, 2 August 1963. Clayton's interpretation of the design characteristics specified by the BTC Technical Committee for a 'Standard' Type 1. Clayton's tender for this design was successful and went into production to cover BR's Type 1 diesel-electric locomotive requirements commencing 1962. (David Dippie)

This proposal became the subject of a separate submission to the BTC. The other Regions were asked for their views on the suitability of such a locomotive for their own traffic purposes, but most subscribed to the Bo-Bo single-cab proposal. The Eastern Region suggested a small number of 0-6-2 locomotives (17) of 500hp with a train heating boiler for empty coach stock working; however, in the absence of any similar requirements on other Regions, they agreed to the Bo-Bo option.

The subject of ten Bo-Bo single-cab diesel-electric locomotives, with an altered gear ratio for hump shunting purposes on Eastern Region at Wath, Temple Mills, Ripple Lane and Tinsley, was resurrected as part of the whole Type 1 debate. In the event, the use of 'master-and-slave' 700hp standard diesel-shunter twins was adopted at Tinsley only (D4500-2).

The Technical Committee Meeting of 23 September 1960 (Min.866) accepted the findings of the 15 September memorandum and recommended that for future standardization on British Railways there should be two Type 1 locomotives, as follows:

- 900hp Bo-Bo type diesel-electrics with provision for train heating, steam or electric, and,
- 650hp 0-6-0 diesel-hydraulic without train heating (Western Region)

On the 'ancillary' items, the Committee were not satisfied that there was a case for small numbers of alternative designs.

D9519, 86A Cardiff Canton, 18 August 1970. The Western Region opted for a cheaper alternative to the Bo-Bo design in the form of an 0-6-0 diesel-hydraulic built at Swindon Works. Fifty-six were built although anything up to 400 had been intimated as replacements for the ubiquitous 0-6-0T steam fleet! (Anthony Sayer)

2.3 The Slow Transition to 'Standard'.

A Memorandum to the W&EC dated 5 October 1959 indicated that submissions had been made by the respective General Managers for a further eleven Main-Line Diesel Locomotives Area Schemes, the basis for justifying the ordering of new diesel locomotives, including 'notional schemes' for Glasgow North, Glasgow South, and Fife. The identified Type 1 requirements were 47 locomotives for the Glasgow North scheme, 65 for Glasgow South and 34 for Fife, a total of 146. The Memorandum authors recommended the provisional approval of these schemes subject to the later submission of detailed proposals.

This memorandum was considered by the Modernisation Committee of the General Staff on 5 November and agreed as the basis for the submission of Building Programmes for 1961 and 1962.

A further Memorandum submission to the W&EC dated 16 November 1959 detailed the locomotive requirements to be included as part of the 1961 and 1962 Building Programmes. Thus the 1961 Programme totalled 301 main-line diesel locomotives including 58 Type 1s (47 for Glasgow North and 11 for Glasgow South at a cost of £3,364,000); similarly the 1962 Programme included 171 locomotives, of which 88 were Type 1s (54 for Glasgow South and 34 for Fife at a cost of £5,104,000). At this time, English Electric locomotives were envisaged for all of the Type 1 requirements.

The W&EC on 3 December 1959 (Meeting Min.1611/7) agreed to recommend to the BTC that approval be given to, inter alia, the Glasgow North, Glasgow South and Fife 'notional' Area Schemes and their associated locomotive requirements. On 17 December 1959, the BTC (Meeting Min.12/510) approved the recommendation. It was noted that the Supply Committee should consider the proposed allocation of these locomotives between manufacturers.

In parallel, a Memorandum to the Supply Committee dated 15 December 1959 proposed the allocation of orders for manufacture recognizing that, whilst for Type 1 requirements the English Electric 1000hp locomotives was preferred type, there now existed a requirement for a new design with a centre-cab to improve all-round vision.

It was thought possible that the new design of locomotives could be available from 1962 onwards and, therefore, it was proposed to order only the 78 Type 1 locomotives associated with the 1961 Programme from English Electric i.e. fifty-eight for the Scottish Region (D8070-99, D8100-16 to 65A Eastfield, and D8117-27 to 66A Polmadie, the latter eleven to operate the Glasgow General Terminus Quay to Ravenscraig iron-ore trains) plus an additional twenty for the Eastern Region (Sheffield) (D8050-D8069).

A recommendation regarding the balance of eighty-eight (all for the Scottish Region) was delayed pending confirmation regarding availability of the new centre cab Type 1s.

The Supply Committee of 17 December 1959 (Meeting Minute 721) agreed to recommend to the BTC the 1961 Programme allocation (including the 58 English Electric Type 1s for the Scottish Region) and also noted that 'the allocation of orders for the 88 Type 1 locomotives to be constructed under the 1962 Programme had been deferred until more reliable information could be obtained regarding the availability of new designs'. On 31 December, the BTC Meeting (Minute 12/528) approved the recommendation.

Numerous subsequent W&EC Meeting minutes throughout 1960 carried the following comment (or similar):

'Details of allocation of orders for 88 Type 1 locomotives to be constructed under the 1962 Programme to follow.'

Chapter 3
THE 'STANDARD' DIESEL-ELECTRIC TYPE 1: ORDER PLACEMENT

3.1 D8500-D8587.
During 1960, invitations to tender for a centre-cab Bo-Bo Type 1 design were issued to seven contractors (English Electric, NBL, Brush, AEI, Clayton Equipment and two unknown).

BTH had acted as main contractor for the Paxman-engined D8200-9 series of 'Pilot Scheme' locomotives and whilst BTH supplied the electrical equipment for these locomotives, they subcontracted a substantial amount of the work i.e. to Paxman (engine), Clayton Equipment (bogies and superstructure) and Yorkshire Engine Company (underframe and locomotive erection). Three repeat orders saw the responsibility for underframe and locomotive erection transferred to Clayton Equipment.

BTH and Metropolitan-Vickers were amalgamated in 1928 to form Associated Electrical Industries (AEI), but the two companies operated autonomously until January 1960, when the two brand identities were discontinued. D8237-43 were, therefore, introduced under the AEI identity.

Whilst Clayton Equipment had not previously been invited to tender for locomotives as main contractor, the work undertaken in the construction of D8210-43 put them in a position to be considered for such a role. Whether AEI would have looked to Clayton to provide locomotive underframes, superstructure components, bogies and final erection had they submitted a successful bid, is unknown. However, unlike AEI, Clayton was in a position to be able to offer alternative electrical equipment options, a factor which ultimately proved crucial.

The Memorandum to the Technical Committee dated 15 September 1960 entitled *Standardisation of Main Line Diesel Electric Locomotives – Types 1 and 4* (already mentioned above in terms of the broad Type 1 debate) provided the results of the various bids, together with a recommendation as to the preferred supplier. Overall, five contractors out of seven invited responded to the invitation to tender for the Type 1 requirement with prices ranging from £50,612 to £59,490 for a quantity of seventy-five. Quotations for other quantities were supplied by the Contractors (giving an idea of potential volume discounts available) but all BR archive material indicates analysis based around the supply of seventy-five locomotives.

The Memorandum, virtually in its entirety, as it relates to the Type 1 subject, is given below:

'In reply to (*a*) recent enquiry, and in the 800-1000hp type category, five firms offered a total of 23 variations with Paxman, Maybach, English Electric and Rolls-Royce engines, and electrical equipment by AEI, GEC, Brush and English Electric. The attached table (Appendix 'A') summarises the prices of the lowest tenders from each firm.

'Two firms, namely AEI and Clayton, are considerably cheaper than their nearest competitors. Offers by Brush, EE Co. and NB Loco. Co., having no technical features of superior worth which would justify their higher price, are ruled out on cost.

'The lowest priced AEI and Clayton offers have the following features:

AEI/Clayton offers (1)

	AEI	Clayton with AEI Equipment
Single vertical 900hp Paxman engine Type	YHXL	YHXL
Height of bonnet above rail level	10' 4"	10' 3"
Length, cab to end of bonnet		
Front end of locomotive	31' 6"	26' 0"
Back end of locomotive	4' 6"	9' 6"
Price (75) with Spanner steam heating boiler	£52,502	£53,120
Price (75) without steam heating boiler	£50,612	£50,790
Deliveries commence	Dec. 1961	Dec. 1961

'The YHXL engine, of which we have 54 in service or on order on Type 1 locomotives (D8200-43, D8400-9), by virtue of a medium high speed of 1250rpm, is of moderate height and the level of the bonnet top as indicated above, satisfies that part of the explanatory notes issued with the enquiry which specified that "the driver in both directions of travel shall be able to see all signals on right-hand curves from his seated position". The arrangement of the cab off-centre does not, however, satisfy the second requirement advised to the manufacturers, that "the maximum possible field of vision in all directions is required from the single cab".

'A truly central cab is offered as an alternative by both AEI and Clayton, at prices also lower than their competitors. This involves the use of two flat Paxman engines of 450hp each of Type 6ZHXL. These engines of which we have only had a limited experience.....cost £2,380 per locomotive more, but the operating requirement is fully met. These alternative offers have the following features:-

AEI/Clayton offers (2)

	AEI	Clayton	
		AEI Electrics	GEC Electrics
Two horizontal Paxman engines:			
Type	6ZHXL	6ZHXL	6ZHXL
Total hp	900	900	900
Height of bonnet above rail level	9' 10"	9' 10"	9' 10"
Width of bonnet	7' 0"	5' 9"	5' 9"
Length of bonnet on either side of cab	Equal	Equal	Equal
Position of steam boiler	Under bonnet	Inside cab	Inside cab
Price (75) with Spanner boiler	£54,882	£55,480	£55,400
Price (75) without boiler	£52,992	£53,150	£53,070
Delivery	Dec. 1961	Dec. 1961	Feb. 1962

'The above mentioned prices . . . include a new lighter traction motor, the outcome of the new collaboration between AEI and the ALCO Co. of America. We see no reason why this latest technical development should not be accepted under guarantee, in place of the motor used on the existing BTH Type 1 locomotives, which would cost £2,210 more per locomotive. A Rolls-Royce Vee engine of the same power (i.e. 2 x 450hp) has been quoted at approximately the same price, but this is a new design with no experience, and it will not be available until the end of 1962. Provision is made for electric heating, if and when this is required in substitution for steam.

'The Chief Traffic Officer has intimated that he much prefers the proposals having twin flat engines and central cab. His reasons are:

(1) Better visibility in either direction, the central cab position making for more satisfactory judgement when carrying out shunting duties.
(2) As two engines are provided, it will be possible during the time the locomotive is in traffic, and when full power is not required, for one engine only to be used, thereby saving fuel.
(3) When a steam heating boiler is required, a position inside the cab enables its functioning to be under the control of the driver, thereby avoiding the necessity of a second man for this purpose (i.e. empty stock working into terminals).

'It is the opinion of the Chief Traffic Officer that the additional cost is justified to obtain a locomotive fully suitable for all his requirements.

'Of the three alternative offers for this kind of locomotive outlined above, the AEI offer, which is the cheapest, sites the boiler under the bonnet and not in the cab, which does not satisfy the Chief Traffic Officer's point (3) above. Moreover, its greater bonnet width somewhat impedes clear vision of the buffer beams. Of the Clayton offers, that with GEC equipment is marginally cheaper (£80) but the alternative offer with AEI equipment is preferred because:

(i) Commencing delivery date is better by two months.
(ii) We already have good experience with AEI equipment on existing Type 1 locomotives.
(iii) The Paxman flat engines of which we already have experience . . . were associated with AEI equipment.

'We recommend, therefore, that the order for 88 Type 1 locomotives authorised for Scotland under Works & Equipment Minute 1611/7 of 3 December 1959, be placed with Messrs. Clayton Equipment Co., and have AEI electrics at a cost of:

£53,150 without boiler
£55,480 with boiler

subject to the following considerations:

(a) The above prices are for 75 locomotives and the Chief Contracts Officer should be asked to negotiate some reduction in cost for the higher number required.
(b) Although the Glasgow South and Fife schemes for which these locomotives are required, indicate freight duties only, the General Manager, Scottish Region, should be asked to indicate which, if any, of these locomotives should have boilers for empty stock working.
(c) Although the Scottish Region have hitherto been supplied with English Electric Type 1 locomotives, it is of interest to note from Appendix 'A' that the English Electric Co's. quotation for a locomotive on the lines of the present recommendation is £6,340 more per locomotive, or approximately £½ million more for the 88 locomotives required.

Signed: Chief Electrical Engineer, Chief Mechanical Engineer.

Appendix 'A'. Summary of Prices

Type 1.

	Prices for 75 locomotives (£)				Paxman Engine
	Without boiler		With Boiler		
	Existing Traction Motors	New Traction Motors	Existing Traction Motors	New Traction Motors	
AEI (900h.p.) (AEI Electric)	£52,822	£50,612	£54,712	£52,502	1 Vertical
	£55,202	£52,992	£57,092	£54,882	2 Flat
Clayton (900h.p.) (AEI Electric)		£50,790		£53,120	1 Vertical
		£53,150		£55,480	2 Flat
Clayton (900h.p.) (GEC Electric)	£53,070		£55,400		2 Flat
NBL/GEC/Paxman (1000h.p.)	£57,000		£59,480		1 Vertical
NBL/GEC/Paxman (900h.p.)	£59,000		£61,480		2 Flat
Brush/Paxman (900h.p.)	£57,250		£59,450		2 Flat
EE Co (1000h.p.)	£58,770		£61,256		1 Vertical (EE engine)
EE Co (900h.p.)	£59,490		£61,850		2 Flat

The Technical Committee Meeting of 23 September 1960 (Min.866) approved the recommendation stating that 'the order for 88 Type 1 locomotives authorised for the Scottish Region by Works & Equipment Committee Minute No.1611/7 - 3 December 1959, should be placed with Messrs. Clayton Equipment Co., subject to the considerations mentioned in the Memorandum, the locomotives to have Paxman flat engines type 6ZHXL'. It will be noted, however, that no specific reference was made to the type of electrical equipment to be installed even though the technical team recommended AEI equipment.

With locomotive provision apparently sorted out with respect to the Glasgow North, Glasgow South and Fife Area Schemes, all that was now required was the upgrading of the Area Schemes themselves from 'notional' to 'firm' by incorporating all necessary justification information and associated quantified evidence (e.g. benefits of single-manning, re-diagramming of services to take advantage of the characteristics of diesel locomotives, etc). In addition, the Fife scheme required the mileages proposed for the diesel locomotives to be quantified, split between the thirty-four Type 1s and twenty Type 2s.

A Memorandum to the W&EC dated 10 October 1960 attempted to provide the additional information required:

'Main Line Diesel Locomotive Area Schemes: Glasgow North District, Glasgow South District, Fife Area'.
1. The main line diesel locomotive schemes considered by the Committee at their meeting on 3rd December, 1959 included three Scottish Region 'notional' schemes to which provisional approval was given and the "firm" details were to be submitted later. The broad details submitted for the schemes concerned were:

Main Line Diesel Locomotive Area Schemes

Area Scheme	Diesel Locomotives			Net Revenue Improvement	Excluding Interest Return on	
	No.	Gross Outlay £000	Net Outlay £000	£000	Gross Outlay %	Net Outlay %
Glasgow North	47	2726	1072	142	5.2	13.3

Main Line Diesel Locomotive Area Schemes

Area Scheme	Diesel Locomotives			Net Revenue Improvement	Excluding Interest Return on	
	No.	Gross Outlay £000	Net Outlay £000	£000	Gross Outlay %	Net Outlay %
Glasgow South	65	3770	2063	258	6.9	12.5
Fife	54	3472	1515	262	7.6	16.0

Note:
Net Outlay: Additional outlay in excess of replacement cost of displaced steam locomotives.

2. The Scottish Region has reported that a further examination has been made of the above schemes. The pattern of the freight services for which the locomotives are required is constantly changing in the heavy industrial areas concerned and diagrams are subject to extensive alterations, to conform to the changing traffic flows. This situation is expected to continue and it is considered that there would be little purpose at this stage in preparing for the diesel locomotives detailed diagrams that are certain to be unsuitable for the conditions when the locomotives are delivered. It is intended, therefore, that immediately prior to the delivery of the locomotives, the diagrams will be entirely re-cast to take full advantage of the potential of the locomotives having regard to the traffic conditions at that time.

3. Regarding single-manning the Region considers that, at present, no material saving from this source can be expected pending the fitting of vacuum brakes to all mineral wagons and the elimination of the obsolescent merchandise wagon stock that is to remain unfitted until replaced by new construction.

4. In the Fife area the principal freight services are provided to meet the requirements of the National Coal Board and because of the closures and developments under the modernization plans of the Board train workings have to be frequently varied to suit the changing traffic conditions. The type of locomotive to be provided on any particular diagram can vary from day to day depending on the number of wagons to be moved, the shunting to be performed and power availability. In these circumstances the Region does not consider it practicable to estimate separately the annual mileage for each type of diesel locomotive and expect that it will be approximately the same for both types.

5. In the above circumstances, it is the view of the Scottish Region that the estimates originally given represent the best estimates that can be provided at the present time of the financial results expected from the schemes.

6. It is recommended that, having regard to these further explanation, firm approval be given to the three schemes concerned.

Signed: Chief Traffic Officer,
Chief Mechanical Engineer,
Chief Electrical Engineer,
Director of Costings.

The contents of the Memorandum were discussed at the W&EC Meeting on 8 November 1960. The discussion was recorded in Minute 1912/9 in unusual length which suggested some difficulty in getting the submission approved, perhaps not surprising given the vague and unquantified nature of the evidence provided:

'Dieselisation of Freight Working in the Glasgow North & South Districts and the Fife Area, involving the use of 166* Diesel-Electric Main-Line Locomotives to replace 297 Steam Locomotives. (*146 Type 1's and 20 Type 2's) 'Mr Crabtree explained that the constantly changing pattern of freight services in the areas concerned entailed extensive alterations to the diagrams. The Region, which had gone into the matter very carefully, felt that no useful purpose would be served at this stage in preparing detailed diagrams which would not be applicable at the time the diesel locomotives were delivered. In the case of the Glasgow districts, for example, there were alterations to one or other of the existing diagrams every two days of so. It was proposed, therefore, that immediately prior to the delivery of the locomotives the diagrams should be re-cast to take full advantage of the potential of the locomotives, having regard to the traffic conditions at that time. The Region was satisfied that if the number of locomotives at present envisaged was then found to be in excess of requirements, the balance could be absorbed in other area schemes which had not yet been put forward. He pointed out that this was the first scheme to be submitted for the elimination of steam locomotives on freight services.

'The Committee commented on the low annual mileage which would be worked by the locomotives and MrCrabtree explained that this was due to the fact that they would operate almost exclusively on low classified short trip working, with a high percentage of terminal working and shunting. He pointed out, however, that notwithstanding this, they would still work twice the mileage of the present steam locomotives. In answer to an enquiry as to whether certain of the services could not be worked by diesel shunting engines, he said that the Region had already gone to the limit with 350hp shunters in trip working and there was no further scope for the use of such engines on the services and diagrams covered by these three schemes.

'The Committee were not satisfied that there was no possibility of effecting savings through the manning arrangements and after full discussion they decided to do no more than note the scheme. They agreed, however, that orders should now be placed for 88 Type 1 locomotives for which tenders have been received. These would not necessarily be allocated to the Glasgow and Fife schemes, but would ensure that the Region were covered on submission of the final schemes, which should contain full financial justification for the proposals.'

In parallel, work continued on the order placement. A Memorandum to the Supply Committee (from the Chief Contracts Officer) dated 17 January 1961 summarised the bid assessment undertaken via the Technical Committee and how they reached the decision to endorse the design offered by the Clayton Equipment Co. (as discussed above). The Memorandum included the following paragraphs:

'The Technical Committee considered the various designs offered and by Minute No.866 endorsed a recommendation to adopt the design offered by Clayton Equipment Co. Clayton quoted alternative prices for locomotives with AEI, GEC, or Brush electrical equipments, the cheapest being that with GEC equipment at £53,070 for 75 and £52,568 for 100. All tenders were subject to price variation from 11th July 1960. The delivery offered by Clayton with GEC equipment is to commence in 17 months after receipt of the order and to continue at the rate of 5 locomotives per month. In view of the long period between the tender date and completion of delivery, it is not proposed to ask for a fixed price but an endeavour will be made to agree a price lower than that offered for 75…..'

'It is recommended that, subject to the views of the Ministry, an order be placed with The Clayton Equipment Co. Ltd. for the supply of 88 Type 1 Diesel-Electric Main Line Locomotives with GEC electrical equipments at a price not exceeding £53,070 each, giving a total cost of £4,670,160 at mid-1960 price levels. Where train-heating boilers are required the price will be increased by approximately £2,330 per locomotive.'

It will be noted that the authors of the Memorandum to the Supply Committee accepted the Technical Committee recommendations with respect to lead manufacturer (i.e. Clayton Equipment Co.) but changed the electrical equipment supplier from AEI to GEC, on the basis of the lower price offered by GEC.

The Supply Committee Meeting Minute 891 of 19 January 1961 (Min.891) records that the following contract was approved:

'With the Clayton Equipment Co. Ltd. for the supply of 88 Type 1 Diesel-Electric main-line locomotives with GEC electrical equipment at a price not exceeding £53,070 each, plus £2,330 per locomotive where train-heating boilers are required.

'Subject to confirmation of the choice of electrical equipment by MrRatter and to clearance of the expenditure by the Ministry of Transport.'

A number of observations are pertinent here:

- The eighty-eight locomotives were ultimately delivered with GEC electrical equipment.
- In the event, no Claytons were delivered with train-heating equipment. Had this been known in advance, the alternative AEI sponsored centre-cab offering would have met the required design criteria with an associated cost saving of £78 per locomotive or £6,864 for the full order.
- The total all-up cost of £4,670,160 compared very favourably with the authorised estimated cost level of £5,104,000 already signed-off by the BTC on 17 December 1959 (Meeting Minute 12/510).
- The Original Cost per locomotive recorded on the withdrawal documentation for the first eighty-eight locomotives was £57571/£57572, reflecting the application of the price variation clause between July 1960 and delivery.

The W&EC Meeting Minutes of 21 February 1961 (Min. 1999/9) with regard to the order for eighty-eight Claytons finally changed from 'Details to follow' to 'These locomotives to be purchased from contractors', fourteen months after authorization.

The order for the eighty-eight locomotives was placed on the Clayton Equipment Co. during April 1961:

W&EC Meeting 03/12/59 (Min.1611/7) / BTC Meeting 17/12/59 (Min.12/510(b))

| 88 Clayton/Paxman | D8500-D8587 | Ordered 04/61 | Delivered 09/62-02/65 |

It will have been noted above that the North British Locomotive Co. Ltd. (NBL) quoted for the first batch of the centre-cab Type 1s. The award of the contract to Clayton caused considerable consternation in Scotland as shown by the following extract from *The Daily Telegraph and Morning Post*, dated Thursday, 18 May 1961:

'Protest Over Rail Order.'
'The British Transport Commission was strongly criticized by the general council of the Scottish TUC yesterday for placing a £5 million order for 88 diesel locomotives with an English firm instead of North British Locomotive Company in Glasgow.

'The diesels are for service in the Scottish area. The council argued that the North British Company is so desperately in need of orders that it is contemplating a run-down and eventual closure at its Springburn works.

'Yesterday, Mr George Middleton, general secretary of the Scottish TUC, said what disturbed them was that North

D8500, Cricklewood (en route Marylebone Goods Yard for exhibition), 26 July 1962. (Transport Treasury)

View from a Clayton driving seat, Millerhill Yard, 18 August 1972.
(Anthony Sayer)

British has been completely passed over and the contract awarded to Clayton Equipment Company….. All this must seem to those interested in the preservation of industry in Scotland like the economics of the madhouse.

'A spokesman for the British Transport Commission said the design produced by Clayton Equipment most satisfactorily met the Commission's specification, and the price was 'right'. The Commission had nothing against the North British Locomotive Company.'

3.2 D8588-D8616.

A Memorandum to the W&EC dated 1 December 1961 implicitly included a recommendation for the purchase of twenty-nine Type 1 main line diesel-locomotives as part of the original Holbeck, Hull and Sheffield (II) Area Schemes. In August 1961, the method for approving diesel orders was changed complicating the locomotive ordering process during the transition period. The content of the Memorandum explained the situation:

'Hitherto, expenditure on diesel locomotives has been approved by the Commission in annual (*Building*) programmes. The programmes have had regard to the investment likely to be available in the year to which they have referred and, also, to capacity for manufacturing the types required.

'The annual programmes have been derived from diesel traction schemes (*Area Schemes*) prepared by Regions to a prescribed code and approved by the Works and Equipment Committee.

'The programme procedure has now been abolished, with effect from 15th August 1961, and, instead, individual diesel schemes will be treated as single comprehensive projects.'

Six Area Schemes prepared on the basis of the annual Building Programme methodology fell foul of the new arrangements. These Schemes involved an overall requirement for 691 locomotives. At the time of the system changeover, 374 of the 691 locomotives had already been authorised by the BTC and with some additional information had been signed off by the Ministry of Transport. The remaining 317 locomotives required approval under the more exacting requirements of the new system. Of the 317, 41 were associated with North Eastern schemes (16 Holbeck (including 3 Type 1 locomotives), and 25 Hull (13 Type 1s)) and 62 with the Eastern Region Sheffield (II) scheme (13 Type 1s).

The 1 December 1961 Memorandum recommended that the Commission approve the procurement of the 317 locomotives based around the new individual dieselisation project methodology. The W&EC of 19 December (Minute 2193/3) authorised the procurement of the outstanding 317 locomotives across the six Area Schemes. Comments recorded in the minute included three key paragraphs:

'The Technical Adviser pointed out that although there were substantial variations in the proposed annual mileages for the diesel locomotives as between the different schemes, these should be viewed in the knowledge that the mileage for the steam locomotives to be replaced averaged only 22,000 per year and that the schemes revealed that the replacement rate of steam by diesel locomotives was rather more than two to one.

'The Committee agreed that, as a result of the traffic studies which were now taking place, a different pattern of train service might well emerge, which would probably have the effect of improving the financial returns on diesel locomotive schemes. Approximately 1,900 diesel locomotives had been authorised to date and many more would be required before the changeover from steam traction was completed. Even if a re-examination of the six schemes in question revealed that a reduction could be made in the number of locomotives required there was ample scope in connection with further schemes which the Regions were developing to absorb any locomotives which might become surplus to requirements.

'The Committee, in recommending the procurement of the outstanding 317 locomotives, accepted the fact that the financial improvement for each scheme had been calculated on a conservative basis and that this improvement would undoubtedly be enhanced following the traffic studies now in place.'

The BTC Meeting of 11 January 1962 (Min.15/7) agreed that tenders should be sought for the 317 locomotives and recommendations for purchase should be made by the Supply Committee to the Commission.

An invitation to tender process followed for the Type 1 requirement of twenty-nine locomotives (closing date May 1962).

As part of the *original* contract with Clayton, BR included a clause that allowed for any further locomotives of the Clayton design to be built under license in either BR workshops or by other external manufacturers, these believed to be based on the following terms agreed with LAMA:

- £7,500 lump sum to Clayton Equipment at the signing of the licensing agreement.
- A fee of £300 per locomotive to Clayton.
- The licensor (i.e. Clayton) to be offered contracts after the initial order for half of each of the Commission's subsequent requirements from Industry for locomotives to such a design.

A Memorandum to the Supply Committee dated 22 June 1962 detailed the resulting tender submissions:

'Type 1 900hp Diesel-Electric Locomotives.'

1. The Commission by Minute No.15/7 agreed that tenders should be obtained for the construction of 29 Type 1 Diesel-Electric Locomotives . . .

2. Tenders were invited for locomotives to the 'Clayton' design incorporating Paxman engines from eight contractors, five of whom responded. The lowest tenders received are as follows:

Tenders

Contractor	Electrics	Price each (subject to price variation (£))				
		30	40	65	80	130
Beyer Peacock	CP	£53,470*	£52,930	£52,440	£52,240	£51,750
Clayton Equipment Co.	CP	£53,382	£53,072	£52,664	£52,403	£51,817
	GEC	£55,076	£54,912	£54,594	£52,240	£53,464
	AEI	£55,169	£54,765	£54,378	£54,136	£53,521
	Brush	£56,476	£56,276	£55,608	£55,316	£54,492
AEI	AEI	£53,978	£52,592	£51,637	£51,140	£50,344

* Fixed for 1963.

3. Beyer Peacock have offered delivery of 25 in 1963 at a fixed price of £53,470 with the balance subject to price variation, whereas the next lowest price of £53,382 by Clayton is subject to price variation on the whole quantity. As wages and prices of materials in the remainder of 1962 and in 1963 are likely to rise by more than £88 - the difference between Beyer Peacock's and Clayton's offers - Beyer Peacock's offer is the more attractive for the 29 locomotives required. Their price of £53,470 compares with an estimated price payable by Clayton under a current contract, allowing for modifications and price variation, of £55,325.
4. It was a condition of the current contract with Clayton that in the event of our requiring further locomotives utilising their design they would be offered at least 50% of our requirements from the trade provided their price and delivery were satisfactory. In view of the small number currently required, Clayton have however agreed to waive this proviso on the understanding that subject to their price and delivery being reasonable they will be offered a share of any further requirements to give continuity of production in 1964 at their current rate.
5. Prices for larger quantities have been obtained to ascertain the saving which would result from continuity of production, but whilst there is the possibility of further Type 1 locomotives being required the position at present is not sufficiently clear to justify forward ordering.
6. It is therefore proposed to place the order with Beyer Peacock for 29 Type 1 900h.p. diesel-electric locomotives at £53,470 each, subject to price variation on deliveries in 1964, making a total estimated cost of £1,551,000 which compares with an estimate of £1,670,000 included in the Area Schemes.

It will be noted that Tenders were requested for the supply of differing numbers of locomotives, well beyond the required twenty-nine, to permit any financial benefits from long-term construction continuity to be quantified. Quite how the license fees were incorporated into the calculations is unclear.

The Supply Committee Meeting of 28 June 1962 (Min.1123) agreed to recommend to the Commission that contracts be placed as follows:

'Type 1 900hp Diesel-Electric Locomotives.'
'With Beyer Peacock & Co. Ltd. for supply of 29 locomotives at a price of £53,520 each, subject to price variation in 1964, at total estimated cost of £1,552,080.'

It will be noted that the price per locomotive between the memorandum dated 22 June 62 and the Meeting of 28 June had increased by £50; the reason for this unknown.

A Memorandum from the Supply Committee to the BTC dated 11 July 1962 requested, inter alia, authorisation for the construction of twenty-nine Type 1s by Beyer Peacock (delivery September 1963 to March 1964), making the comment that 'Beyer Peacock's favourable offer for the Type 1 locomotives is probably due to their having a gap to fill following completion in September 1963 of the Type 3 hydraulics for which it is not proposed to place any follow-on order'.

The BTC Meeting of 12 July 1962 (Min.15/190) approved the purchase, subject to clearance with the Ministry of Transport.

The Supply Committee Meeting Minutes of 26 July recorded that the Contracts Officer 'hoped to place the contract [for the twenty-nine Type 1s] within the next few days'.

It is interesting to note that the BTC Meeting Minute of 11 January 1962 (Min.15/7) in approving the orders for the twenty-nine Type 1s (i.e. three to Holbeck, thirteen to Hull and thirteen to Sheffield), also recorded the following caveat:

'This approval in no way commits the Commission to the precise allocation of locomotives shown in the individual schemes although the Regions could plan on the assurance that they would receive an appropriate number of locomotives of the Types required. The schemes are to be reassessed by the Regions and Area Boards and the requirements of each fully justified, regard being had to the necessity for achieving better utilisation than is at present the case with diesel locomotives and to current views about the future pattern of railway business.'

On delivery, the twenty-nine Claytons were actually delivered to Thornaby (four), Gateshead (twelve) and Tinsley/Barrow Hill (thirteen).

The original cost per locomotive recorded on the withdrawal documentation for D8588-D8616 was £56088/£56089, presumably reflecting licencing fees and the application of the price variation clause for the 1964 deliveries.

W&EC Meeting 19/12/61 (Min.2193/3) / BTC Meeting 11/01/62 (Min.15/7)
29 Type 1

Subject Supply Committee (SC) allocation of construction contractor

SC Meeting 28/08/62 (Min.1123) / BTC Meeting 12/07/62 (Min.15/90)
29 Beyer Peacock/Paxman   D8588-D8616   Ordered Circa 07/62   Delivery 03/64-04/64

Chapter 4

THE NEXT 100 TYPE 1s: D8617-D8716 OR BACK TO PLAN A?

4.1 Ultimate Requirements for Diesel Locomotives.

A Memorandum to the British Railways Board (BRB, successor to the BTC) from the Planning Committee dated 13 December 1963 and entitled *Diesel Locomotives* included a Section 'Ultimate Requirements' which discussed the ultimate size of the BR diesel fleet and the fact that this depended on numerous factors which were not yet finally decided, including:

- Shape and size of system;
- Ultimate level and character of traffic;
- Extent to which unremunerative passenger services would be withdrawn;
- Final effect of freight and coal concentration;
- Size of Liner Train network;
- Extent to which 'Merry-go-Round' working would be adopted;
- Eventual motive power maintenance standards;
- The degree to which train loading could be improved, itself dependent on wagon brake system upgrading;
- The future of electrification.

The Memorandum went on to state:

'So far as can be judged at present, the effect which the above factors could have on the ultimate traction fleet indicates that even if they all applied in the way most favourable to a reduction in locomotive requirements, the final demand would still exceed 3,100.'

To bring the fleet close to this level, the Memorandum proposed the building of 430 diesel locomotives at a total estimated gross cost of £37.35m (or £39m including associated maintenance facilities), including 100 Type 1s (circa £5.6m). The Memorandum recommended the construction of the 430 locomotives in 1964 and 1965 to meet 'ultimate requirements'.

On 19 December 1963, the BRB (as recorded in the Meeting Minutes (Min.63/382) approved the recommendation, subject to the number of Type 3 locomotives being reduced from eighty to seventy. The paper was used as the basis of a submission to the Ministry of Transport seeking blanket approval for 420 locomotives (with a revised overall estimated cost of £36.45m).

When a Memorandum was sent to the Supply Committee on 6 January 1964, approval from the Ministry of Transport was still awaited; however, the Supply Committed progressed matters to ensure a state of readiness on receipt of MoT approval.

With respect to the 100 Type 1 locomotives, the Memorandum stated:

'Locomotives of this type are currently in production by Clayton Equipment, the original designers, and Beyer Peacock. Workshops have not built this type. Both contractors should complete delivery of their existing orders early in 1964, and both are in a position to take on additional orders. It was agreed with Clayton Equipment at the time of their initial order that the Board should be free to use the design for building in their own workshops and to have locomotives built by other contractors on payment of a royalty by the building contractor, on the understanding that Clayton Equipment would receive orders for at least 50% of the Board's requirements from the trade. So far, Clayton Equipment have received an order for 88 locomotives for which they quoted the lowest price, and although strictly they were entitled to share the next order of 29, the whole quantity was allocated to Beyer Peacock on the understanding that Clayton Equipment would receive an appropriate share of the Board's next requirement. The Board now require 100 Type 1 and Clayton Equipment have offered, provided an early order is placed, to deliver 14 in 1964 and 84 in 1965. Beyer Peacock have offered, again subject to an early order, to deliver 10/12 in 1964 and 60 in 1965. Both such offers incorporate Crompton Parkinson equipment, the approximate prices being Clayton £54,000 and Beyer Peacock £55,000.

'Clayton have also offered an alternative price of approximately £57,000 for locomotives with GEC equipment. The Chief Mechanical Engineer considered that Hunslet might be in a position to offer an attractive price and they have been invited to tender for 25, which is their estimated capacity in 1965. The Workshops have indicated that they are interested in building Type 1 locomotives at Swindon and say they could deliver 60 in 1965, but we have not yet received their price.

'There is ample capacity available for Type 1 locomotives given the delivery required and, subject to technical views on the type of electrical equipment to be fitted, it is proposed to proceed on the basis of the lowest cost to the Board having regard to our commitment to Clayton. If this should mean that Swindon obtain orders approximating 60, the balance, which would then be relatively small, should all go to Clayton Equipment, subject to their price bearing a reasonable comparison with any Hunslet may submit. If, on the other hand, Swindon's price is too high, the total requirement would be allocated to Clayton Equipment, Beyer Peacock and possibly Hunslet at the lowest overall cost.'

4.2 The Clayton/ English Electric Conundrum.

Technical issues with the Clayton fleet were highlighted by the Chief Mechanical Engineer, J.F. Harrison, in a Memorandum to the British Railways Management Committee dated, 21 February 1964 entitled *Diesel Availability* in which reservations were expressed with respect to the placement of further orders:

'.....until the cause [of the piston seizures] is found and rectified, it would be unwise to order more locomotives of this type.'

Following further deliberations, a Memorandum from the Supply Committee to the BRB dated 10 March 1964 stated that:

'There is a technical problem with the engines for this type and until it is resolved no orders will be placed.
'The intention is to order in due course:-

Clayton Engineering	40
Beyer Peacock	10
Workshops	50

'. . .a full submission will be made before the orders are placed.'

The 'technical problem' persisted and the intended order for 100 Type 1s was ultimately split to ensure continuity of supply. Whilst there is an obvious gap in available archive material, a Memorandum from the Chief Operating Officer and the Chief Mechanical Engineer to the Supply Committee dated 15 April 1964 explained the situation:

'At the meeting held on 13th April . . . it was agreed in conversation that the 100 Type 1 locomotives authorised by Board Minute 63/382 of the 19th December 1963, should be divided into two parts:-
'50 Type 1 centre-cab Clayton design . . .
'50 Type 1 end-cab English Electric design . . .

'The reason for dividing this order into two parts is that considerable trouble and difficulties have occurred with the engines and mechanical parts of the Clayton/Paxman Type 1's. The Chief Mechanical Engineer does not consider it wise, at this stage, until the modifications which he has agreed with the firms concerned are implemented and proved satisfactory, to order further locomotives of this type. Therefore if 100 locomotives are to be delivered to British Railways by the end of 1965 in accordance with the programme, it is vital that we obtain some locomotives, for certain, which are satisfactory, and these can be the English Electric Type 1's, 25 of which can be delivered for certain by the end of 1965, and the further 25 follow in 6/8 weeks afterwards.

'It has been confirmed with Claytons that they could deliver 50 of their Type 1's by the end of 1965 provided they received an order as late as February 1965.

'By February 1965 the modifications referred to above to the Clayton/Paxman Type 1's will have been proved, or not, and it will then be possible either to order the 50 Clayton/Paxman and complete the picture, or to order them with Rolls-Royce engines which by that time will also have been proved in service (two such locomotives of existing orders are being fitted with Rolls-Royce engines). Or, failing both, a further 50 Type 1 English Electric could then be placed . . . for delivery during the early part of 1966.

'We recommend, therefore, that an order should be placed with English Electric very quickly (the delivery of 25 in 1965 is dependent on this order being placed within the next few days), so as to ensure delivery in 1965 of 25 of the order of 50 Type 1 English Electric end-cab locomotives at a price of £59,500 each and a total cost of £2,975,000.'

Min.64/111 of the BRB Meeting of 23 April 1964 recorded:

'The Board considered a Memorandum, dated 15th April 1964, by the Chief Operating Officer and the Chief Mechanical Engineer to the Supply Committee, recommending that an order be placed with the English Electric Company for 50 Type 1 diesel main line locomotives at a cost of £2,975,000. This would leave open the question of the allocation of the order for the remaining 50 of the 100 Type 1 locomotives authorised by minute 63/382 on 19th December 1963, until the proving or otherwise of the Clayton Type locomotive.

'Mr Shirley explained the urgency of making this recommendation on behalf of the Supply Committee, in advance of their next meeting, in order to ensure delivery of 25 of the English Electric locomotives in 1965. The Board approved the recommendation, leaving in abeyance the question of the ordering of the remaining 50 Type 1 locomotives.'

The Clayton technical issues continued. J.F. Harrison in Appendix C of his 'Report on the Working of the CME Department' dated December 1964 stated:

' . . .this type of locomotive cannot be repeated with present equipment until more experience has been gained with the modified equipment.'

A Memorandum to the Supply Committee dated 7 December 1964 addressed the issue of the allocation for construction of the outstanding 50 Type 1 locomotives. The Memorandum explained:

'. . . At that time [April 1964] it was anticipated that by February 1965 sufficient experience would have been gained with the modified Paxman engines and two Clayton locomotives fitted with Rolls-Royce engines to enable a decision to be made regarding the remaining 50. The Rolls-Royce engine locomotives are only just entering service and other difficulties have arisen so that it will not be known until at least mid-1965 whether the Clayton locomotive will be satisfactory, at which time it would not be possible to obtain completion of an order before early 1967.

'English Electric have offered to supply 50 locomotives commencing in April 1966 with completion in September 1966 at £59,500 each, subject to price variation, which is the same price as that under the current contract. Workshops were invited to tender for the 50 locomotives to the English Electric design, and although originally they decided not to quote they have recently decided they would like to build the bogies, underframes and bodywork, leaving English Electric to supply and install the power equipments. This is being explored.

'In the meantime, if delivery is to be maintained by September 1966, and in order to obtain the benefit of continuity of production, it is necessary to place an order with English Electric for the power equipments for which they have quoted £32,440 per set, subject to price variation, with delivery commencing January 1966 and finishing July 1966.

'In the circumstances the Committee is requested to approve the placing of an order with English Electric Co. Ltd., for 50 power equipments for Type 1 locomotives at a total cost of £1,622,000.'

This request was approved by the Supply Committee on 10 December 1964 (Min.292). However, manufacturing arrangements were altered as a further Memorandum to the Supply Committee, dated 23 December, explains:

'English Electric have offered to supply [an additional] 50 complete locomotives commencing April 1966 with completion in September 1966 at £59,500 each, subject to price variation, which is the same price as under the present contract. Workshops will have surplus labour at this time suitable for undertaking the construction of mechanical parts, and, as a result of discussions between English Electric, the Chief Mechanical Engineer and Workshops, it has been agreed that Workshops should manufacture the bogies. On this basis English Electric are prepared to reduce their price by £6,208, which the Chief Mechanical Engineer considers reasonable. There has not been sufficient time for Workshops to submit an estimate for the bogies.

'To ensure continuity of production of the locomotives, English Electric require an order by the end of this month, and the Committee is requested:

(a) to recommend to the Board that an order be placed with English Electric Co. Ltd. for the supply of 50 Type 1 diesel electric locomotives, at a price of £53,292 each, subject to price variation; the order already placed for 50 power equipments will be cancelled, and,

(b) to approve the placing of an order on BR Workshops for 100 bogies, at a price to be agreed.'

On 19 January 1965, a Memorandum to the BRB requested approval for the construction of 50 Type 1s by English Electric, excluding bogies, at a price of £53,292 each, subject to price variation. The BRB approved the request at their Meeting on 28 January (Minute 65/30).

The delivery of the Type 1s by English Electric was delayed as a consequence of issues with the construction of AL6 electric locomotives at Crewe Works. To maintain the delivery deadline of the electric locomotives, twenty of the Crewe order were transferred to English Electric with a consequent short delay to the Type 1s.

Thus, delivery of the so-called 'Standard Diesel-Electric Type 1' ceased after 117 locomotives, with the title effectively reverting to the English Electric product first produced in 1957. The transfer back to the English Electric design incurred an extra initial cost although this will have been recouped many times over by improved reliability and availability. Appendix A to the W&EC meeting of 17 May 1966 entitled *Overspending of Construction of Locomotives and Rolling Stock* included the following entry:

Overspending of Construction of Locomotives and Rolling Stock

Original Authority		Construction Project	Authorised Outlay	Estimated Final Cost	Overspending
Minute	Date		(£)	(£)	(£)
BRB 63/382	19/12/63	100 Type 1 D/E Locomotives	5,600,000	6,401,000	801,200

BRB Meeting 19/12/63 (Min.63/382)			
100 Type 1s authorised.			
BRB Meeting 23/04/64 (Min.64/111)			
50 English Electric Type 1s authorised	D8128-D8177	Ordered circa 04-05/64	Delivered 01/66-11/66
BTC Meeting 28/01/65 (Min.65/30)			
50 English Electric Type 1s authorised	D8178-D8199, D8300-D8327	Ordered circa 01-02/65	Delivered 12/66-02/68

D8190 and D8076, Nottingham Midland, 7 September 1973. English Electric Type 1s back in the ascendency! D8190 was one of the final batch of 100 Type 1s built for BR and as can be seen the visibility problem has gone away by the simple expedient of double-heading in nose-to-nose formation. Instant Type 4 power, and no requirement for a brake-tender on this sizeable Class 9 unfitted freight. (Rail-Online)

Chapter 5
SOME TECHNICAL ASPECTS

5.1 History of Clayton Equipment.

A brief historical summary of Clayton Equipment is given below (based on information supplied on their website (claytonequipment.co.uk)):

- Founded in 1931 by Stanley Reid Devlin as The Clayton Equipment Company Ltd.
- Operated from International Combustion Ltd until the outbreak of World War Two and trading as Clayton Equipment, Sinfin Lane, Derby.
- Produced industrial goods including structural steelwork, farm buildings, conveyors and elevators.
- Acquired new premises at Record Works, Hatton, Derbyshire in 1946. Facilities installed for the building of locomotives and industrial equipment for home and export markets.
- Supplied locomotive components for the BR modernisation programme (BTH Type 1).
- International Combustion Holdings Ltd acquired a 100 per cent shareholding of Clayton in 1957, but the company continued to operate autonomously.
- £5million contract from BR to supply eighty-eight locomotives and £1.75million order for ten 2,500hp locomotives for Cuba.
- Concentration on the sales of mining locomotives and tunnelling equipment, increasingly for the export market as the UK domestic market declined.
- International Combustion was acquired by Clarke Chapman in 1974, merging with Reyroll Parsons in 1979 to form Northern Engineering Industries (NEI). Ten years later NEI was acquired by Rolls-Royce with Clayton Equipment becoming part of the Rolls-Royce Industrial Power Group.
- In March 2005, after fifty years as a subsidiary of major British organisations, Clayton Equipment once again became an independent company.
- Moved to new premises at Burton-on-Trent in 2006.

5.2 Locomotive Leading Particulars.

5.2.1 D8500-85

Engine: Two Paxman 6ZHXL, 6-cylinders (firing order 1/5/3/6/2/4 with No.1 cylinder at the Free End and No.6 at the Generator End), 4-stroke, pressure-charged. Maximum continuous rated output: 450hp at 1500rpm each. 6ZHXL: 6 (No. of Cylinders), Z (in-line, horizontal/'flat'), H (7in cylinder bore size), X (turbo-charged), L (locomotive).

Turbo Charger: Napier MS 90.

Engine Fuel Tank Capacity: 500gal.

Main Generator: Two GEC WT800, 6-pole, self-ventilated. Continuous rating: 400V, 700 A at 1500rpm.

Auxiliary Generator: Two GEC WT763, self-ventilated. Continuous rating: 110V, 318A at 1500rpm.

Traction Motors: Four GEC WT421, force-ventilated, 4-pole, nose-suspended, single-reduction gear drive, 15:66 gear ratio.
Continuous rating: 158hp, 397V, 350A at 480rpm.

Performance:
Maximum Tractive Effort: 40,000lbs at 26.2 per cent adhesion, at 1,400A each main generator.
Continuous Tractive Effort: 18,000lb at 12.8mph at 700A each main generator.
Rail hp at Continuous Rating: 602hp
Full engine output available between 3 and 60mph.

Braking: Air for locomotive, vacuum for train giving a brake force of 86.7per cent of locomotive weight in working order.

Maximum Permitted Speed: 60mph.

Curve: 3½ chains (230ft) minimum radius, without gauge widening, dead slow speed.

Minimum Hump Radius: 660ft.

Weight in Working Order: 68 tons.

Axle Weight in Working Order: 17 tons (all axles).

Length overall: 50ft 7½in

Length over buffer beams: 47ft 0in

Width overall: 8ft 9½in

Height overall: 12ft 8in

Bogie Wheelbase: 8ft 6in

Distance between bogie pivot centres: 28ft 0in

Total Wheelbase: 36ft 6in

Wheel Diameter: 3ft 3½in

Train Heating Boiler: None fitted.

Wheel Arrangement: Bo-Bo.

Multiple Working (as built): Red Diamond.

Official photograph of D8500. No 1 end closest to camera. (Rail-Online)

Some Technical Aspects • 39

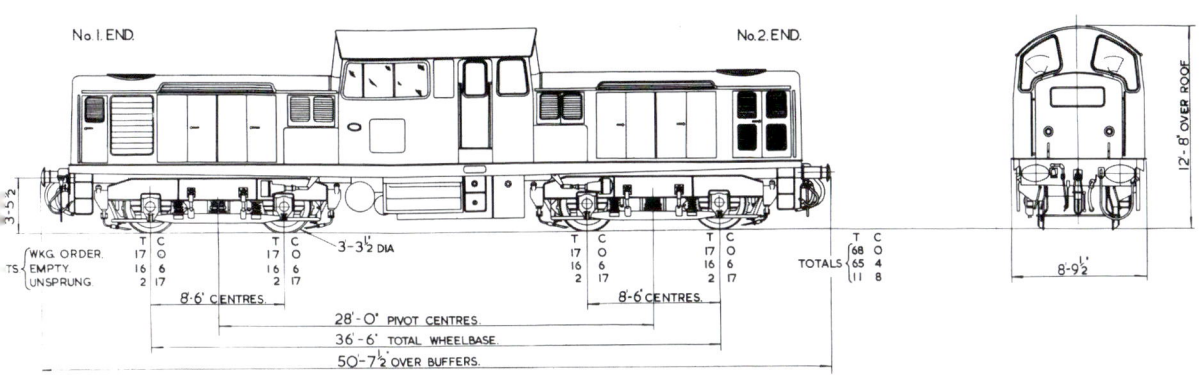

Line Drawing DE/1004/1 for D8500-85. Paxman engines, GEC electrical equipment.
(BR *Main-Line Diesel Locomotive Diagrams*, September 1961)

5.2.2 D8586/7 (where different from D8500-85)
Engine: Two Rolls-Royce DV8T, 8-cylinders, four stroke.
Turbo Charger: Holset.

D8587, International Combustion Ltd, Derby, 4 December 1964. The '88' on the buffer indicates that this was the last locomotive of the first order for Claytons. Note the 1500hp prototype DHP1 behind D8587. (Brian Lee)

D8587, International Combustion Ltd, Derby, 2 February 1965. The raised sections above the Rolls-Royce engines are clearly evident. Even allowing for the raised sections, the lack of headroom between the top of the engine and the bonnet roof was such that roof hatches had to be fitted to allow maintenance access to the top end of the engine. Such roof hatches, well illustrated in this photograph, were specific to only D8586 and D8587. (Brian Lee)

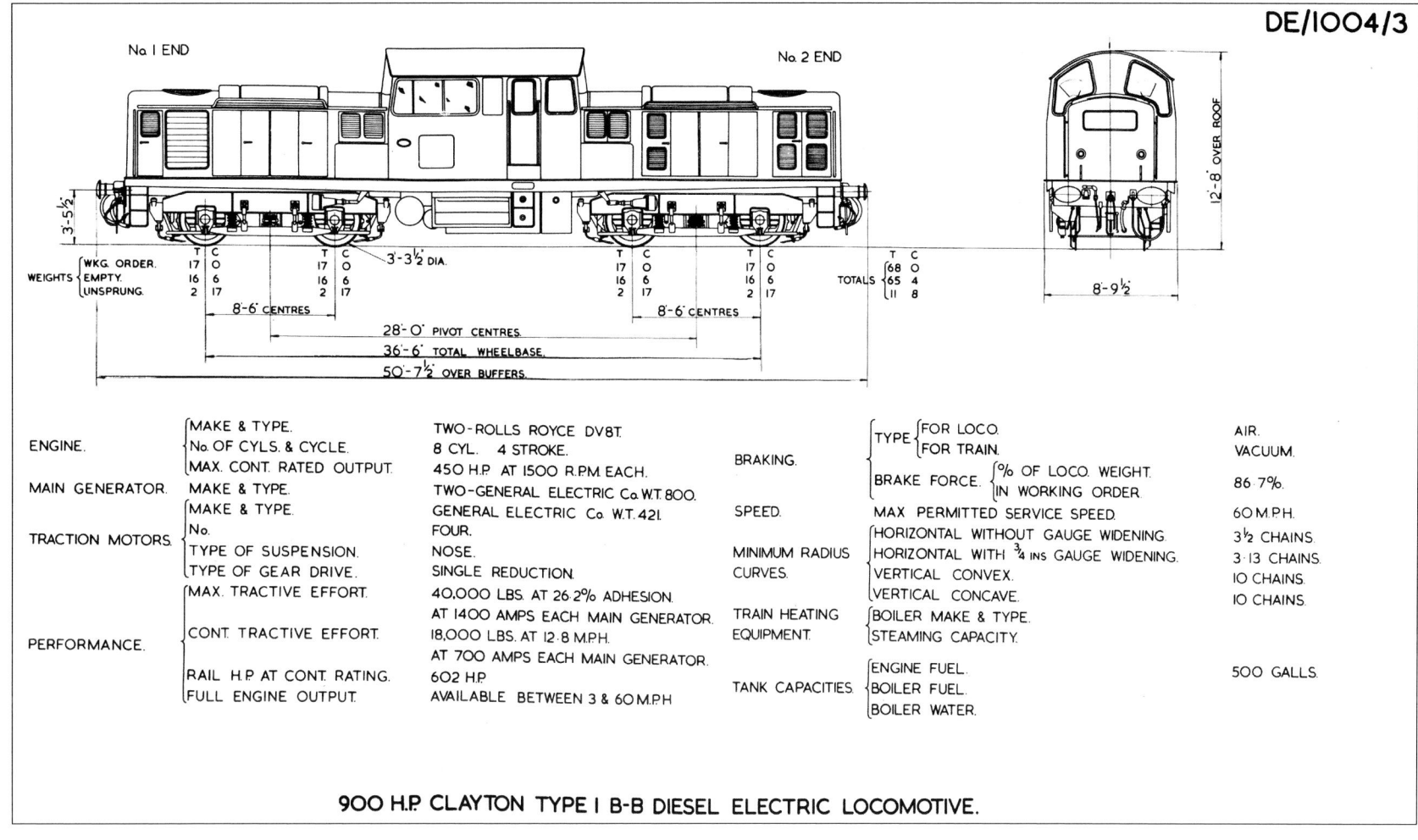

Line Drawing DE/1004/3 for D8586/7. Rolls-Royce engines, GEC electrical equipment. (BR *Main-Line Diesel Locomotive Diagrams, September 1961*)

5.2.3 D8588–D8616 (where different from D8500-85)

Main Generator: Two Crompton Parkinson CG 1086A1, 4-pole, self-ventilated.
Continuous rating: 305V, 900A at 1500rpm.

Auxiliary Generator: Two Crompton Parkinson CAG 1087A1, self-ventilated.
Continuous rating: 110V, 310A at 1500rpm.

Traction Motors: Four Crompton Parkinson C 1066A1, force-ventilated, 4-pole, nose-suspended, single-reduction gear drive, 13:81 gear ratio.
Continuous rating: 160hp, 305V, 350A at 670rpm.

Performance: Maximum Tractive Effort: 43,000lbs at 28.2 per cent adhesion, at 1,740A each main generator.
Continuous Tractive Effort: 18,400lb at 12.6mph at 900A each main generator.
Rail hp at Continuous Rating: 618hp
Full engine output available between 3 and 60mph.

Braking: Air for locomotive, vacuum for train giving a brake force of 88 per cent of locomotive weight in working order.

Multiple Working: Blue Star.

D8616, Marylebone, 30 April 1965.
The two-tone green livery is shown to good effect. Western Region diesel-hydraulic D9541 is positioned behind the Clayton. (Colour-Rail)

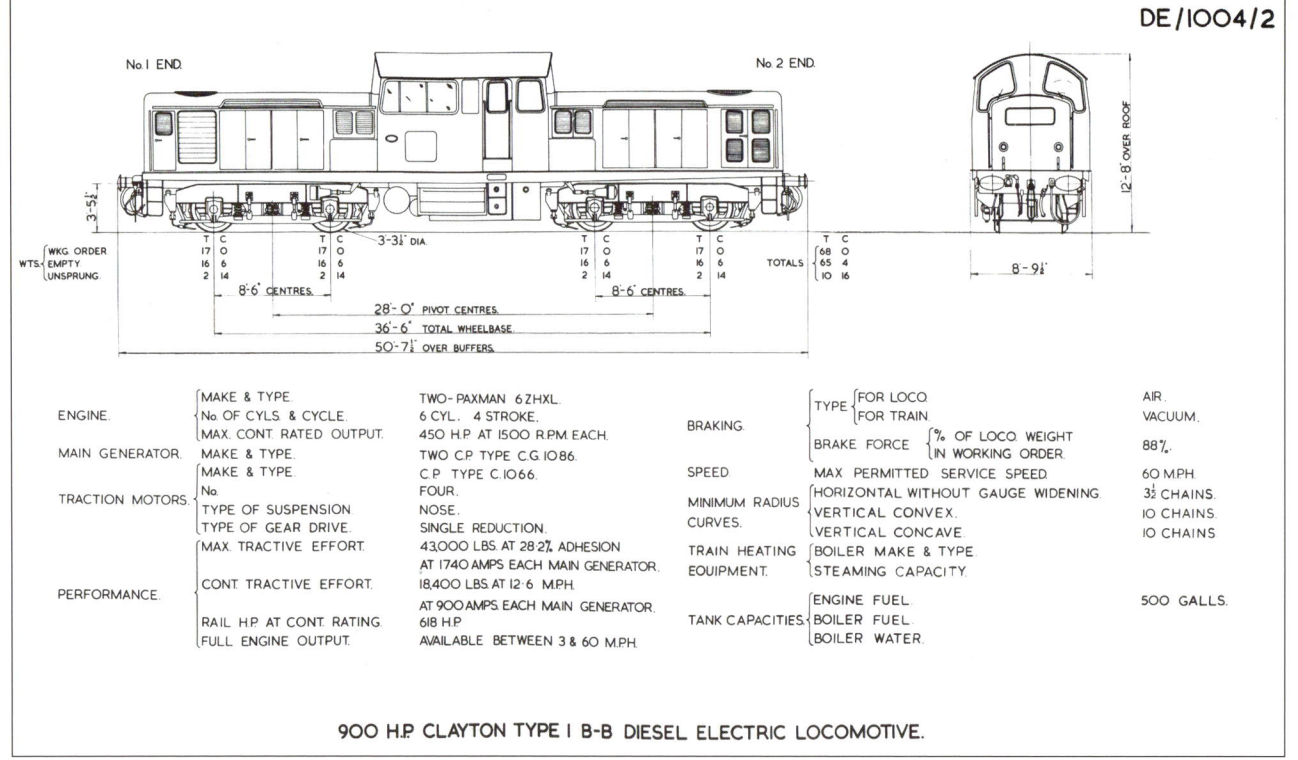

Line Drawing DE/1004/2 for D8588-D8616. Paxman engines, Crompton Parkinson electrical equipment. (BR *Main-Line Diesel Locomotive Diagrams, September 1961*)

ENGINE.	MAKE & TYPE.	TWO-PAXMAN 6ZHXL.		BRAKING.	TYPE {FOR LOCO. / FOR TRAIN.	AIR. / VACUUM.
	No. OF CYLS. & CYCLE.	6 CYL. 4 STROKE.			BRAKE FORCE {% OF LOCO. WEIGHT IN WORKING ORDER	88%.
	MAX. CONT. RATED OUTPUT.	450 H.P. AT 1500 R.P.M. EACH.		SPEED.	MAX PERMITTED SERVICE SPEED.	60 M.P.H.
MAIN GENERATOR.	MAKE & TYPE.	TWO C.P. TYPE C.G. 1086.		MINIMUM RADIUS CURVES.	HORIZONTAL WITHOUT GAUGE WIDENING.	3½ CHAINS.
TRACTION MOTORS.	MAKE & TYPE.	C.P. TYPE C.1066.			VERTICAL CONVEX.	10 CHAINS.
	No.	FOUR.			VERTICAL CONCAVE.	10 CHAINS.
	TYPE OF SUSPENSION	NOSE.		TRAIN HEATING EQUIPMENT.	BOILER MAKE & TYPE.	
	TYPE OF GEAR DRIVE.	SINGLE REDUCTION.			STEAMING CAPACITY.	
	MAX. TRACTIVE EFFORT.	43,000 LBS. AT 28·2% ADHESION AT 1740 AMPS EACH MAIN GENERATOR.		TANK CAPACITIES.	ENGINE FUEL.	500 GALLS.
PERFORMANCE.	CONT. TRACTIVE EFFORT.	18,400 LBS. AT 12·6 M.P.H. AT 900 AMPS EACH MAIN GENERATOR.			BOILER FUEL.	
	RAIL H.P. AT CONT. RATING.	618 H.P.			BOILER WATER.	
	FULL ENGINE OUTPUT.	AVAILABLE BETWEEN 3 & 60 M.P.H.				

900 H.P. CLAYTON TYPE I B-B DIESEL ELECTRIC LOCOMOTIVE.

Some Technical Aspects • 43

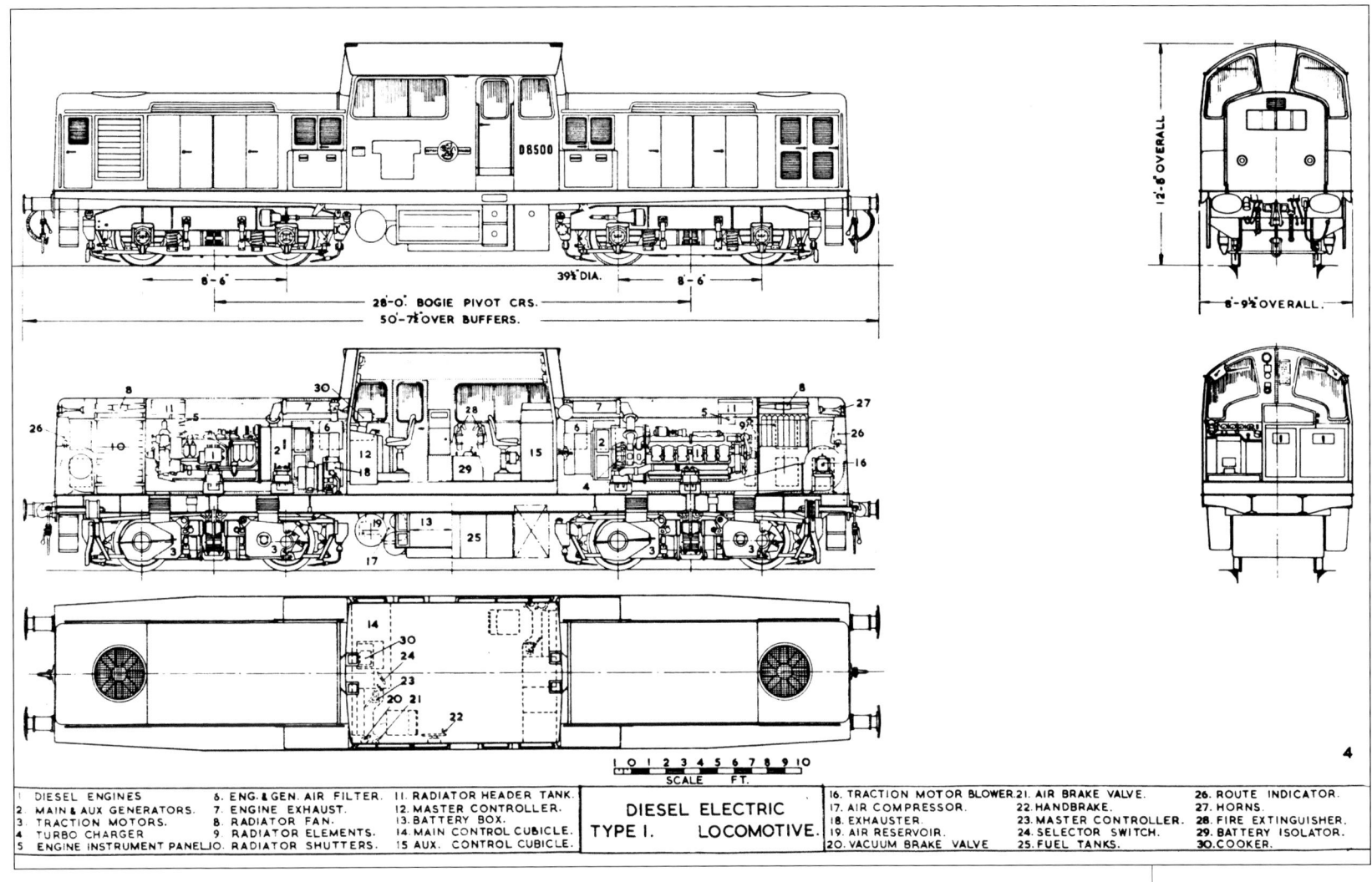

Layout Drawing for D8500-85. (B.R. *Main-Line Locomotive Layout Diagrams*)

5.3 Engines.
5.3.1 Paxman 6ZHXL Prototypes.
The website 'Richard Carr's Paxman History Pages' provides a succinct history of the Paxman 6ZHXL engine:

'Introduced in 1954, this was a horizontal or "flat" 6 in-line cylinder engine. Evolved from the YH, it was essentially half (i.e. one bank) of a Vee 12YH. The ZH was designed for British Railways primarily as an under-floor mounted power unit for railcars. Paxman was unsuccessful in persuading BR to adopt the engine for this application but large numbers of the pressure-charged version were supplied for the Class 17 "Clayton" Type 1 diesel-electric locomotive.'

Appendix B to a memorandum to the Technical Committee dated 15 September 1960 entitled *Standardisation of Main Line Diesel Electric Locomotives – Types 1 and 4* provided additional background:

<u>Experience with Paxman Engines.</u>
'Two prototype flat 6ZHXL engines of 450hp each were mounted under floor on two standard LMR coaches driving the road wheels through BTH electric transmission.

'Between September 1956 and August 1957, a distance of 38,000 miles was run in 900 engine hours, without trouble, and a complete dismantling and inspection at the end of the period indicated very good internal condition. Following this, further satisfactory running took place on the L.M. and E. Regions, bringing the total mileage to 43,000.

'In January 1958, the engines were dismantled from the vehicles and were returned to the firm, because Commission policy was to standardise BUT and Rolls-Royce engines for under-floor railcar propulsion, and it was not felt that an additional type of engine for this purpose could be justified.

'The above flat engine, and the same engines now offered for the Type 1 [Clayton] locomotives, have the same bore and stroke and belong to the same design series as the 16YHXL, modified only to permit horizontal operation. This latter engine of vertical type, is already in use on 42 [sic 54] Type 1 locomotives allocated to the Eastern Region . . .'

Diesel-Electric DMU fitted with two experimental under-floor Paxman 'flat' ZH engines, Derby, 14 October 1956. LMS coaches 9828 and 9821 built by BRCW in 1926. Whether these engines had cast-iron or aluminium cylinder heads and crankcases is unknown. (Courtesy and copyright Paxman Archive Trust)

5.3.2 Paxman Engines (D8500-85, D8588-D8616).

Each locomotive deployed two Paxman 6ZHXL turbo-charged 6-cylinder diesel engines rated at 450hp at 1,500rpm. This was a lightweight in-line, four-cycle, direct-injection, horizontal engine with a height of 33 inches. The engine was developed from Paxman's YHXL conventional upright engine as deployed in 16-cylinder form in the D82xx and D84xx locomotives; the ZHXL design was originally intended for under-floor installation in diesel railcars but was offered by Clayton as an above-frame solution to BR's centre-cab locomotive specification.

Cast-aluminium was used for the 6-cylinder monobloc; the cylinders were 7in bore with the pistons having 7¾in stroke. The cylinder heads were cast in light alloy.

Turbocharging was achieved by Napier equipment (one charger per engine)

The engine auxiliary drives were at the free (i.e. non-generator) end of the engine and taken through a torsionally flexible coupling. Engine speed was controlled by an Ardleigh type 303 pneumatic control governor. An independent electrically-driven pump was fitted for priming the lubricating system before starting.

Each engine/generator group was mounted on four low-frequency resilient mountings and

Paxman 6ZHXL diesel engine (with generator attached). From a brochure entitled 'Paxman High-Powered Flat Diesels for Rail Traction', dated June 1961. (Courtesy Paxman Archive Trust)

Drive end view of 6ZHXL engine.

PAXMAN HORIZONTAL DIESELS

The flat ZH engine belongs to the well established Paxman 7" bore range which in all parts of the world has for Rail Traction, Industrial and Marine applications a record of long life and economical trouble free operation borne out by an aggregate of millions of running hours.

Paxman 7" bore diesels are employed by British Railways in their 800 b.h.p. Type 1 locomotives for passenger and mixed traffic service and are also extensively used for industrial shunting duties. Both types of locomotives have a high earning capacity since little time is spent in the sheds on maintenance work, the running period recommended by this Company before a major overhaul being 18,000 hours under average conditions; in many cases this has been extended in practice to as much as 40,000 hours.

The ZH engine was specifically designed for Rail Traction purposes and other applications demanding a minimum overall height and incorporates experience gained in powering diesel-electric and hydraulic railcars and locomotives for British Railways and in over 30 countries overseas.

RATINGS AT 1,500 R.P.M.

6ZHL	330 b.h.p.
6ZHXL (Pressure-charged)	450 b.h.p.

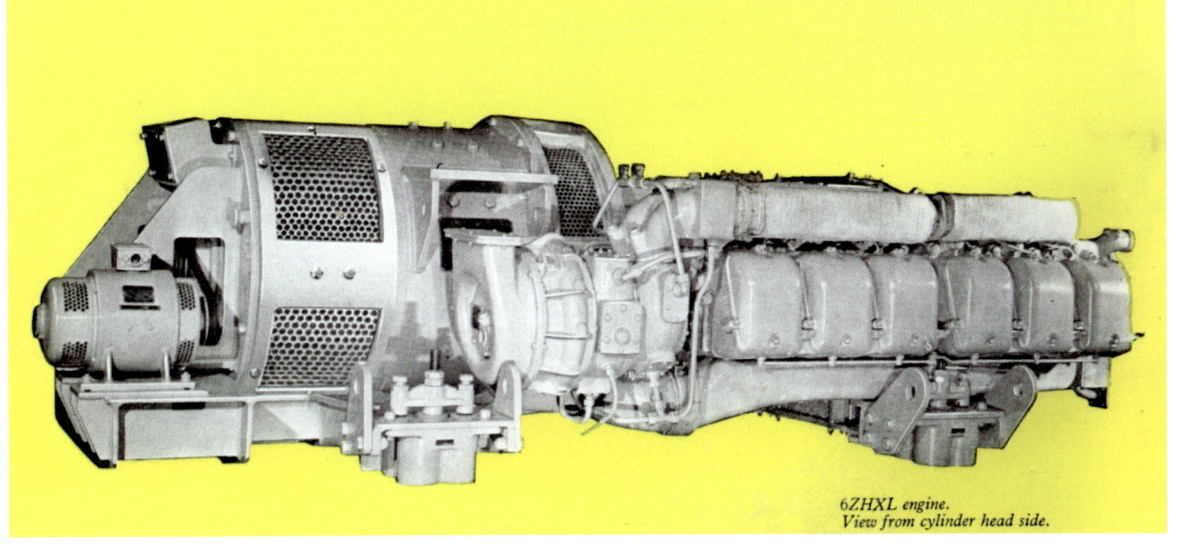

6ZHXL engine. View from cylinder head side.

Paxman ZH Publicity Photo Record Card dated March 1961.
(Courtesy and copyright Paxman Archive Trust)

had its own radiator and cooling equipment with a single radiator panel mounted in the bodyside. Each cooling group was of the Serck-Behr type in which the fan was driven hydrostatically and its speed regulated under thermostatic control. Under similar thermostatic control, shutters outside the radiator elements were opened and closed pneumatically. Air was drawn in over the radiator blocks through the left-side of the bonnet and expelled upwards through the bonnet roof.

Combustion air for the diesel engine was drawn from the generator compartment through oil-wetted filter panels mounted in the bonnet side doors.

The engine exhaust passed through stainless steel flexible bellows and mild steel ducting and was expelled through bifurcated outlets attached to the outer cab bulkheads.

One fuel tank was provided to supply both engines, but separate fuel lines were installed.

A 20-gallon radiator header tank was fitted together with a low-level switch which shut down the engine if loss of coolant occurred.

Some Technical Aspects • 47

(E1239) 6ZHXL ENGINES CONTRACT 56594 - 56769
Before leaving the works each Paxman engine is carefully inspected and tested in one of these test shops. A test certificate is issued with every engine.

ZHXL engines on test at Paxman's Works, Colchester. The hand-written comment 'Contract 56594-56769' refers to 176 engines built for D8500-87. (Courtesy and copyright Paxman Archive Trust)

5.3.3 Rolls-Royce Engines (D8586/7). Archive information indicates that the initial order for 88 Clayton locomotives when authorised in December 1959 were all to be built with Paxman engines.

The only subsequent archive material which sheds any light on the substitution of Rolls-Royce engines on the final two locomotives of the original order is a document which recorded *New Locomotives from Contractors*, produced by the CM&EE Department, Glasgow. The report for four weeks ending 13 June 1964 carried the comment 'Delivery of last two locomotives now altered to September 1964 due to fitting of alternative power unit'. This change was presumably to test an alternative to the Paxman engine which was already exhibiting serious problems in 1963 (see Section 11).

When first introduced in 1962, the Rolls-Royce DV8 8-cylinder V-form engine was applied to marine auxiliaries; D8586 and D8587 were fitted with the turbocharged DV8T version. To comply with the traction equipment, the engine was set at 450hp at 1500rpm, compared with a potential maximum rating of 534hp at 1800rpm.

The engine housing was a monobloc cast-iron structure combining the crankcase and cylinder blocks with the gear case at the rear as an integral part of the crankcase. Side cover plates were fitted to both sides of the crankcase to facilitate inspection and the removal of the connecting rod big-ends.

The separate cylinder heads each incorporated two inlet and two exhaust valves, operated by pushrods from a single camshaft. Aluminium alloy pistons were cooled by a metered supply of oil through the connecting rods. Turbocharging was achieved by a Holset turbocharger mounted at the rear of each engine.

To accommodate this V-engine the engine compartment had to be raised slightly, but otherwise the mechanical and electrical parts, together with traction characteristics, remained unchanged.

5.4 Electrical Equipment.
5.4.1 GEC (D8500-D8587).
The main generator (two per locomotive) was a self-ventilated GEC Type WT800 single bearing machine with a continuous rating of 274kW. The two generator shells were flange-mounted on to each engine crankcase. The two auxiliary generators, GEC Type WT763 with a continuous rating of 35kW, were flange-mounted on the end of each main generator. Ventilation of both the main and auxiliary generators at each end of the locomotive was by fans mounted at the driving end of the armature.

The four force-ventilated series-wound traction motors were GEC Type WT421 with a continuous rating of 157hp at 480rpm. These motors were axle-hung on white-metal sleeve bearings with 15:66 reduction gearing. The traction motors were arranged permanently in parallel with two weak-field steps. Transition from full to weak-field was initiated automatically in steps at any engine load setting without increasing engine speed. Backward transition from weak-field to full-field was one step only.

5.4.2 Crompton Parkinson (D8588-D8616).
The second order for the Clayton design incorporated Crompton Parkinson electric equipment, as follows:

- Main Generator: 2 x Type CG1086
- Auxiliary Generator: 2 x Type CAG1087
- Traction Motors: 4x Type C1066, nose-suspended, single-reduction.

The Crompton Parkinson electrical equipment resulted in slightly different traction performance characteristics as illustrated on the Line Drawing DE/1004/2 above.

Paxman ZH Publicity Photo Record Card dated October 1963. Crompton Parkinson generator attached. The Contract Nos. 57349-57406 relate to the 58 engines built for D8588-D8616. Note the incorrect spelling of Beyer Peacock. (Courtesy and copyright Paxman Archive Trust)

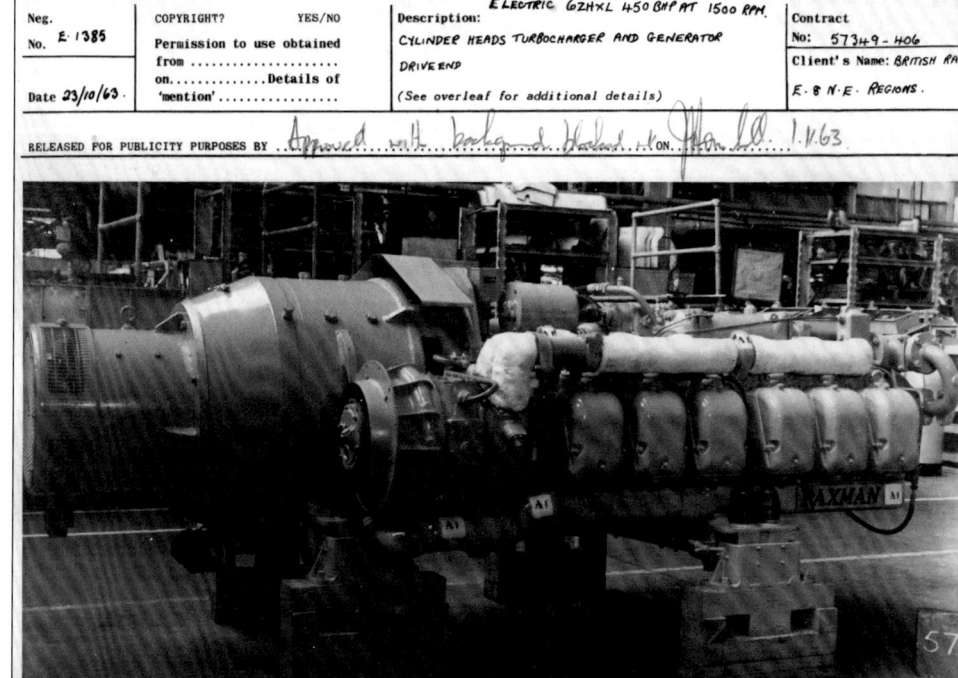

5.4.3 Power Control.

The Clayton fleet offered several generator/traction motor combinations for the driver; this situation applied irrespective of the type of equipment installed.

Both engines ran independently and could be operated singly or in tandem. When running as a pair, both engines were controlled simultaneously. Each engine drove a generator, which delivered power to two traction motors. Under 'normal' conditions, therefore, the two engine/generator sets combined to supply the total power of the locomotive. Either set could be shut down and isolated, leaving the locomotive to run on half power. In addition, provision was made for switching all four traction motors onto either generator; the consequent series-parallel grouping of the four motors enabled the maximum possible starting tractive effort. The choice of the various power combinations was made with a Traction Power Selector switch located on top of the No.1 end control cubicle.

Power selector positions:

- Both engine/generator sets shut down; traction motors isolated. No power.
- No.2 engine/generator set shut down; Nos.3 & 4 traction motors isolated. Nos.1 & 2 traction motors connected in parallel across No.1 generator. Locomotive on half power.
- No.1 engine/generator set shut down; Nos.1 & 2 traction motors isolated. Nos.3 & 4 traction motors connected in parallel across No.2 generator. Locomotive on half power.
- Nos.1 & 2 traction motors connected in parallel across No.1 generator, and, Nos.3 & 4 traction motors connected in parallel across No.2 generator. Normal operating configuration; full power available.
- No.2 engine generator shut down; all 4 traction motors connected in series-parallel across No.1 generator. Locomotive on half power, but higher starting tractive effort; fuel saving during shunting operations.
- No.1 engine generator shut down; all 4 traction motors connected in series-parallel across No.2 generator. Locomotive on half power, but higher starting tractive effort; fuel saving during shunting operations.

The switch positions could be used to allow the locomotive to travel under its own power in the event of the failure of an engine, a generator or a traction motor.

At the No.1 end of the cab is the main power control cubicle and at No.2 end the auxiliary control cubicle. Doors across the full extent of the front panels provided access from the cab to the equipment in the auxiliary control cubicle and to the majority of the equipment in the main control cubicle. Access to the other equipment in the main power control cubicle was through the adjacent generator compartment.

5.5 Underframe.

The underframe was designed to fully support the equipment above and below footplate level and to withstand an end buffing load of 200 tons without permanent deformation.

The frame was of all-welded steel construction. The main longitudinals were deep-section beams extending the full length of the locomotive and braced with cross-stretchers of plate and channel. The sole bars were 'fish-bellied' for appearance only giving very limited additional strength to frame structure. The underframe top was completely covered with continuously welded deck-plate, preventing oil penetration to traction motors, cables, etc. Any leaking or spilt oil was designed to drain into large trays under the engines and sumps built into the under-frame.

Jacking and lifting brackets were built into the under-frame slightly forward of the bogie pivot centres and the whole of the locomotive, including bogies, could be lifted from these points.

A large proportion of the brake piping was mounted in the under-frame together with the carriage-heating through steam pipe. Provision was made on the under-frame for the attachment of a standard BR snowplough.

The 500-gallon fuel tank and lead-acid batteries were under-slung from the underframe beneath the cab.

5.6 Equipment Layout and Superstructure.

The locomotive was symmetrically arranged about the transverse centre line with the two power units positioned fore and aft of the centre-cab.

The superstructure was divided into three main sections: No.1 equipment casing, cab, and No.2 equipment casing. Each equipment casing was divided into three compartments: the radiator section; the engine compartment; and the generator compartment.

Each radiator section was a sealed compartment and housed a radiator, fan, fan ducting (for engine cooling, as described above), route indicator and a traction motor blower. Air for the traction motor blower was drawn in through oil-wetted filter panels in the radiator compartment side doors (on the right-hand side as viewed from each cab) and blown through steel ducting and flexible moulded rubber bellows into the traction motors. This section was separated from the engine compartment by a bulkhead.

Each engine compartment contained either a Paxman or Rolls-Royce diesel engine.

Each generator compartment was sealed from the engine room to prevent recirculation of warm air and the ingress of oil mist and dirt. The generator cooling air was drawn in through body side grilles positioned on each side of the bonnet immediately ahead of the cab and discharged into the engine compartment, thereby pressurising it, and the engine air flow then passing to atmosphere through the long thin louvres fitted in the engine compartment bonnet roof.

A three-way 'pneuphonic' horn was mounted in each nose compartment; this enabled the driver to sound high and low notes on his front facing horn and a high note on the rear facing horn. A four-character route indicator was fitted at each end of the locomotive, access to the winding handles being via the radiator compartment.

The single centre-cab was built of steel sections and plates and lined with fibreglass panels and hardboard. The space between the interior and exterior was filled with heat and sound absorbing material. Large armour-plate glass occupied most of the front and rear cab walls above bonnet level, and large sliding side windows were provided adjacent to the driving position.

The cab had two identical diagonally-opposed driving positions. No secondman seats were provided.

5.7 Bogies.

The bogie design was of conventional construction fabricated from rolled steel sections and plate, with SKF roller bearing axle boxes working in guides and linked by under-slung equalising beams which fitted into stirrups cast integrally with the axle boxes. A dished centre pivot enabled the bogies to adjust themselves to the changing gradients when working over humps in marshalling yards. The bogies were to all intents and purposes identical to those provided by Clayton Equipment for the BTH D82xx locomotives

No.1 End Driving Position. Note the Traction Power Selector switch located on top of the main control cubicle (see Section 5.4.3). (Rail-Online)

Clayton bogie as deployed on both the BTH and Clayton Type 1s. (Rail-Online)

5.8 Train Heating.

The design provided for the installation of a Spanner train-heating boiler and although it was proposed that about twenty locomotives were to receive boilers, in practice no locomotive ever carried such equipment. A section of the cab roof was removable to allow the boiler to be installed or removed if so required.

The space allotted for the boiler in the centre of the cab explains the large dimensions of the centre-cab and the extra length of the Claytons (50ft 7½in, compared with 42ft 0in and 46ft 9⅜in for the BTH and English Electric Type 1 locomotives respectively).

The first batch of Claytons, D8500-87, were fitted with through train heating steam pipes. Of the second batch only the Yorkshire-allocated locomotives were fitted with through steam pipes (D8604-16). The Beyer Peacock 'Maintenance Instructions' specifically state: 'Carriage Heating Train Pipe fitted to Locomotive Nos. D8604 to D8616 only'.

5.9 Multiple Working.

Connections for multiple-working were fitted to the buffer beam enabling up to three locomotives to be controlled from the lead locomotive.

The first 88 Claytons had GEC electrical equipment with stepped control of engine speed; with a 'Red Diamond' multiple working restriction, they could not work in multiple with other classes. The later Beyer Peacock batch of 29 locomotives had Crompton Parkinson electrical equipment with continuous pneumatic control of engine speed replacing the earlier 'notched' system; their 'Blue Star' multiple-working system allowed coupling with several other diesel classes.

5.10 Locomotive Erection.

M. Alden, in his article in *Classic Diesels & Electrics* (Issue No.2) described the locomotive erection process:

'The Record Works at Hatton consisted of three main shops. No.1 Shop was responsible for manufacturing of the main frame; No.2 for the bogies and running gear and No.3 for the superstructures forming the cab and both bonnets.

'Once all three sections were completed, they were then transferred by road to the works of International Combustion at Derby, where a purpose-built erecting shop with cranes and storage areas had been made ready to assemble the locos. This shop was entirely under the management of Clayton. Here the locomotives were built and the fitting-out with major components, engines, generators and other electrical equipment was carried out.'

5.11 The Clayton Conundrum: Which is No.1 End?

On the footplate area immediately ahead of each corner of the cab were some 'boxes' (one in each corner roughly the size of a large tool-box, four in total).

The two boxes adjacent to No.1 end cab both contained Northey Type 125RE vacuum-brake exhausters, whereas at the No.2 end the left-hand box (from the cab looking forward) contained the Control Reservoir Tank. The right-hand box containing the Weak-Field Resistance equipment; uniquely, the side of this latter box included a grille to facilitate cooling of the resistances.

So, when viewed from the side, if the right box had a grille, then the locomotive No.1 end was to the left and No.2 to the right. If there was no grille visible on either box, then the No.1 end was to the right and No.2 to the left.

The Claytons bucked convention with respect to the rule that No.1 end always contained the radiator compartment. In the sense that the Claytons were effectively two locomotives in one, each locomotive had a radiator compartment at each end, albeit on one side only. Thus, whichever side of the locomotive was viewed, the large radiator grille (with ten rotating slats) could be seen at the left-hand end; the four small fixed grilles at the right-hand end were the air intake grilles for the traction motor blowers (again applicable to both sides of the locomotive).

Quite whether the No.1 end of the Claytons was officially determined by the positioning of the Northey exhausters is highly doubtful! More likely is the positioning of the main and auxiliary control cubicles located in the cab, with the main control cubicle (with the Traction Power Selector Switch) defining the No.1 end and the auxiliary cubicle No.2 end; or, alternatively, No.1 end was defined by the position of the parking brake wheel in the cab.

Above left: **D8568, 11 October 2019.** No grille on box on footplate. No.2 end to left, No.1 to right.

Abovce right: **D8568, 14 October 2019.** Grille on box on footplate. No.1 end to left, No.2 to right.

D8568, Nene Valley Railway, Wansford, 11 October 2019. No.1 end nearest. Compare this picture with the one below; note that the large radiator grilles are at diametrically opposite corners of the locomotive. (Anthony Sayer)

D8568, Nene Valley Railway, Wansford, 14 October 2019. No.2 end nearest. (Anthony Sayer)

Chapter 6
APPEARANCE, DESIGN AND STYLING

There is remarkably little archive material on the design and styling of the Clayton locomotives, particularly when compared with other classes. It is certainly clear that a very strong working relationship developed between the nominated design consultant and the BR and Clayton Equipment management teams which undoubtedly made at least this part of the design workload a relatively slick and straightforward process.

Work on the appearance design of the Clayton design commencing in May 1961. As with the 'Pilot Scheme' designs, BR employed an external designer to oversee the work; such was the considered success of the BTH Type 1 that BR looked to recruit Mr John Barnes

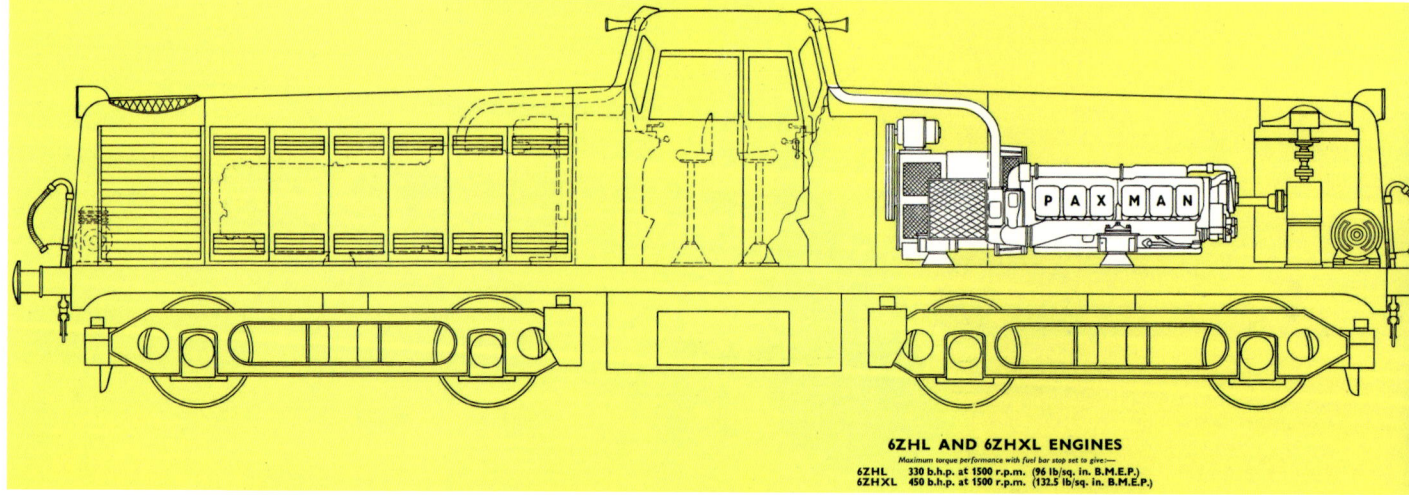

Illustration from 'Paxman High-Powered Flat Diesels for Rail Traction' brochure dated June 1961. The locomotive diagram has presumably been supplied to Paxman from Clayton Equipment and clearly pre-dates the employment of John Barnes, the Design Consultant. Points of interest:

- **Short** 'steeple' cab, prior to the stipulation to include a train-heating boiler within the cab.
- **Radiator** fan driven off the free end of the Paxman engine.
- **Exhausters** in the radiator compartment and no 'boxes' on the footplate in front of the cab ends.
- **Slightly** sloping bonnets.
- **Headcode** boxes perched on top of the bonnet ends.

(Courtesy Paxman Archive Trust)

Appearance, Design and Styling • 55

from Messrs. Allen-Bowden Ltd once again for the styling work on the Clayton fleet. As Mr G. Williams (BTC Design Officer) indicated in a letter to Mr F.D. Hollister of Allen-Bowden Ltd dated 09 May 1961:

'The styling of the Type 1 BTH locomotives was considered one of the more successful achievements so far as single cab locomotives are concerned, and we consider it appropriate and desirable that Mr John Barnes should deal with the Clayton locomotives.'

Discussions were successful and Barnes became the Design Consultant. Barnes had his first meeting with Clayton Equipment on 30 June 1961. On the basis of engineering design, information obtained and the high-level design parameters (i.e. centre-cab, low bonnets, etc) Barnes commenced work producing initial design drawings.

A Paxman brochure produced in June 1961 included two schematics of the Clayton design, one illustrating the positioning of the flat engines and other equipment within an early body design and another (almost invisible) sketch on the back cover of the brochure showing an interim stage of appearance design development. The two schematics are reproduced on page 54 and below.

Barnes' initial sketches were discussed with Mr G. Williams on

Artist's impression of a Clayton locomotive; heavily manipulated in Photoshop to bring out the almost invisible sketch on the back cover of the Paxman brochure dated June 1961. Key features:

- Large cab capable of accommodating a train-heating boiler.
- Cab with very small hoods over the enlarged driving windows.
- Single (un-bifurcated) exhaust pipe positioned to one side of a small central cab window.
- Horizontal bonnet tops.
- Radiator grilles on all corners
- Grilles on all bonnet side access doors.
- High-level diminutive head-code boxes.
- No footplate 'boxes' ahead of the cab fronts suggesting retention of the exhausters within the radiator compartment.

(Courtesy Paxman Archive Trust)

18 August 1961. Williams discussed Barnes' proposals with Mr E.S. Cox (Assistant Chief Mechanical Engineer) on 25 August who, in turn, debated the subject with Clayton Equipment personnel on 1 September.

The outcome of these meetings, notified to Barnes on 7 September, was general acceptance of his ideas including some of what became the more 'iconic' elements of the Clayton design i.e:

- An enlarged roof 'peak' covering the top of the exhaust casing.
- The exhaust casing (now containing bifurcated exhaust pipes) between two cab windows obviating the need for the small centre window.
- Cab windows equipped with bright metal frames.
- Long thin ventilators above engine compartment side access doors, which replaced most of the access door grilles.

However, Barnes' suggestion to modify the 'curious' shape of the cab front windows was not accepted and the original Clayton proposal for a lowering of the glass between the bonnet casing and the screen pillar was upheld by Cox. As Williams later explained in a letter to Barnes on 25 September 1961:

'The windscreen is I am afraid *fait d'accompli*, the firm [Clayton Equipment] seems to have convinced Mr Cox that the drop ends are necessary for better visibility.'

To improve the general appearance of the locomotive, Barnes wanted to decrease the height of the cab by 5 inches i.e.

D8500, Cricklewood (en route to Marylebone), 26 July 1962. Note the lack of Red Diamond coupling code signs on the yellow panel. (Transport Treasury)

to decrease the rather excessive depth of the driving window at the centre. Cox responded to Barnes' suggestion in a letter to Williams dated 2 October 1961 as follows:

'There is a technical reason why the roof height of the above locomotives should be made as high as possible. Because the [train-heating] boiler is enclosed in the casing inside the cab, as much space above it as can be arranged is necessary for the adequate operation of the ventilation arrangements, and for clearing away the pocket of hot air which will tend to linger above the boiler during operation.

'This matter was considered very carefully during the inspection of the mock-up at the Clayton Co's Works [on 1 September 1961], and the decision then given that the cab should remain at 12' 8" high.'

Barnes was notified accordingly. A train-heating boiler was never actually fitted to any locomotive, but this component still managed to exert a very strong influence the final design.

A Clayton Equipment Co. Ltd. Drawing No. WL-01-263 dated 13 December 1961 included the following specific modifications:

Mod.A   10/08/61 'Bottom line of cab (*front*) windows lowered.'
Mod.B   05/09/61 'Cab roof altered. Louvres in engine comp't roof added. Louvres in doors removed.'
Mod.C   27/11/61 'Cab handrails mod'd. Cab door kicking strip added. Solebar cut-out mod'd.'

In a 'final' letter from Williams to Barnes on the subject, dated 27 July 1962, Williams commented:

'I saw the "Clayton" locomotive today for the first time at its official viewing at Marylebone station. I do congratulate you on the work which you have put into this. The whole unit was very much admired by those present and it is clear that the very careful detailing over which you have taken so much care has been well worthwhile.'

D8574, 64B Haymarket, 7 January 1968. In my humble opinion the Claytons looked as attractive in blue livery is it did in two-tone green. The Red-Diamond coupling code is still retained at this point. (John Grey Turner)

Chapter 7
DELIVERY AND ACCEPTANCE TESTING

7.1 Delivery Dates.
See Section 10.

7.2 Deliveries: Plan v. Actual.
7.2.1 Memorandum to the Works & Equipment and Supply Committees, 13 February 1963.
Progress of Rolling Stock Construction - Year 1962.
Deliveries – 52 weeks ended 29/12/62:
Estimated in January 1962   32
Estimated in September 1962   32
Actual   14

Remark: 'Engine/generator coupling problems necessitated placing an embargo on delivery of locomotives on 8th December pending solution of problems.'

7.2.2 Summary of Reports from the Sponsoring Region.
Output of New Locomotives Built or Sponsored by L.M. Region (from CM&EE Department, Derby).
See table on page 59.

Reasons given for shortfalls:

4we14/07/62 and
4we11/08/62 reports:-

'Firm (*Clayton Equipment*) anticipate delivering first locomotive on 25/8. There has been a delay at Paxman's with type testing of the engines and generators, also when Paxman sent first (*engine/generator*) coupling discs to GEC they were found to be out of dynamic balance and convex on one face and had to be balanced and re-machined . . .'

4we08/09/62 report:-

'Firm state difficulty is being experienced by Paxman with over-speed trips . . .'

4we06/10/62 and
4we03/11/62 reports:-

'Difficulty experienced by Paxman's with over-speed trips has been solved . . .'

4we01/12/62 report:-

'Difficulty with engines and generators . . .'

4we29/12/62 report:-

'Difficulty with engines and generators . . .

'At a meeting here on 7/1 between LMR and Clayton's representatives it was agreed that LMR should wait for Clayton's proposals for modifications following meeting to be held between Clayton, GEC and Paxman representatives on 10/1 before taking any action to lift the embargo on deliveries. There are now 8 locos complete, which if released would give 22 revenue-earning whilst modifications to be agreed, following meeting referred to, were put into effect commencing on the 23rd locomotive. No further locomotives would be delivered without modifications and a programme for the return of the 22 unmodified would be worked out later.'

4we26/01/63 report:-
As 4we 29/12/62, plus:-

'A further meeting between Clayton and their sub-contractors was held on the 25th January; modifications were put forward but agreement has not yet been reached and the embargo on deliveries still stands.'

Delivery and Acceptance Testing • 59

Output of New Locomotives Built or Sponsored by L.M.Region (from CM&EE Department, Derby)

Period Ending	Delivered Actual	Delivered Cumulative	Target Cumulative	Locomotives Delivered
4we 14/07/62	0	0	2	-
4we 11/08/62	0	0	6	-
4we 08/09/62	0	0	11	-
4we 06/10/62	2	2	16	D8500/2
4we 03/11/62	7	9	17	D8501/3-8
4we 01/12/62	3	12	21	D8509-11
4we 29/12/62	2	14	25	D8512/3
4we 26/01/63	0	14	30	-
4we 23/02/63	0	14	35	-
4we 23/03/63	0	14	40	-
4we 20/04/63	2	16	44	D8516/8
4we 18/05/63	4	20	49	D8520/9-31
4we 15/06/63	3	23	53	D8534/7/9
4we 13/07/63	4	27	58	D8517/22/38/41, plus D8514/23 added in pencil.

4we23/02/63 report:-

'A meeting between CM&EE representatives and Paxman was held at firm's works 13/3. Difficulty appears to have been resolved and it is anticipated that deliveries of locomotives will re-commence shortly.'

4we23/03/63 report:-

'Proposed by Clayton Equipment Company to change engines and generators on all locomotives. Engines put up for test are being rejected because of high oil lubricating consumption . . .'

4we20/04/63 (*Deliveries recommenced*), 4we18/05/63 and 4we15/06/63 reports:-

'Please see . . .(*correspondence*) . . . concerning cracked crankshafts and the decision to return all the engine/generator sets for modification in all the locomotives so far completed.'

7.2.3 Summary of Reports from the Receiving Region.
New Locomotives from Contractors (from CM&EE Dept, Glasgow).
See tables on page 60. N.B. Reports for 1965 (and, therefore, D8587) not seen.

Reasons given for shortfalls:

4we06/10/62 to 4we26/01/63:

'Production difficulties experienced by Contractors.'

4we23/02/63, 4we23/03/63:

'Delivery of locomotives suspended pending Contractors investigating vibration in engine/generator power units and also crankshaft fracture in certain locomotives.'

4we20/04/63 to 4we16/05/64:

'Shortfall due to delays in Maker's Works caused mainly by investigations into excessive vibration in engine/ generator power units and engine crankshaft fractures.'

4we13/06/64

'Delivery of last two locomotives now altered to September 1964 due to fitting of alternative power unit.'

4we11/07/64 to 4we28/11/64

'Delivery of last two locomotives delayed due to fitting of alternative power unit.'

4we26/12/64

'Delivery of final locomotive delayed due to fitting of alternative power unit.'

New Locomotives from Contractors (from CM&EE Department, Glasgow)

Period Ending	Delivered Actual	Delivered Cumulative (1962)	Target Cumulative (To date)	Full Year Target (1962)	Locomotives Delivered	Comments
4we 08/09/62	0	0	0	32	-	
4we 06/10/62	2	2	11	32	D8500/2	
4we 03/11/62	7	9	16	32	D8501/3-8	
4we 01/12/62	3	12	20	25	D8509-11	Cumulative/Full Year targets reset.
4we 29/12/62	2	14	27	27	D8512/3	Cumulative/Full Year targets reset.

Period Ending	Delivered Actual	Delivered Cumulative (1963)	Target Cumulative (To date)	Full Year Target (1963)	Locomotives	Comments
4we 26/01/63	0	0	30	74	-	Full Year Target set to complete 88 loco order
4we 23/02/63	0	0	35	74	-	
4we 23/03/63	0	0	40	74	-	
4we 20/04/63	2	2	45	74	D8516/8	
4we 18/05/63	4	6	49	74	D8520/9-31	
4we 15/06/63	3	9	53	74	D8534/7/9	
4we 13/07/63	6	15	58	74	D8514/7/22/3/38/41	
4we 10/08/63	8	23	63	74	D8515/9/21/4/8/32/3/5	
4we 07/09/63	6	29	67	74	D8525-7/36/40/2	
4we 05/10/63	6	35	46	59	D8543-8	Cumulative/Full Year targets reset.
4we 02/11/63	7	42	51	59	D8549/51-6	
4we 30/11/63	7	49	55	59	D8550/7-62	
4we 28/12/63	4	53	59	59	D8563-6	

Period Ending	Delivered Actual	Delivered Cumulative (1964)	Target Cumulative (To date)	Full Year Target (1964)	Locomotives	Comments
4we 25/01/64	6	6	88	21	D8567-72	Full Year Target set to complete 88 loco order. Target Cumulative (To Date) reset to original target.
4we 22/02/64	5	11	88	21	D8573-7	
4we 21/03/64	3	14	88	21	D8578-80	
4we 18/04/64	3	17	88	21	D8581-3	
4we 16/05/64	2	19	88	21	D8584/5	
4we 26/12/64	1	20	88	21	D8586	

7.2.4 'Diesel Train Locomotive Performance' (Report by Chief Mechanical Engineer), 20/08/63.

Appendix D. B.R. Standard Main-Line Diesel Locomotives.

Type	hp	Engine	Transmission	Total no. ordered	No. delivered as at 15/06/63
Type 1 D/E	900	Paxman Flat	Electric	117	23

Remark: 'Worked satisfactorily after initial trouble with whirling armature shaft causing engine crankshaft failures – now cured.'

7.2.5 Memorandum from Chief Mechanical Engineer to Chief Operating Officer, 18/09/64.

Diesel Train Locomotive Deliveries.

'Clayton Type 1's: Two locomotives for the Scottish Region are being fitted with Rolls-Royce engines and will be delivered before the end of the year.

'Beyer Peacock Type 1's: A strike at the firms works delayed this contract initially, but all 16 locomotives for the North Eastern have now been completed and delivery of the Eastern Region's 13 locomotives has now commenced.

'From the above it will be seen that the only factors which could be said to have an effect on steam locomotive condemnations are the Type 4's and Type 1's for the Eastern Region . . .'

7.2.6 Claim for Late Delivery.

Late delivery of the early Clayton locomotives had a financial impact on BR both from the point of view of not realising the operational benefits of a diesel fleet compared with steam, and, having to maintain essentially worn-out steam locomotives to ensure motive power continuity whilst the Claytons were out use pending modification.

Discussions between BR, Clayton Equipment and Paxman regarding delivery delays and associated issues started in 1965 and were concluded in June1966 on the basis that BR's original claim for liquidated damages caused by late delivery of £45,000 was reduced to the actual loss of £25,000. It was recognised that:

'Clayton were not dilatory in carrying out their part of the work and did everything possible to expedite completion with the result that all the locomotives were (*ultimately*) delivered before the completion of delivery date.'

7.3 Acceptance Testing.

D8500-87: Once a locomotive was completed it undertook static load bank testing at the International Combustion Works, followed by main-line trials which involved hauling a 200-ton train from Derby to Chinley and return.

D8588-D8616: Main-line tested between Gorton and Derby via Chinley.

7.4 Delivery Route.

The D8500-87 batch were frequently noted passing Yorkshire on delivery from Derby to Scotland, often in pairs, on a Thursday lunchtime.

D8579, Location and date unknown.
(Author's Collection)

Chapter 8
ALLOCATIONS

8.1 Allocations – General Comments.

Compiling the allocation history of the Clayton fleet proved to be far from straightforward and it has been necessary to combine elements from various sources. By far the best sources were 'Diesel Locomotive Record Cards' (DLRC) and the official Stock Alterations published by the four Regions which operated the fleet.

However, as far the DLRCs are concerned, only 106 of the 117 have been unearthed. This, combined with the fact that the cards only covered allocations up to about 1966, meant that alternative sources were required to obtain a complete picture. Similarly, the regional Stock Alterations, whilst providing weekly information, were only found for the London Midland and Scottish Regions.

Additional sources perused include the Stock Alteration reports provided by the RCTS, SLS and LCGB. Whilst the information provided in these magazines was sourced from the official BR regional sources mentioned above, it is very clear that significant amounts of weekly information were aggregated into several weeks, and even months in some cases, leading to a significant loss of accuracy. In some cases, particularly for D8500-34 when allocated to the LMR, this aggregation caused the complete loss of some vital pieces of information; luckily the primary source LMR weekly stock alteration reports were available to ascertain the real picture in appropriate detail. On the plus side, however, the various societies aggregated their information into different time periods; as a consequence it has been possible to quote the overlapping period as the re-allocation time period, thereby reducing considerably the number of weeks associated with a particular re-allocation. Magazine reporting of Inter-Regional transfers of locomotive was frequently confusing.

Additional sources of information included Internal BR memos listing weekly storage of locomotives (these never being released into the public domain) and LMR 'Engine History Card' (EHC) information.

It is believed that the allocation history provided below is as detailed as can now be obtained. Certainly by combining the various available sources, a very comprehensive picture has been obtained, down to individual week or even actual date levels. Some caution is required, however, as it is clear in some cases that what appears to be an actual date is in fact a week-ending date. Inevitably, there were some reporting discrepancies between the various sources, usually regarding the actual date within a week-ending period; where this was the case (and where DLRCs are available) the DLRC information has been preferentially used.

8.2 Allocations – Numerical Order.

A full listing of locomotive allocations in numerical order are provided in Section 10. Depot codes used in this and subsequent Sections are as follows:

ScR	64B	Haymarket (Edinburgh),
	66A	Polmadie (Glasgow)
NER	51L	Thornaby (Teesside)
	52A	Gateshead
	52H	Tyne Dock
	56B	Ardsley (Leeds)
ER	41A	Tinsley (Sheffield)
	41E	Barrow Hill (Staveley) (diesel depot from 05/04/65)
LMR	12A	Carlisle Kingmoor (up to 16/06/68)
	D10	Preston Division (from 17/06/68)
o/l	On loan	
perm	Permanent	

New depot codes were applicable from 6 May 1973 (e.g. ED Eastfield, PO Polmadie, etc.); however, for clarity, the earlier codes have been used throughout.

8.3 Scottish Region Allocation History.

Year	Locomotives	Re-Allocations	Date	Net ScR Stock
1962/63	D8500-53	To 66A from New		
1963-65	D8554-87	To 64B from New		88
1966	D8604-16	41E to 64B	22/05/1966	101
1967	D8604/5	64B to 52A (on-loan)	12/06/1967	99
		64B to 52A (permanent)	03/12/1967	
	D8552/3	66A to 64B	12/06/1967	
	D8500-3	66A to 12A	13/08/1967	95
	D8552-4/6	64B to 66A	19/08/1967	
	D8555	64B to 66A	20/08/1967	
	D8556	66A to 64B	20/08/1967	
	D8556-9	64B to 66A	02/09/1967	
	D8504-6	66A to 12A	10/09/1967	92
	D8507-9	66A to 12A	08/10/1967	89
	D8512-6	66A to 12A	05/11/1967	84
	D8560	64B to 66A	03/11/1967	
	D8510/1/7-9	66A to 12A	03/12/1967	79
	D8520	66A to 12A	31/12/1967	78
1968	D8561	64B to 66A	16/01/1968	
	D8521-5	66A to 12A	28/01/1968	73
	D8562	64B to 66A	16/03/1968	
	D8563	64B to 66A	18/03/1968	
	D8526-32	66A to 12A	24/03/1968	66
	D8564/5/7	64B to 66A	04/05/1968	
	D8533/4	66A to 12A	19/05/1968	64
	D8566	64B to 66A	26/05/1968	
	D8537	Withdrawn (ex-66A (store))	21/07/1968	63
	D8568/9	64B to 66A	02/09/1968	
	D8570	64B to 66A	06/09/1968	
	D8502/15/25	12A to 64B	12/10/1968	66
	D8609/11	Withdrawn (ex-64B)	13/10/1968	64
	D8553	Withdrawn (ex-66A (store))	27/10/1968	
	D8575	Withdrawn (ex-64B (store))	27/10/1968	62
	D8584/5	Withdrawn (ex-64B (store))	29/10/1968	60
	D8502/15/25	64B to 66A	11/11/1968	
	D8570	Withdrawn (ex-66A (store))	24/11/1968	59
	D8566/9	Withdrawn (ex-66A)	04/12/1968	57
1969	D8582	Withdrawn (ex-64B)	05/01/1969	56
	D8571-3	64B to 66A	14/01/1969	

Year	Locomotives	Re-Allocations	Date	Net ScR Stock
	D8574/6	64B to 66A	27/01/1969	
	D8547/56/60/4	Withdrawn (ex-66A (store))	16/02/1969	
	D8577	Withdrawn (ex-64B (store))	16/02/1969	51
	D8578/9	64B to 66A	18-19/02/69	
	D8576	Withdrawn (ex-66A)	28/02/1969	50
	D8500/4/5/8/13	Reinstated to 66A	18/05/1969	
	D8544	Withdrawn (ex-66A)	18/05/1969	
	D8554/71/2/8	Withdrawn (ex-66A (store))	18/05/1969	50
	D8505	66A to 64B	27/07/1969	
	D8530	Reinstated to 64B	10/05/1969	51
	D8503/6	Reinstated to 66A	02/11/1969	
	D8507/10/6/28/9/31	Reinstated to 64B	02/11/1969	59
	D8516/31	64B to 66A	14/11/1969	
	D8561/3	66A to 64B	14/11/1969	
	D8565/79	66A to 64B	23/11/1969	
1970	Nil			
1971	D8530	Withdrawn (ex-64B)	03/03/1971	58
	D8500/35	Withdrawn (ex-66A)	27/03/1971	
	D8510, D8606	Withdrawn (ex-64B)	27/03/1971	54
	D8588/99, D8600/2/3	52A to 64B	25/04/1971	59
	D8607/8/10/2-6	64B to 66A	05/05/1971	
	D8594/7/8	52A to 64B	23/05/1971	62
	D8592/3, D8601/4	52A to 64B	06/06/1971	66
	D8506/31/43	Withdrawn (ex-66A (store))	21/09/1971	
	D8513/49/55/7, D8616	Withdrawn (ex-66A)	21/09/1971	
	D8583/6	Withdrawn (ex-64B (store))	21/09/1971	
	D8592/4	Withdrawn (ex-64B)	21/09/1971	54
	D8579/88/93, D8604	Withdrawn (ex-64B)	04/10/1971	50
	D8502/15/40/1/50/1, D8607/8/10/2/4/5	Withdrawn (ex-66A)	05/10/1971	
	D8528/99	Withdrawn (ex-64B)	05/10/1971	36
	D8503/16/25/38/9/42/5/6/59/62/7/8/73/4, D8613	Withdrawn (ex-66A)	06/10/1971	
	D8505/29/80/1/7/97/8, D8600-3	Withdrawn (ex-64B)	06/10/1971	10
	D8504	Withdrawn (ex-66A)	07/11/1971	
	D8561/5	Withdrawn (ex-64B)	07/11/1971	
	D8529/98	Reinstated to 64B	07/11/1971	
	D8574	Reinstated to 66A	07/11/1971	10

Year	Locomotives	Re-Allocations	Date	Net ScR Stock
	D8507/29/63/98	64B to 66A	06/12/1971	
	D8507/8/29/36/48/52/ 8/63/74/98	Withdrawn (ex-66A)	31/12/1971	0

Notes:-
1. Short-term trial/training re-allocations are excluded (e.g. D8500/1/36/45).
2. LMR (12A) transfers from Scottish Region in 1967/68 are based on the permanent transfer dates.
3. During the period when D8500-34 (35 locomotives) were transferred to 12A, D8552-67 (16 locomotives) were transferred from 64B to 66A.
4. Locomotives stored, then subsequently reinstated, in 1968/69 are excluded from the above analysis.
5. Table excludes the reinstatement and immediate withdrawal of D8506/30 in May 1969.
6. D8592 was possibly also withdrawn from stored status as implied by the Performance &Service Problems Meeting (8 September 1971).
7. Excluding periods in store and the short-term transfers of D8536 and D8545 to the NER and Leith respectively, D8535-51 were only ever allocated to 66A Polmadie.
8. John Hooper's Clayton book (2016) lists D8547 as re-allocated to Haymarket on 10/10/63 (on loan) reverting to Polmadie on 25/10/63, although I have found no 'official' evidence of these two re-allocations.
9. Excluding periods in store, D8575/7/80-7 were similarly only ever allocated to 64B Haymarket.

8.4 London Midland Region Allocation History.

Year	Locomotives	Re-Allocations	Date	Net LMR Stock
1967	D8500-3	66A to 12A	13/08/1967	4
	D8504-6	66A to 12A	10/09/1967	7
	D8507-9	66A to 12A	08/10/1967	10
	D8512-6	66A to 12A	05/11/1967	15
	D8510/1/7-9	66A to 12A	03/12/1967	20
	D8520	66A to 12A	31/12/1967	21
1968	D8521-5	66A to 12A	28/01/1968	26
	D8526-32	66A to 12A	24/03/1968	33
	D8533/4	66A to 12A	19/05/1968	35
	D8500-34	12A to D10	17/06/1968	
	D8500/1/4-9/11-3/6-24/6/8-31/3/4	Withdrawn (ex-D10)	12/10/1968	
	D8502/15/25	D10 to 64B	12/10/1968	5
	D8503/14	Withdrawn (ex-D10 (store))	02/11/1968	
	D8512/29	Reinstated to D10 (store)	02/11/1968	5
	D8512/29	Reinstated to D10	04/11/1968	
	D8510/2/27/9/32	Withdrawn (ex-D10)	28/12/1968	0

Notes:
1. Period on LMR: 13/08/67-28/12/68
2. LMR (12A) transfers from Scottish Region in 1967/68 are based on permanent transfer dates only.

8.5 Eastern Region Allocation History.

Year	Locomotives	Re-Allocations	Date	Net ER Stock
1964/65	D8604-15	To 41A from New		
1965	D8604-15	41A to 41E	05/04/1965	
	D8616	To 41E from New	23/04/1965	13
1966	D8604-16	41E to 64B	22/05/1966	0

Note:
1. Period on ER: 15/09/64-22/05/66.

8.6 North Eastern Region Allocation History.

Year	Locomotives	Re-Allocations	Date	Net NER Stock
1964	D8588-91	To 51L from New		
	D8592-9, D8600-3	To 52A from New		16
1965	Nil			
1966	D8588-91	51L to 52A	19/06/1966	
1967	D8604/5	64B to 52A (on-loan)	12/06/1967	18
	D8604/5	64B to 52A (permanent)	03/12/1967	
1968	D8601/2/4/5	52A to 51L	28/01/1968	
	D8599, D8600/3	52A to 51L	31/03/1968	
	D8605	Withdrawn (ex-51L)	28/10/1968	17
	D8591/5/6	Withdrawn (ex-52A)	22/12/1968	14
1969	D8599, D8600	51L to 52A	02/03/1969	
1970	D8601	51L to 52A	15/03/1970	
	D8602-4	51L to 52A	03/05/1970	
	D8589	Withdrawn (ex-52A)	06/07/1970	13
1971	D8590	Withdrawn (ex-52A)	27/03/1971	12
	D8588/99, D8600/2/3	52A to 64B	25/04/1971	7
	D8594/7/8	52A to 64B	23/05/1971	4
	D8592/3, D8601/4	52A to 64B	06/06/1971	0

Notes:
1. Period on NER: 20/03/64-06/06/71.
2. Periods at 51L: 20/03/64-19/06/66, 28/01/68-03/05/70.
3. Period at 52A: 12/05/64-06/06/71.
4. D8595/6 were only ever allocated to 52A Gateshead.
5. From 1 January 1967 the North Eastern Region was merged into the Eastern Region.

D8544, 66A Polmadie, 23 January 1968. (TOPticl Digital Memories)

D8558, 66B Motherwell, 9 September 1971. (Rail-Online)

D8587 and D8586, 64A St Margarets, 9 July 1966. Both Rolls-Royce Claytons together. (RCTS Archive)

D8587, 64B Haymarket, 1967.
(Rail-Online)

D8573, 64C Dalry Road, 7 April 1964.
(Colour-Rail)

Allocations • 71

D8584, Millerhill s.p., 7 June 1965. (Author's Collection)

Three Claytons and a Class 21, 61A Kittybrewster, Undated. Probably soon after May 1966 when D8604-16 were re-allocated to Scotland from Barrow Hill. Although allocated to 64B Haymarket, many of this batch gravitated to the Aberdeen area during the summer of 1966 (see Section 14.4). (Author's Collection)

D8598, 52A Gateshead, circa 1968. (Rail-Online)

D8603, 51L Thornaby, 15 June 1969. (Colour-Rail)

Allocations • 73

D8614, 41E Barrow Hill, 22 August 1965.
(Colour-Rail)

D8516, 12A Carlisle Kingmoor Diesel, 19 April 1968.
(TOPtical Digital Memories (Roger Hateley))

Chapter 9
OVERHAUL AND REPAIRS

9.1 Workshops - Allocation of Work.

In the early years, St Rollox Works was responsible for both Classified and Unclassified repairs for the Scottish fleet of Claytons (D8500-87), with Doncaster Works undertaking similar responsibilities for the Eastern and North-Eastern locomotives (D8588-D8616).

The idea of concentrating all Clayton repairs at St Rollox Works was discussed at the BRB Mechanical Engineering Committee on 23 July 1963 but was not pursued further at that time.

Based on the DLRCs, St Rollox Works started to carry out work on the ER/NER Claytons from mid-1966 with D8602 receiving Unclassified repairs between 06/07/66 and 19/07/66 and D8593 receiving similar repairs between 01/08/66 and 14/08/66.

The last three DLRC recorded Clayton repairs undertaken at Doncaster were D8603 (Classified 23/03/66-14/05/66), D8593 (Classified 05/05/66-03/06/66) and D8592 (Unclassified 12/05/66-30/07/66). A DLRC listed visit to Doncaster of D8592 during the period 15/06/67-23/06/67 is believed to be an error and should have been recorded as St Rollox.

With the transfer of D8604-16 from the Eastern Region to the Scottish Region en masse on May 1966, only 16 Claytons remained south of the Scottish border, as against 101 in Scotland. The decision to concentrate all Clayton shopping work in Scotland from mid-1966 onwards was therefore entirely logical.

Inverurie Works undertook some AWS fitment work during 1964/65.

The Polmadie and Perth Repair Bays also provided heavy repair facilities in support of the Main Works, although reference to this never appeared on the DLRCs.

9.2 Workshops - Repair Categories.

The BR Workshop Schedule for the full Clayton fleet (D8500-D8616) provides the following insights:

Amendment No.19, January 1968

'LIGHT attentions to these locomotives is no longer considered necessary.'

Amendment Nos.65-67, January 1970

'Classified repairs (INTERMEDIATE and GENERAL) are those which are carried out on a regular basis with the object of enabling the locomotive to remain in service for a pre-determined number of engine hours until the next Classified repair.

'Unclassified repairs are those which arise in consequence of a collision or accident or failure for any reason of a component which necessitates shopping the locomotive to rectify the defect.

'The periodicity of Workshops repair will vary from one Region to another, depending on various factors. Basically, however, shopping will be a function of the condition of the locomotive and the responsibility for deciding the periodicity of shopping should rest the Regional CM&EE. The following table is offered as a guide . . . .

Periods (engine hrs)	Body	Engines/Generators	Bogies	Traction Motors
12,000	Intermediate	General	Intermediate	Intermediate
24,000	General	General	General	Intermediate
36,000	Intermediate	General	Intermediate	General

Allocations • 75

'For Workshop Planning purposes, these periodicities should be converted into calendar months dependent on utilisation of the locomotive class, which may vary between Regions.'

9.3 Locomotive Works Histories.
Works history information for each of the 117 Claytons is provided in Section 10.

D8600, St Rollox Works (Erecting Shop), 1970. D8600 spent a prolonged period of time in St Rollox Works during 1970 for Unclassified repairs. Note the BRCW Type 2 (D5402) and NBL Type 2 locomotives behind D8600, both with newly applied full yellow front ends. (Transport Topics)

D8607, Doncaster Works, 9 January 1966. Just over four months after this photograph was taken, D8607 was reallocated to 64B Haymarket and, as a consequence, all subsequent major repair work was undertaken by St Rollox Works. (Keith Long)

Chapter 10
LOCOMOTIVE HISTORIES

10.1 Explanation of Information Provided.

10.1.1 Depot Allocations and Withdrawals.
Sections 8.1 and 8.2 explain the sourcing of depot allocation information and the depot code abbreviations used.

Occasionally two dates have been found for some Inter-Regional transfers; this reflects situations where the providing and receiving Regions report different dates. Such instances are presented as '2we090967 (010967/030967)-12A'; in all cases, the first date in the brackets is that given by the providing Region, and the latter date by the receiving Region.

10.1.2 Sightings: Primary and Secondary Sources.
Locomotive histories have been built up for all 117 locomotives developed from primary sources (personal observations and photographs) and secondary sources (magazine reports, archive depot/works visit sightings and listings which provide fully dated information). However, due to severe space constraints only significant sightings are reproduced here; however, information is provided to cover all Works visit sightings, and most post-withdrawal observations.

Whilst Primary information is undoubtedly the best source of sighting details and associated anecdotal information, it is not without shortcomings; for example, for St Rollox Works certain areas were not always visited (most notably the Scrapping Area) which inevitably led to some incomplete reporting (see Section 20).

10.1.3 DLRC Works Information.
Works information from the DLRC (B.R. 9215/1) is provided in full. Ian Sixsmith, author of many of the Irwell Press locomotive histories, explains:

'Dates of leaving and entry to works were of course to some extent nominal and a day or two . . .either side should always be assumed. The works were not above 'fiddling' dates at the beginning or the end of a month to enhance the monthly figures . . .It was thus not entirely unknown for a locomotive to be out on the road with the figures still showing it still in works and vice versa - for a few days at least . . .'

Cards or card-sets covering 106 locomotives of the 117-strong Clayton fleet have been seen (i.e. D8503/5-43/5-53/5-70/73-7/9-88/91-9 and D8600-16) although the card-sets for D8503 (post-ScR 11/69 reinstatement only) and D8505/13/29/31 (pre-LMR withdrawal only) were clearly incomplete.

Unfortunately no DLRC information has been found for eleven locomotives i.e. D8500-2/4/44/54/71/2/8/89/90, although some Classified/ Unclassified information for D8500/72/8 has been derived from the St Rollox Shopping Summaries (SRSS) for 1967/68.

The overhaul categories provided on the DLRCs are as follows:-

G	General.
I	Intermediate.
C	Classified (level unspecified, actual levels dependent on specific components (see Section 9.2)).
Rect	Rectification (i.e. re-call to works to correct identified faults).
U	Unclassified (i.e. unscheduled).

A small number of locomotives entered Works prior to end-December 1963 and received LC and HC repairs; these were 'Light Casual' and 'Heavy Casual' respectively and were equivalent to the subsequent Unclassified repairs.

10.1.4 BR *Fires on Diesel Train Locomotives* Reports (1961-68).
All reported incidents are included (see Section 13.3).

10.1.5 Other Secondary information.
On certain occasions other secondary information has been included provided that it is fully dated and provides additional commentary around already established primary sources.

10.2 Data Presentation.
The date format in the locomotives' logs is 'mmyy' or 'ddmmyy', as opposed to 'mm/yy' or 'dd/mm/yy', for clarity. The more conventional 'dd/mm/yy' format is, however, used in the locomotive 'Notes' (excepting quotes where the original published format is retained).

Text colour coding used is as follows:

Blue	Key dates (e.g. New, depot transfers, Works data (prefixed by 'DLRC'), withdrawal and disposal). Disposal information is sourced from *Diesel & Electric Locomotives for Scrap* (A. Butlin) (D&ELfS) and the *Railway Observer*. It should be noted, however, that some of the disposal information conflicts with the sightings; this is discussed further in Sections 19-21.
Black	Locomotive sightings, and, fire damage reports (suffixed by 'FTDL').
Purple	Dates when locomotives were NOT seen at a particular depot or works.
Red	Sightings 'conflicting' with works information, or, duplicated sightings on specified dates.

The DLRC references to Works visits in the following logs (e.g. DLRC: St Rollox: 180866-291066 I) specify 'Works Location', 'Date in Works' to 'Date out of works' and 'Class of Repair'. There are many instances in the logs where the early part of the date range includes at period awaiting works attention prior to actually arriving in Works. So, for the example Works visit quoted for D8506 above, this locomotive was seen at 66A Polmadie on 210866 and has been deliberately highlighted in red to emphasise the date overlap.

The following Works sighting abbreviations are used: E.Shop: Erecting Shop, CS: Carriage Sidings.

'Nil' refers to no observations at the specified location.

Abbreviations of the sources used in the 'Notes' at the end of many Locomotive Histories are explained in the 'References & Sources' section.

10.3 Locomotive Histories.
Abridged histories are provided on the following pages.

D8500   Serial No.: 4365/U 1   N.B. No DLRC information.

D8500, Cricklewood, 26 July 1962. En route to Marylebone for inspection. (Transport Treasury)

Whatstandwell: 180762 (1T48 test train)
Midland Main-Line: 250762 (l/e to Marylebone)
Cricklewood: 260762
Marylebone: 260762 (BTC inspection)
Mill Hill/St Albans: 270762 (l/e to Derby)

New (ScR): 3we290962 (100962)-66A

66A Polmadie: 100962 (day of arrival)
55A Leeds Holbeck: 171062 (en route Polmadie with D8504/5)

66A Polmadie: 170263
65C Parkhead: 210363/ 040463/ 120463/ 200463
63A Perth: 120563/ 010663. Not listed 220663.

Transfer NER: 1we070963 (020963)-NER(52H/56B) (on loan)

66A Polmadie: 020963 (departed for Tyne Dock)
Tyne Dock: 050963 (Tyne Dock-Consett iron-ore trial, with D8501)

Transfer ScR: 1we051063 (300963)-66A (*see note 1*)

St Rollox Works: 201064 (E.Shop)

66A Polmadie: 200567 (outside Repair Shop)
St Rollox Works: 280567. SRSS:St Rollox: xxxxxx-we240667 I

Transfer LMR: 1we190867 (130867)-12A

St Rollox Works: 110568 (minor repairs)

Transfer: 1we220668 (170668)-D10

12A Carlisle Kingmoor (Diesel): 080968
65A Eastfield: 100968 (en route to St Rollox Works?)

Withdrawn (ex-LMR): 1we121068 (121068) (ex-D10).

12A Carlisle Kingmoor (Steam): 241068/ 071268/ 311268/ 030169/ 110169
Kingmoor Yard: xx0269. Not listed 120569.

Reinstated (ScR): 1we240569 (180569)-66A

Kingmoor Yard: Nil. Not listed 260569.

Withdrawn (ex-ScR): 4we030471 (270371) (ex-66A).

66A Polmadie: 250471/ 280571/ 070871/ 160871/ 210871/ 290871
St Rollox Works: 270971. Not listed 161071/ 211171.

Disposal: St Rollox Works: 1171 (D&ELfS).
Commencing month of cutting-up: 1171 (RO0776).

Note:
1. Official transfer dates from NER back to ScR in September 1963 for D8500 and D8501 seem to have been inter-changed. On the basis of sightings, D8500 should show return to ScR as 1we070963 and D8501 as 1we051063.

D8501 Serial No.: 4365/U 2 N.B. No DLRC information.

Official photograph of D8501. Note the light-grey painted bogie equalising beams, axle boxes, springs, sand boxes, brake cylinders and pipe work for publicity purposes. 0T44 in headcode box (see below). (Courtesy and copyright Paxman Archive Trust)

Belper: 230862 (on test, 0T44 Derby-Millers Dale & return, primer)
Shipley: 101062 (with D8503)

New (ScR): 1we131062 (101062)-66A

66A Polmadie: 170263
65C Parkhead: 210363/040463/ 120463/ 200463
63A Perth: 120563/010663. Not listed 220663.

Transfer NER: 1we070963 (020963)-NER (52H) (on loan)

66A Polmadie: 020963 (departed for Tyne Dock)
Tyne Dock: 050963 (Tyne Dock-Consett iron-ore trial, with D8500)
65B St Rollox: 080963
56B Ardsley: 140963 (moved to Ardsley for trials (E&W Yorkshire Union Railway route), with D8536)
Neville Hill West: 250963 (test train from Hunslet East Branch)
Heckmondwike: 260963 (Healey Mills-Low Moor (Bradford) freight)

Transfer ScR: 1we070963 (040963)-66A (*see note 1*)

Milnwood Junction: 260466 (LE to 66B Motherwell, 'serious' fire) (FDTL)
66A Polmadie: 040566/290566 (fire-damage)/ 030766/ 170766

Transfer LMR: 1we190867 (130867)-12A

St Rollox Works: 241067 (E.Shop)/ 041167 (under repair)
P&SP:Ex-Works 141167

Transfer: 1we220668 (170668)-D10

Withdrawn (ex-LMR): 1we121068 (121068) (ex-D10)

12A Carlisle Kingmoor (Steam): 071268/ 311268/ 030169/ 110169. Not listed 241068.
Kingmoor Yard: xx0269/120569/ 260569/ 060769/ 200869/ 240869/ 040969/ 200969/ 280370/ 300370/ xx0470
Abington: 270470 (8Z42 Carlisle Kingmoor-Polmadie, hauled by D1739, with D8509/14)
66A Polmadie: 240570/130670/ 140670/ 030770/ 020870/ 040870. Not listed 300770/ 310770.
St Rollox Works: 220870/031070 (Old Paint Shop sidings)/ 041070 (beyond old Carriage Shop)/ 101070/ 100171/ 200271 (½mile from Works)/ 140371/ 170471/ 190571 (dumped near old Carriage Shop)/ 310571/ 190671 (¼mile north of Works)/ 070871/ 100871/ 160871/ 210871/ 300871/ 270971/ 161071 (Yard)/ 211171/ 281171. Not listed 090172.

Disposal: St Rollox Works: 1271 (D&ELfS).
Commencing month of cutting-up: 1271 (RO0776).

Notes:
1. See Note 1 for D8500.
2. D8501: '...one week at 52H until 14/09/63 and three weeks at 56B until 05/10/1963 when it did tests on local trips.' (K.Long)

D8502　Serial No.: 4365/U 3　N.B. No DLRC information.

55A Leeds Holbeck: 051062

New (ScR): 1we061062 (061062)-66A

66A Polmadie: 170263
65C Parkhead: 210363/040463/ 120463/ 200463
XX: xxxx63 (location of post-storage repairs unknown)

Transfer LMR: 1we190867 (130867)-12A

St Rollox Works: 241067 (Works Yard)/ 041167 (E.Shop)

12A Carlisle Kingmoor (Steam): 101267

Transfer: 1we220668 (170668)-D10

Transfer ScR: 1we121068 (121068/061068)-64B

Transfer: 1we161168 (111168)-66A

Withdrawn (ex-ScR): 4we091071 (051071) (ex-66A)

66A Polmadie: 271171
Ex-67D Ardrossan: Arrived mid-1271/030172/ 090172/ 260272/ 300372/ 250472/ 140572/ 170772/ 170872/ 060972
66A Polmadie: 061172/091172/ 191172. Not listed 141072.
St Rollox Works: 100273/110273. Not listed 080473/ 210473/ 120573.

Disposal: St Rollox Works: 0473 (D&ELfS).

D8503　Serial No.: 4365/U 4　N.B. Incomplete DLRC information.

D8503, St Rollox Works (Old Paint Shop Sidings), undated. The loco behind D8503 is believed to be D8555. Due to the lack of DLRC information it has not been possible to precisely pin down the date that this locomotive received blue livery; however, the large sized 'double-arrow' emblems indicates August 1967 or later; D8503 was a Carlisle Kingmoor-allocated locomotive at this time, so the sighting of it at Polmadie on 31 December 1967 may be indicative of a move to or from St Rollox for Classified repair and repaint. (Colour-Rail)

Shipley: 101062 (with D8503)

New (ScR): 1we131062 (101062)-66A

66A Polmadie: 170263
65C Parkhead: 210363/040463
63A Perth: 120463/190463/ 120563/ 010663/ 230763. Not listed 070463/ 220663.

Transfer LMR: 1we190867 (130867)-12A

66A Polmadie: 311267 (en route to/from St Rollox Works?)

Transfer: 1we220668 (170668)-D10

Stored ( ): 1we121068 (121068) (ex-D10)

12A Carlisle Kingmoor (Steam): 241068

Withdrawn (ex-LMR): 1we021168 (021168) (ex-D10 (store))

12A Carlisle Kingmoor (Steam): 071268/ 311268/ 030169/ 110169

Kingmoor Yard: xx0269/120569/ 260569/ 060769/ 200869/ 240869/ 040969/ 200969
66A Polmadie: arrived ex-Carlisle 151069 (with D8506)

Reinstated (ScR): 1we081169 (021169)-66A

St Rollox Works: Nil. DLRC: Glasgow: 031169-251169 U (Power unit)

St Rollox Works: Nil. DLRC: Glasgow: 081269-060170 U (One power unit (General repair))

66A Polmadie: 310770/020870/ 040870 (store)/ 160870/ 220870

Withdrawn (ex-ScR): 4we091071 (061071) (ex-66A)

St Rollox Works: 281171/090172/ 120272/ 270272/ 010472/ 020472 (Works Yard)/ 090472 (Old Paint Shop Yard)/ 140472 (Top Yard)/ 250472/ 260472/ 140572/ 280572. Not listed 211171 and 100672/ 130872.

Disposal: St Rollox Works: 0672 (D&ELfS). Commencing month of cutting-up: 0672 (RO0776).

D8504   Serial No.: 4365/U 5   N.B. No DLRC information.

D8504, 66A Polmadie, 28 September 1974.
(Anthony Sayer)

55A Leeds Holbeck: 171062 (crew-change) (with D8500/5)

New (ScR): 2we271062 (171062)-66A

66A Polmadie: 170263
65C Parkhead: 210363/040463

63A Perth: 120463/190463/120563/010663. Not listed 070463 and 220663.

Inverurie Works: xx0665

Transfer LMR(o/l): 2we090967 (010967/030967)-12A
Transfer LMR(perm): 1we 160967 (100967)-12A

Transfer: 1we220668 (170668)-D10

9C Reddish: Departed 180768 after repairs

Withdrawn (ex-LMR): 1we121068 (121068) (ex-D10)

12A Carlisle Kingmoor (Steam): 241068/ 071268/ 311268/ 030169/ 110169
Kingmoor Yard: xx0269. Not listed 120569.

Reinstated (ScR): 1we240569 (180569)-66A

Kingmoor Yard: 260569
66A Polmadie: 130769

St Rollox Works: 100171. Not listed 200271.
P&SP: Repair extended due to waiting Main Generator Armature Bearings (Classified Repair).

Withdrawn (ex-ScR): 4we271171 (071171) (ex-66A)

66A Polmadie: 271171

Ex-67D Ardrossan: Arrived mid-1271/ 030172/ 090172/ 260272/ 300372/ 250472/ 140572/ 170772/ 170872/ 060972
66A Polmadie: 091172/ 191172/ 100273/ 250273/ 100473/ 120573/ 270573/ 090673/ 080773/ 170873/ 150973/ 131073/ 181173/ 130174/ 260174/ 170274/ 230374/ 140474/ 280574/ 150674/ 290674/ 030874/ 100874/ 250874/ 280974/ 051074/ 161174/ 260175/ 080375/ 160375/ 290375/ 050475/ 270475. Not listed 141072.
St Rollox Works: 170575 (West of Old Scrapping Area)/ 270575/ 070675 (Dump)/ mid-0775 (Works Yard)/ 120875
Transfer to J. Cashmore, Great Bridge: 17-xx0975 (D8504/46/63, 9X20 via Carlisle/Hellifield)
Hunslet South Sidings: 180975 (with D8546/63). Not listed 260975.
Chesterfield: 110975 (with D8546/63, hauled by 40019)
J. Cashmore, Great Bridge: Nil.

Disposal: J. Cashmore, Great Bridge: 0975-1175 (D&ELfS). '...scrapped about 11/75'. (AHBRD&E5)
Disposal not proven.

D8505   Serial No.: 4365/U 6   N.B. Incomplete DLRC information.

D8505, Location and date unknown.
(Colour-Rail)

55A Leeds Holbeck: 171062 (crew-change) (with D8500/4)

New (ScR): 2we271062 (171062)-66A

66A Polmadie: 170263

65C Parkhead: 210363/040463/ 120463/ 200463
63A Perth: 120563
66A Polmadie: 020663

St Rollox Works: Nil. DLRC: St Rollox: 200663-200663 NC

66A Polmadie: 201264 / 241264/ 030165
St Rollox Works: Nil. DLRC: St Rollox: 221264-090165 U

St Rollox Works: Nil. DLRC: St Rollox: 281166-280167 I

66B Motherwell: 200867
66A Polmadie: 200867/020967
St Rollox Works: Nil. Not listed 170867/ 210867.
DLRC: St Rollox: 180867-130967* U * Date modified on card to 110967.

Transfer LMR(o/l): 1we090967 (060967/030967)-12A
Transfer LMR(perm): 1we160967 (100967)-12A

Transfer: 1we220668 (170668)-D10

12A Carlisle Kingmoor (Diesel): 250868
St Rollox Works: August BH weekend (310868-020968). Not listed 260868 and 210968.

12A Carlisle Kingmoor (Diesel): 091068

Withdrawn (ex-LMR): 1we121068 (121068) (ex-D10)

12A Carlisle Kingmoor (Steam): 311268/ 030169/ 110169. Not listed 241068/ 071268.

Kingmoor Yard: xx0269. Not listed 120569.

Reinstated (ScR): 1we240569 (180569)-66A

Kingmoor Yard: 260569

Transfers: 1we020869 (270769)-64B

Withdrawn (ex-ScR): 4we091071 (061071) (ex-64B)

Millerhill s.p.: 271171
64B Haymarket: 121271/030172/ 080172
Millerhill Yard: 270272/ 310372/ 230472/ 130572/ 030672/ 230772/ 180872/ 260872/ 020972
St Rollox Works: 141072 (Dump)/221072/ 281072 (Top Yard)/ 051172/ 191172/ 100273/ 110273/ 080473/ 210473/ 220473/ 050573/ 120573 (Top Yard)/ 130573/ 280573/ 290573/ 090673 (Yard)/ 170673/ 050873/ 110873/ 120873/ 140873/ 160873/ 150973/ 131073/ 041173/ 181173/ 130174/ 100374/ 230374 (Scrap Road)/ 130474/ 140474/ xx0474 (being-cut)/ 010574 (Scrapping Area, part cut-up)/ 150674 (unidentified part cut-up remains)/ 070675 (unidentified part cut-up remains). Not listed 280574.

Disposal: St Rollox Works: 0574 (D&ELfS).

Note:
1. St Rollox Works, 150674: 'frames and bogies of two others were seen on the scrap-road' (RO0974). Assumed to be D8505/59; see Section 20.4.

D8506   Serial No.: 4365/U 7

D8506, 10A Carnforth, 24 May 1968.
(Author's Collection)

New (ScR): 2we271062 (241062)-66A

66A Polmadie: 170263
65C Parkhead: 210363/ 040463/ 120463/ 130463/ 140463/ 200463
63A Perth: 120563

St Rollox Works: Nil. DLRC: St Rollox: 181163-071263 LC

St Rollox Works: 170564. DLRC: St Rollox: 180364-220564 U

St Rollox Works: Nil. DLRC: St Rollox: 231164-121264 U

66A Polmadie: 210866
St Rollox Works: Nil. DLRC: St Rollox: 180866-291066 I

Transfer LMR(o/l): 1we090967 (060967/030967)-12A
Transfer LMR(perm): 1we 160967 (100967)-12A

Transfer: 1we220668 (170668)-D10

12A Carlisle Kingmoor (Diesel): 250868
St Rollox Works: August BH weekend (*310868-020968*).
DLRC: St Rollox: 260868-310868 U

Withdrawn (ex-LMR): 1we121068 (121068) (ex-D10)

12A Carlisle Kingmoor (Steam): 241068/ 071268/ 311268/ 030169/ 110169
Kingmoor Yard: xx0269/ 120569

Reinstated (ScR): 1we240569 (180569)-66A
Withdrawn: 1we310569 (180569) (ex-66A/D10)
(late entry)

Kingmoor Yard: 260569/200869/ 240869/ 040969/ 200969. Not listed 060769.
66A Polmadie: arrived ex-Carlisle 151069 (with D8503)

Reinstated (ScR): 1we081169 (021169)-66A

St Rollox Works: xx0370. DLRC: Glasgow: 160270-140370 U
XX: xx0370 (ex-Works)
66A Polmadie: 290370 (Repair Bay)/040470

Stored ( ): 5we110971 (250771) (ex-66A) (late entry)

66A Polmadie: 070871/080871/ 160871/ 210871/ 290871

Withdrawn (ex-ScR): 4we091071 (210971) (ex-66A (store))

66A Polmadie: 260971
St Rollox Works: 281171. Not listed 161071/ 211171.

Disposal: St Rollox Works: 1271 (D&ELfS).
Commencing month of cutting-up: 1271 (RO0776).

D8507   Serial No.: 4365/U 8

D8507, 12A Carlisle Kingmoor (Steam), October 1967. Four months after its repaint into blue livery. As a pre-August 1967 recipient of blue livery, D8507 received a small 'double-arrow' emblem. (Colour-Rail)

Derby (Sidings opposite shed): 100363 (stored)/ 230363 (stored)

New (ScR): 1we031162 (301062)-66A

Derby (Sidings opposite Shed): 070463
66A Polmadie: 310563. Not listed 140463.

66A Polmadie: 050164

St Rollox Works: Nil. DLRC: St Rollox: 271263-240164 LC

St Rollox Works: Nil. DLRC: St Rollox: 250264-130364 U

St Rollox Works: 230965. DLRC: St Rollox: 070965-081065 U

St Rollox Works: 280567. DLRC: St Rollox: 020567-090667 I

Transfer LMR: 1we141067 (081067)-12A

Transfer: 1we220668 (170668)-D10

Withdrawn (ex-LMR): 1we121068 (121068) (ex-D10)

12A Carlisle Kingmoor (Steam): 241068/ 071268/ 311268/ 030169/ 110169
Kingmoor Yard: xx0269/120569/ 260569/ 200869/ 240869/ 040969/ 200969. Not listed 060769.
64B Haymarket: arrived ex-Carlisle on 111069 (with D8510/6)

Reinstated (ScR): 1we081169 (021169)-64B

64B Haymarket: 160570 (damage to one end)
St Rollox Works: 250570. DLRC: Glasgow: 050570-060670 U

64B Haymarket: 130670/140670

St Rollox Works: 140371. Not listed 200271 and 170471.

Transfers: 5we010172 (061271)-66A

64B Haymarket: 121271
66A Polmadie: 171271

Withdrawn (ex ScR): 2we010172 (311271) (ex-66A)

66A Polmadie: 030172/090172/ 270272/ 300372/ 230472/ 140572/ 100672/ 080772/ 170872/ 290872/ 060972/ 160972
St Rollox Works: 141072 (Dump)/221072/ 051172/ 191172 (outside)/ 100273/ 110273/ 080473/ 210473/ 220473/ 230473 (CS)/ 050573/ 120573 (CS)/ 130573/ 280573/ 290573/ 090673 (Yard)/ 170673/ 050873 (CS)/ 110873/ 120873/ 140873/ 160873/ 150973/ 131073/ 041173/ 181173/ 130174/ 100374/ 230374/ 130474/ 140474/ 010574 (CS)/ 110574 (CS)/ 280574/ 150674 (Dump)/ 200674/ 030874/ 240874/ 051074 (Dump)/ 260175/ 170575 (CS)/ 270575/ 070675 (Dump)/ 120875. Not listed 260974/ 080375/ 200375/ 290375/ mid-0775.
Transfer to R.A. King, Norwich: 04-080975 (D8507/8/16/25/9/31/6/52/74, 9X20 dep. Glasgow 03.35 040975, arr. March Up Yard 17.55 080975; dep. March 19.45 080975, arr. Norwich Trowse Yard 23.30 080975)
Tyne Yard: 060975
Doncaster: 070975 (with D8508/16/25/9/31/6/52/74)
March: 080975 (hauled by 47360, with D8508/16/25/9/31/6/52/74)
Norwich mpd: 140975
R.A. King, Norwich (Trowse Upper Junction): 270975 (Entrance)/ xx1075 (whole). Not listed 241075.

Disposal: A. King & Son, Norwich: 1075 (D&ELfS).

D8508　　Serial No.: 4365/U 9

D8508, St Rollox Work, 26 September 1974. Note the dark green cab front and exhaust stack cowling which replaced the Sherwood Green at some point. D8508 is believed to have been the last Clayton to receive a Classified Repair (released 26 June 1971, or, 14 August 1971 if the subsequent rectification work is included). (Anthony Sayer)

New (ScR): 1we031162 (301062)-66A

Derby St Andrews Goods Yard: 2we020363 (stored)/ 100363
66A Polmadie: Not listed 140463.

St Rollox Works: Nil. DLRC :St Rollox: 300964-101064 U

66A Polmadie: 270365/ 180465 (No.1 end accident damage)/ 190465
St Rollox Works: 270565/290565/ 060665 (E.Shop).
DLRC: St Rollox: 120465-260665 U

XX: 150866 (fire, location unknown) (FDTL)
66A Polmadie: 210866

St Rollox Works: Nil. DLRC: St Rollox: 281166-140167 I

Transfer LMR: 1we141067 (081067)-12A

Transfer: 1we220668 (170668)-D10

12A Carlisle Kingmoor (Steam): 091068

Withdrawn (ex-LMR): 1we121068 (121068) (ex-D10)

12A Carlisle Kingmoor (Steam): 241068/ 311268/ 030169/ 110169. Not listed 071268.
Kingmoor Yard: xx0269. Not listed 120569.

Reinstated (ScR): 1we180569 (180569)-66A

Kingmoor Yard: Nil. Not listed 260569.

St Rollox Works: 170471 (E.Shop)/190571 (Works Yard)/ 310571/ 190671 (E.Shop). DLRC: Glasgow: 310371-260671 C

St Rollox Works: Nil. DLRC: Glasgow: 020871-140871 Rect

Withdrawn (ex-ScR): 2we010172 (311271) (ex-66A)

66A Polmadie: 030172/090172/ 270272/ 020472/ 230472/ 140572/ 100672/ 080772/ 170872/ 290872/ 060972/ 160972/ 141072

St Rollox Works: 221072/051172/ 191172/ 100273/ 110273/ 180373 (CS)/ 080473/ 210473/ 220473/ 230473 (CS)/ 050573/ 120573 (CS)/ 130573/ 280573/ 290573/ 090673 (Yard)/ 170673/ 040773/ 050873 (CS)/ 110873/ 120873/ 140873/ 160873/ 150973/ 131073/ 231073 (CS)/ 041173/ 181173/ 130174/ 100374/ 230374/ 130474/ 140474/ 010574 (CS)/ 280574/ 150674 (Dump)/ 200674/ 030874/ 240874/ 260974 (N of Traverser)/ 051074 (N of Traverser)/ 121074 (N of Traverser)/ 221274/ 260175/ 080375/ 200375/ 290375/ 170575 (N of Traverser)/ 270575/ 070675 (Dump)/ 120675 (N of Traverser)/ mid-0775 (Works Yard)/ 120875. Not listed 141072.
Transfer to R.A. King, Norwich: 04-080975 (D8507/8/16/25/9/31/6/52/74, 9X20 dep. Glasgow 03.35 040975, arr March Up Yard 17.55 080975; dep. March 19.45 080975, arr Norwich Trowse Yard 23.30 080975)
Tyne Yard: 060975
Doncaster: 070975 (with 8507/16/25/9/31/6/52/74)
March: 080975 (hauled by 47360, with D8507/16/25/9/31/6/52/74)
Wensum Yard, Norwich: 270975/241075
R.A. King, Norwich: 311075/011175. Not listed 271275.

Disposal: A. King & Son, Norwich: 1175 (D&ELfS).

Note:
1. See Section 13.1 for additional details regarding accident damage sustained by D8508 in April 1965.

D8509 Serial No.: 4365/U 10

New (ScR): 2we241162 (131162)-66A

66A Polmadie: 170263
65C Parkhead: 210363/040463/ 120463/ 200463
63A Perth: 120563

St Rollox Works: Nil. DLRC: St Rollox: 260264-190364 U

St Rollox Works: Nil. DLRC: St Rollox: 300165-300165 U

St Rollox Works: 100466

Transfer LMR: 1we141067 (081067)-12A

Transfer: 1we220668 (170668)-D10

12A Carlisle Kingmoor (Diesel): 041068

Withdrawn (ex-LMR): 1we121068 (121068) (ex-D10)

12A Carlisle Kingmoor (Steam): 241068/ 071268/ 311268/ 030169/ 110169
Kingmoor Yard: xx0269/120569/ 260569/ 060769/ 200869/ 240869/ 040969/ 200969/ 280370/ 300370/ xx0470
Abington: 270470 (8Z42 Carlisle Kingmoor-Polmadie, hauled by D1739, with D8501/14)
66A Polmadie: 240570/140670/ 030770/ 020870/ 040870/ 160870/ 220870/ 041070/ 071170/ 140271/ 110471/ 250471. Not listed 300770/ 310770 and 100171.
St Rollox Works: 170471/190571 (dumped near old Carriage Shop)/ 310571/ 190671 (¼mile north of Works)/ 070871/ 100871/ 160871/ 210871/ 300871/ 270971/ 161071 (Yard)/ 211171/ 281171. Not listed 090172/ 120272.

Disposal: St Rollox Works: 0172 (D&ELfS). Commencing month of cutting-up: 0172 (RO0776).

D8510 Serial No.: 4365/U 11

D8510, 64B Haymarket, 1970. (Transport Topics)

New (ScR): 2we241162 (131162)-66A

66A Polmadie: 170263
65C Parkhead: 210363/040463/ 120463/ 200463
63A Perth: 120563/010663/ 220663. Not listed 230663.

St Rollox Works: 271063/291063 (E.Shop). **DLRC:** St Rollox: 101063-081163 HC

St Rollox Works: Nil. **DLRC:** St Rollox: 180366-020466 U

St Rollox Works: 241067 (E.Shop)/041167 (E.Shop). **DLRC:** St Rollox: 121067-181167 I

Transfer LMR(o/l): 1we251167 (241167/251167)-12A
Transfer LMR(perm): 1we091267 (031267)-12A

Carlisle: 020368 (SLS/MLS "West Cumberland Rail Tour" (1T90 Brake-Van tour to Buckhill (RNAD Gates), Workington Moss Bay Iron Works, Ullcoats Mine, Beckermet Mine)

Transfer: 1we220668 (170668)-D10

Stored ( ): 1we121068 (121068) (ex-D10)

Reinstated: 1we301168 (141068)-D10 (late entry)

Note:
1. 'Held at Kingmoor Diesel Depot 12/68-5/69' (AHBRD&EFU)

Withdrawn (ex-LMR): 2we281268 (281268) (ex-D10)

12A Carlisle Kingmoor (Diesel): 030169/ 110169
Kingmoor Yard: 120569/260569/ 060769/ 200869/ 240869/ 040969/ 200969
64B Haymarket: arrived ex-Carlisle on 111069 (with D8507/16)

Reinstated (ScR): 1we081169 (021169)-64B

St Rollox Works: Nil. **DLRC:** Glasgow: 230270-070370 U

St Rollox Works: Nil. **DLRC:** Glasgow: 080570-200570 U

Withdrawn (ex-ScR): 4we030471 (270371) (ex-64B)

64B Haymarket: 250471
66A Polmadie: 280571/070871/ 100871/ 160871/ 210871/ 290871
St Rollox Works: 270971/161071 (Yard)/ 211171 (being cut-up). Not listed 090172.

Disposal: St Rollox Works: 1171 (D&ELfS).
'...broken up by 2/11/71.' (AHBRD&E5)
Commencing month of cutting-up: 1171 (RO0776).

D8511　　Serial No.: 4365/U 14

D8511, 66A Polmadie, 6 August 1963.
(David Dippie)

New (ScR): 2we241162 (221162)-66A

66A Polmadie: 170263
65C Parkhead: 210363/ 040463
63A Perth: 120463/ 190463. Not listed 070463 and 120563.

St Rollox Works: Nil. DLRC: St Rollox: 230165-230165 U

St Rollox Works: Nil. DLRC: St Rollox: 240166-120266 I

Transfer LMR(o/l): 1we251167 (231167/241167)-12A
Transfer LMR(perm): 1we091267 (031267)-12A

Transfer: 1we220668 (170668)-D10

St Rollox Works: 220668. Not listed 200668 and 180768.

Withdrawn (ex-LMR): 1we121068 (121068) (ex-D10)

12A Carlisle Kingmoor (Steam): 241068/ 071268/ 311268/ 030169/ 110169
Kingmoor Yard: xx0269/030569/ 120569/ 260569/ 060769/ 200869/ 240869/ 040969/ 200969
66A Polmadie: 141269/301269/ 290370/ 040470/ 110470/ 240570/ 130670/ 140670/ 280670 (out of use, partly cannibalised)/ 030770/ 300770/ 310770/ 020870. Not listed 220870.
J. MacWilliam, Shettleston: 230870 (cab only)

Disposal: J. MacWilliam, Shettleston: 0870 (D&ELfS). '. . .by 23/8/70 only the cab portion of the loco was left . . . disposed of by 3/9/70.' (AHBRD&E5)

D8512   Serial No.: 4365/U 16

D8512, 9A Longsight, undated. Now in Departmental service but still in as-withdrawn green livery. (Colour-Rail)

New (ScR): 4we291262 (07 or 101262)-66A

66A Polmadie: 170263
65C Parkhead: 210363/040463/ 120463/ 200463
63A Perth: 120563

66A Polmadie: 310863
St Rollox Works: Nil. **DLRC: St Rollox:** 280863-120963 HC

St Rollox Works: Nil. **DLRC: St Rollox:** 200164-080264 U

St Rollox Works: Nil. DLRC: St Rollox: 190165-190265 U

66A Polmadie: 241265
St Rollox Works: Nil. DLRC: St Rollox:161265-311265 U

66A Polmadie: 080866
St Rollox Works: 120866/200866 (Erecting Shop). DLRC: St Rollox: 110766-170966 M

Transfer LMR(o/l): 1we141067 (111067/081067)-12A
Transfer LMR(perm): 1we111167 (051167)-12A

St Rollox Works: 080568/110568/ 250568

Transfer: 1we220668 (170668)-D10

12A Carlisle Kingmoor (Diesel): 091068

Withdrawn: 1we121068 (121068) (ex-D10)

Reinstated: 1we021168 (021168)-D10 (stored)
Reinstated: 1we301168 (041168)-D10 (late entry)

Withdrawn (ex-LMR): 2we281268 (281268) (ex-D10)

12A Carlisle Kingmoor (Diesel): 030169/ 110169/ xx0269
Kingmoor Yard: 120569
12A Carlisle Kingmoor (Diesel): 260569

Clay Cross: 230769
Derby RTC: Arrived 230769/100869/ 210969/ 191069/ 080370/ 260470/ 300570/ 130670
Stoke Cockshute Depot: 030770

9A Longsight: 200770/260770/ 030870
St Rollox Works: 220870 (see note 3).
9A Longsight: 120970/211070/ 281070/ 221170
Swindon Works: Nil. Not listed 131270.
Derby RTC: 230571
9A Longsight: 050671 (arrived 050671 from Derby)
Test trains (Styal line): w/c 070671 'until further notice'
Wilmslow: 120671 (with test coach)
Transfer Longsight-Derby RCD: 210671 (for mods)
Test trains (Styal line): ceased by 090871
9A Longsight: 170871
Birmingham New Street: 260871
9A Longsight: 061071 (to Derby for attention)
Derby: 061071
Old Dalby: 141071
Derby RTC/ Derby area: 161071/031171
Test trains (Styal line): 07-091271
9A Longsight: 291271/300172
Test trains (Styal line): w/c 310172
GEC Stafford: Arrived 230272, removed 100372
Stoke: 100372 (GEC Stafford-Cockshute, with D8545/67)
9A Longsight: xx0472/210572/ 030672/ 050772
Transfer Longsight-St Rollox Works: 310772 (9Z12, hauled by D435 to Carlisle, via Ais Gill on 310772, then D6853 from Carlisle to Glasgow)
St Rollox Works: 100872/120872 (Dump)/ 130872/ 170872/ 240872/ 280872/ 290872/ 020972/ 250972/ 141072 (Outer Yard)/ 221072/ 051172/ 191172 (inside). Not listed 060872 and 100273.

Disposal: Departmental S18512; St Rollox Works: 0173 (D&ELfS).
Commencing month of cutting-up: 0173 (RO0776).

Notes:
1. 'D8512/21 . . .Derby Research Centre for experiments with radio-controlled multiple working.' (LCGB1169)
2. '8512 was at Longsight shed on 20th July (*1970*) for crew training on the Styal line. It was due to be sent to Swindon for fitting with visual equipment which will be tested on the Styal line under 25kV conditions.' (RO0970) Did it go to St Rollox instead, or was the 22/08/70 report an error for D8514?
3. "8512, which was seen at Longsight until about September 1970 and then mysteriously disappeared, has now returned.....The loco arrived back on Saturday 5th June (*1971*)." (NCTS No.8)
4. "Stafford. A most unexpected.....occurrence during the national fuel crisis in February/March (*1972*) was the stabling of 3 Clayton Type 17's at the main GEC Works here, apparently for emergency power purposes.....8512, from the Derby Research Dept., arrived on 23/2 and 8545, 8567 (ex 66A) on 24/2. All three were coupled together and placed in a siding adjacent to the GEC Works on the up side of the main line just past the Queensville Curve. Their stay lasted just over two weeks and they were eventually removed on 10/3." (RL0472).
5. RTC Allocations. Reinstated: 23/07/69, Withdrawn: 10/03/72.

D8513 Serial No.: 4365/U 13 N.B. Incomplete DLRC information.

D8513, 66B Motherwell, undated.
(Transport Treasury)

New (ScR): 4we291262 (05 or 071262)-66A

65C Parkhead: 210363/040463/ 120463/ 200463
XX: xxxx63 (location of post-storage repairs unknown)

St Rollox Works: Nil. DLRC: St Rollox: 110264-220264 U

St Rollox Works: Nil. DLRC: St Rollox: 060464-230464 U

St Rollox Works: Nil. DLRC: St Rollox: 081065-091065 U

St Rollox Works: 210467 (under repair)/ 220467 DLRC: St Rollox: 170367-190567 I

Transfer LMR(o/l): 1we211067 (171067/151067)-12A
Transfer LMR(perm): 1we111167 (051167)-12A

Beckermet Colliery: 141267 (OURS brake-van tour)

Transfer: 1we220668 (170668)-D10

St Rollox Works: 210968 (Paint Shop). Not listed 260868.

12A Carlisle Kingmoor (Diesel): 091068

Withdrawn (ex-LMR): 1we121068 (121068) (ex-D10)

12A Carlisle Kingmoor (Steam): 071268/ 311268/ 030169/ 110169. Not listed 241068.
Kingmoor Yard: xx0269. Not listed 120569.

Reinstated (ScR): 1we180569 (180569)-66A

Kingmoor Yard: 260569

Withdrawn (ex-ScR): 4we091071 (210971) (ex-66A)

66A Polmadie: 260971. Not listed 271171.
St Rollox Works: Nil. Not listed 161071/ 211171/ 281171/ 090172.

Disposal: St Rollox Works: 1271 (D&ELfS). Commencing month of cutting-up: 1271 (RO0776). Disposal not proven.

D8514   Serial No.: 4365/U 12

D8514, 66A Polmadie, 1970. Following re-allocation to Carlisle Kingmoor in October 1967, D8514 appears to have spent the first three months in unofficial store. Sightings in October through to December show ongoing residence at the Kingmoor Steam depot and the *Railway Observer* (February 1968) reported: "D8514 spent a period of inactivity in the company of the withdrawn steam locomotives." Ultimately the errant locomotive was transferred to St Rollox Works for repairs. The photograph shows D8514 out of action once again, having been taken out of traffic by the LMR in October 1968. (Transport Topics)

Derby St Andrews Goods Yard: 2we020363 (stored)/ 030363/ 100363 (stored)

New (ScR): 1we130763 (090763)-66A

St Rollox Works: 271063/291063 (Works Yard). **DLRC:** St Rollox: 171063-141163 LC

St Rollox Works: Nil. **DLRC:** St Rollox: 251163-121263 HC

St Rollox Works: Nil. **DLRC:** St Rollox: 090165-110165 U

66A Polmadie: 110465
St Rollox Works: 160465. **DLRC:** St Rollox: 230365-170465 U

Transfer LMR(o/l): 1we141067 (141067/081067)-12A
Transfer LMR(perm): 1we111167 (051167)-12A

12A Carlisle Kingmoor (Steam): 151067/ 201067/ 241067/ 281067/ 191167/ 101267

St Rollox Works: 070168

Transfer: 1we220668 (170668)-D10

Stored: 1we121068 (121068) (ex-D10)

Withdrawn (ex-LMR): 1we021168 (021168) (ex-D10 (store))

12A Carlisle Kingmoor (Steam): 241068/ 071268/ 311268/ 030169/ 110169
Kingmoor Yard: xx0269/120569/ 260569/ 060769/ 200869/ 240869/ 040969/ 200969/ 280370/ 300370/ xx0470
Abington: 270470 (8Z42 Carlisle Kingmoor-Polmadie, hauled by D1739, with D8501/9)
66A Polmadie: 240570/130670/ 140670/ 280670 (out of use, partly cannibalised)/ 030770/ 300770/ 310770/ 020870/ 040870. Not listed 160870/ 220870.
St Rollox Works: 220870 (*sic, incorrectly listed as D8512?*)/ 120970 (*sic, ditto?*)/ 031070 (old Paint Shop sidings)/ 041070 (beyond old Carriage Shop)/ 101070/ 100171/ 200271 (½mile from Works)/ 140371/ 170471/ 190571 (dumped near old Carriage Shop)/ 310571/ 190671 (¼mile north of Works)/ 070871/ 100871/ 160871/ 210871/ 300871/ 270971/ 161071 (Yard)/ 211171/ 281171. Not listed 090172/ 120272.

Disposal: St Rollox Works: 1271 (D&ELfS). Commencing month of cutting-up: 1271 (RO0776).

D8515 Serial No.: 4365/U 20

D8515, 66B Motherwell, 29 May 1966. D8515 was one of three Claytons which were immediately transferred back to the Scottish Region in October 1968 after their short stint on the London Midland Region, without any interim period in store; the other two were D8502/25. Fourteen other ex-Carlisle Claytons were reinstated to traffic in Scotland in 1969 after suffering periods of between 7 and 12½ months in store in the Carlisle area. (Author's Collection)

Derby St Andrews Goods Yard: 2we020363 (stored)/ 030363/ 100363 (stored)/ 070463/ 160663/ 230663/ 070763. Gone by 300763.
Ais Gill: 030863 (with D8533)

New (ScR): 1we030863 (02 or 030863)-66A

63A Perth: 061063/261063 (Works)
St Rollox Works: Nil. DLRC: St Rollox: 191163-231163 NC

St Rollox Works: 170564. DLRC: St Rollox: 180564-210564 U
St Rollox Works: Nil. DLRC: St Rollox: 250564-290564 U

66A Polmadie: 300565
St Rollox Works: 060665 (E.Shop). Not listed 290565. DLRC: St Rollox: 250565-050665 U

XX: 121266 (fire, location unknown) (FDTL)

Transfer LMR(o/l): 1we281067 (261067/221067)-12A
Transfer LMR(perm): 1we111167 (051167)-12A

St.Rollox Works: Nil. DLRC: No entry. P&SP:Ex-Works xx0268 C

Transfer: 1we220668 (170668)-D10

Transfer ScR: 1we121068 (121068/061068)-64B

Transfer: 1we161168 (111168)-66A

Withdrawn (ex-ScR): 4we091071 (051071) (ex-66A)

66A Polmadie: 271171
Ex-67D Ardrossan: Arrived mid-1271/ 030172/ 090172/ 260272/ 300372/ 250472/ 140572
St Rollox Works: 080672 (CS)/ 100672 (CS) / 180672/ 250672 (CS)/ 060772/ 060872/ 100872/ 130872/ 170872 (CS)/ 280872/ 290872/ 020972 (CS)/ 141072/ 281072/ 051172. Not listed 221072/ 191172.

Disposal: St Rollox Works: 1172 (D&ELfS).

Notes:
1. See Section 20.4 for further discussion re. disposal of D8515.

D8516   Serial No.: 4365/U 19

D8516, St Rollox Works, 26 September 1974. Sandwiched between D8531 and D8580. (Anthony Sayer)

Derby St Andrews Goods Yard: 2we020363 (stored)/ 030363/ 100363 (stored)/ 070463

New (ScR): 1we200463 (190463)-66A

66A Polmadie: 201063/ 301063
St Rollox Works: 271063/291063 (Works Yard). DLRC: St Rollox: 171063-221163 LC

66B St Rollox: 231163/241163
St Rollox Works: Nil. DLRC: St Rollox: 021263-041263 NC

St Rollox Works: Nil. DLRC: St Rollox: 121264-121264 U

XX: 110565 (fire, location unknown) (FDTL)

66A Polmadie: 241065

St Rollox Works: Nil. DLRC: St Rollox: 181065-291065 U

St Rollox Works: Nil. DLRC: St Rollox: 010367-080467 G

Transfer LMR(o/l): 1we041167 (291067)-12A
Transfer LMR(perm): 1we111167 (051167)-12A

Transfer: 1we220668 (170668)-D10

Withdrawn (ex-LMR): 1we121068 (121068) (ex-D10)

12A Carlisle Kingmoor (Steam): 241068/ 071268/ 311268/ 030169/ 110169
Kingmoor Yard: xx0269/ 120569/ 260569/ 060769/ 200869/ 240869/ 040969/ 200969
64B Haymarket: arrived ex-Carlisle on 111069 (with D8507/10)

Reinstated (ScR): 1we081169 (021169)-64B

Transfer: 1we151169 (141169)-66A

Withdrawn (ex-ScR): 4we091071 (061071) (ex-66A)

66A Polmadie: 211171/ 271171
Ex-67D Ardrossan: Arrived mid-1271/ 030172/ 090172/ 260272/ 300372/ 110472/ 250472/ 140572

St Rollox Works: 180672/060772/ 060872/ 100872/ 120872 (Dump)/ 130872/ 170872/ 280872/ 290872/ 020972/ 141072 (Dump)/ 221072/ 281072 (Top Yard)/ 051172/ 191172 (outside)/ 100273/ 110273/ 080473/ 210473/ 220473/ 050573/ 120573 (Top Yard)/ 130573/ 280573/ 290573/ 090673 (Yard)/ 170673/ 050873/ 110873/ 120873/ 140873/ 160873/ 150973 131073/ 041173/ 181173/ 130174/ 100374/ 230374/ 130474/ 140474/ 280574/ 150674 (Dump)/ 200674/ 030874/ 240874/ 260974 (N of Traverser)/ 051074 (N of Traverser)/ 221274/ 260175/ 080375/ 200375/ 290375/ 170575 (N of Traverser)/ 270575/ 070675 (Dump)/ mid-0775 (Dump)/ 100875 (N of Traverser)/ 120875. Not listed 100672.
Transfer to R.A. King, Norwich: 04-080975 (D8507/8/16/ 25/9/31/6/52/74, 9X20 dep. Glasgow 03.35 040975, arr. March Up Yard 17.55 080975; dep. March 19.45 080975, arr. Norwich Trowse Yard 23.30 080975)
Tyne Yard: 060975
Doncaster: 070975 (with D8507/8/25/9/31/6/52/74)
March: 080975 (hauled by 47360, with D8507/8/25/9/31/6/52/74)
Wensum Yard, Norwich: 270975/241075
R.A. King, Norwich: 271275. Not listed 311075.

Disposal: A. King & Son, Norwich: 1175-0176 (D&ELfS). '...broken up during 1/76.' (AHBRD&E5)

D8517　　Serial No.: 4365/U 18

D8517 (with D8532), 10A Carnforth, 24 May 1968. (Author's Collection)

Derby St Andrews Goods Yard: 2we020363 (stored)/ 030363/ 100363 (stored)/ 070463/ 130463

New (ScR): 1we290663 (28 or 290663)-66A

66A Polmadie: 170564
St Rollox Works: Nil. DLRC: St Rollox: 130564-280564 U

St Rollox Works: Nil. DLRC: St Rollox: 040265-020465 U

66A Polmadie: 241065
St Rollox Works: Nil. DLRC: St Rollox: 091065-201165 U

XX: 140466 ('serious' fire, location unknown) (FDTL)

St Rollox Works: Nil. DLRC: St Rollox: 160167-110367 I

66A Polmadie: 250667
St Rollox Works: Nil. DLRC: St Rollox: 220667-130767 U

Transfer LMR(o/l): 1we251167 (231167/251167)-12A
Transfer LMR(perm): 1we 091267 (031267)-12A

Transfer: 1we220668 (170668)-D10

Withdrawn (ex-LMR): 1we121068 (121068) (ex-D10)

12A Carlisle Kingmoor (Steam): 241068/ 071268/ 311268/ 030169/ 110169
Kingmoor Yard: xx0269/120569/ 260569/ 060769/ 200869/ 240869/ 040969/ 200969
66A Polmadie: 141269/110170/ 290370/ 040470/ 110470/ 240570/ 130670/ 140670/ 280670 (partly cannibalised)/ 030770/ 300770/ 310770/ 020870. Not listed 220870.
J. MacWilliam, Shettleston: 230870 (intact)

Disposal: J. MacWilliam, Shettleston: 0870 (D&ELfS). '. . .noted still intact on 23/8/70. Scrapping was completed by 25/9/70.' (AHBRD&E5)

D8518   Serial No.: 4365/U 15

D8518 and D8547, Polmadie, 5 August 1966. (Rail-Online)

Derby St Andrews Goods Yard: 2we020363 (stored)/ 030363/ 100363 (stored)/ 070463

New (ScR): 1we200463 (190463)-66A

St Rollox Works: Nil. DLRC: St Rollox: 261264-281264 U

XX: 120166 (fire, location unknown) (FDTL)

66A Polmadie: 290566
St Rollox Works: Nil. DLRC: St Rollox: 260566-020766 I

St Rollox Works: Nil. DLRC: St Rollox: 300567-300667 U

Transfer LMR(o/l): 1we111167 (111167)-12A
Transfer LMR(perm): 1we091267 (031267)-12A

Transfer: 1we220668 (170668)-D10

Withdrawn (ex-LMR): 1we121068 (121068) (ex-D10)

12A Carlisle Kingmoor (Steam): 241068/ 071268/ 311268/ 030169/ 110169
Kingmoor Yard: xx0269/120569/ 260569/ 060769/ 200869/ 240869/ 040969/ 200969/ 280370/ 300370/ xx0470
Abington: 270470 (8Z43 Carlisle Kingmoor-Polmadie, with D8519/20)
66A Polmadie: 240570/130670/ 140670/ 030770/ 300770/ 310770/ 020870/ 040870/ 160870/ 220870/ 041070/ 071170/ 100171/ 140271/ 110471/ 250471/ 280571/ 070871/ 100871/ 160871/ 210871/ 290871
St Rollox Works: 270971/161071 (Works Yard)/ 281171/ 090172/ 120272. Not listed 211171 and 090472.

Disposal: St Rollox Works: 0272 (D&ELfS). Commencing month of cutting-up: 0272 (RO0776).

D8519   Serial No.: 4365/U 22

Derby St Andrews Goods Yard: 2we020363 (stored)/ 030363/ 100363 (stored)/ 070463

New (ScR): 1we200763 (17 or 180763)-66A

St Rollox Works: Nil. DLRC: St Rollox: 210964-260964 U
66A Polmadie: 260964

St Rollox Works: Nil. DLRC: St Rollox: 071264-121264 U

66A Polmadie: 201264 / 241264/ 030165/ 100165
St Rollox Works: Nil. DLRC: St Rollox: 141264-060265 U

St Rollox Works: Nil. DLRC: St Rollox: 160265-270265 U
66A Polmadie: 250265

St Rollox Works: Nil. DLRC: St Rollox: 120365-120365 U

67A Corkerhill: 260367
St Rollox Works: 210467 (E.Shop)/ 220467.
DLRC: St Rollox: 070367-220467 I

St Rollox Works: Nil. DLRC: St Rollox: 260467-280467 U

66A Polmadie: 200567 (Repair Shop)
St Rollox Works: 280567 DLRC: St Rollox: 220567-170667 U

Transfer LMR(o/l): 1we111167 (111167)-12A
Transfer LMR(perm): 1we091267 (031267)-12A

Transfer: 1we220668 (170668)-D10

Withdrawn (ex-LMR): 1we121068 (121068) (ex-D10)

12A Carlisle Kingmoor (Steam): 241068/ 071268/ 311268/ 030169/ 110169
Kingmoor Yard: xx0269/030569/ 120569/ 260569/ 060769/ 200869/ 240869/ 040969/ 200969/ 280370/ 300370/ xx0470
Abington: 270470 (8Z43 Carlisle Kingmoor-Polmadie, with D8518/20)
66A Polmadie: 240570/ 130670/ 140670/ 280670 (out of use, partly cannibalised)/ 030770/ 300770/ 310770/ 020870/ 040870/ 160870/ 220870/ 041070/ 071170/ 100171/ 140271/ 270271)/ 110471/ 250471
St Rollox Works: 170471/190571 (dumped near old Carriage Shop)/ 310571/ 190671 (¼mile north of Works)/ 070871/ 100871/ 160871/ 210871/ 300871/ 270971/ 161071 (Dump)/ 211171/ 281171/ 090172/ 120272/ 270272/ 010472/ 020472 (Works Yard)/ 090472 (Old Paint Shop Yard)/ 140472 (Top Yard)/ 250472/ 260472/ 280572. Not listed 140572/ 100672.

Disposal: St Rollox Works: 0572 (D&ELfS).
Commencing month of cutting-up: 0572 (RO0776).

D8520   Serial No.: 4365/U 23

D8520, 12A Carlisle Kingmoor (Diesel), 19 April 1968. (TOPticl Digital Memories (Roger Hateley))

New (ScR): 1we180563 (170563)-66A

St Rollox Works: Nil. DLRC: St Rollox: 191264-191264 U

XX: 190866 (fire, location unknown) (FDTL)
66A Polmadie: 210866

St Rollox Works: Nil. Not listed 280567. DLRC: St Rollox: 250567-070767 I

66A Polmadie: 200867
St Rollox Works: Nil. DLRC: St Rollox: 170867-250867 U

St Rollox Works: Nil. DLRC: St Rollox: 280867-080967 U

Transfer LMR(o/l): 1we161267 (101267)-12A
Transfer LMR(perm): 1we060168 (311267)-12A

Transfer: 1we220668 (170668)-D10

Withdrawn (ex-LMR): 1we121068 (121068) (ex-D10)

12A Carlisle Kingmoor (Steam): 241068/ 071268/ 311268/ 030169/ 110169
Kingmoor Yard: xx0269/ 120569/ 260569/ 060769/ 200869/ 040969/ 200969/ 280370/ 300370/ xx0470. Not listed 240869.
Abington: 270470 (8Z43 Carlisle Kingmoor-Polmadie, with D8518/9)
66A Polmadie: 240570/ 130670/ 140670/ 030770/ 040870/ 160870/ 220870/ 260870/ 041070/ 071170/ 100171/ 140271/ 110471/ 250471. Not listed 300770/ 310770/ 020870.
St Rollox Works: 170471/ 190571 (dumped near old Carriage Shop)/ 310571/ 190671 (¼mile north of Works)/ 070871/ 100871/ 160871/ 210871/ 300871/ 270971/ 161071 (Dump)/ 211171/ 281171/ 090172/ 120272/ 270272/ 010472/ 020472 (Works Yard)/ 090472 (Old Paint Shop Yard)/ 140472 (Top Yard)/ 250472/ 260472. Not listed 140371 and 140572/ 100672.

Disposal: St Rollox Works: 0572 (D&ELfS). Commencing month of cutting-up: 0572 (RO0776).

D8521 Serial No.: 4365/U 17

D8521, 66A Polmadie, 6 August 1963.
(David Dippie)

Derby St Andrews Goods Yard: 2we020363 (stored)/ 030363/ 100363 (stored)/ 070463/ 140463/ 160663/ 230663

New (ScR): 1we270763 (230763)-66A

St Rollox Works: Nil. DLRC: St Rollox: 260564-030664 U

St Rollox Works: Nil. DLRC: St Rollox: 130267-250367 I

Transfer to LMR(o/l): 2we130168 (190168/070168)-12A

Transfer to LMR(perm): 1we030268 (280168)-12A

Transfer: 1we220668 (170668)-D10

St Rollox Works: 220668

12A Carlisle Kingmoor (Diesel): 091068

Withdrawn (ex LMR): 1we121068 (121068) (ex-D10)

12A Carlisle Kingmoor (Steam): 241068/ 071268/ 311268/ 030169/ 110169
Kingmoor Yard: xx0269/030569/ 120569
12A Carlisle Kingmoor (Diesel): 260569

Derby RCD/Derby area: Arrived 230769/ 120969/ 191069/ 080370/ 220570/ 300570/ 020870/ 130870/ 070970/ 081070

Derby Works: 251070 (Yard)/281070/ 151170/ 201270 (Klondyke Sdgs)/ 070271/ 060371/ 250471/ 020571/230671 (returned to RTC after repaint)

Derby RCD/Derby area: 270771/280871/ 031171/ 230172/ 060272/ 210572/ 040672/ 060872
Transfer Derby RCD-Mickleover: 170872 (hauled by D8598, with D5901)
Egginton Junction/Mickleover: Nil.

Derby Works: 240274 (Works Yard)/030374/ 100374 (Klondyke Sdgs)/ 210474 (Klondyke Sdgs)/ 120574 (Klondyke Sdgs)/ 230674 (Klondyke Sdgs)/ 040874 (Klondyke Sdgs)/ 310874/ 201074 (Klondyke Sdgs)/ 260175 (Klondyke Sdgs)/ 010275/ 160375. Not listed 230275.

Derby area: 120475/100575

Derby Works: 210675

Derby area: 190775
Egginton Junction/Mickleover: 081176/ 050677/ 030777/ 050878
Booked transfer: 9Z10 11.15 Derby RCD-Carlisle 071078, then Carlisle-St Rollox Works 091078 (with D8598)
Chesterfield: 071078 (hauled by 46027, with D8598)
Rotherham: 091078 (with D8598)
Carlisle: xx1078
St Rollox Works: 141078/151078/ 191178/ 251178 (E.Shop Yard)/ 130179/ 210479 (Scrapping Area, baseframe/bogies only). Not listed 250279.

Disposal: Departmental S18521; St Rollox Works: 0479 (D&ELfS).

Notes:
1. 'D8512/21 . . .Derby Research Centre for experiments with radio-controlled multiple working.' (RO1169)
2. 'On 20th June (*1971*) . . .D8521 had been renumbered S18521. It has been painted all over blue including the ends.' (RO0971)
3. Derby Works, 310874: 'S18521 has been converted to a mobile power station..' (RO1174)
4. RTC Allocations. Reinstated: 23/07/69, Withdrawn: 07/10/78.

D8522   Serial No.: 4365/U 24

D8522. St Rollox Works (Erecting Shop Yard), July 1970.
(Grahame Wareham)

Derby St Andrews Goods Yard: 2we020363 (stored, Makers' No.24)/ 030363/ 100363 (stored, Makers' No.24)/ 070463 (primer, Makers' No.24)

New (ScR): 1we060763 (02 or 030763)-66A

66A Polmadie: 050164
St Rollox Works: Nil. DLRC: St Rollox: 040164-300164 U(HC)

St Rollox Works: Nil. DLRC: St Rollox: 030264-050264 U

66A Polmadie: 170967
St Rollox Works: 241067 (E.Shop)/041167 (under repair). DLRC: St Rollox: 130967-041167 I

Transfer LMR(o/l): 2we200168 (190168/070168)-12A
Transfer LMR(perm): 1we030268 (280168)-12A

Transfer: 1we220668 (170668)-D10

Withdrawn (ex-LMR): 1we121068 (121068) (ex-D10)

12A Carlisle Kingmoor (Steam): 241068/ 071268/ 311268/ 030169/ 110169
Kingmoor Yard: xx0269/120569/ 260569/ 060769/ 200869/ 240869/ 040969/ 200969
12A Carlisle Kingmoor (Diesel): 010270
Carlisle Kingmoor Yard: 280370/ 300370/ xx0470
Abington: 280470 (8Z42 Carlisle Kingmoor-Polmadie, hauled by D6839, with D8526/7)
66A Polmadie: Nil. Not listed 240570.
St Rollox Works: 250570/140670/ 110770/ 040870/ 220870/ 120970/ 031070 (old Paint Shop sidings)/ 041070 (beyond old Carriage Shop)/ 101070/ 100171/ 200271 (½mile from Works)/ 140371/ 170471/ 190571 (dumped near old Carriage Shop)/ 310571/ 190671 (¼mile north of Works)/ 070871/ 100871/ 160871/ 210871/ 300871/ 270971/ 161071 (Yard)/ 281171 / 090172/ 120272/ 270272/ 010472/ 020472 (Works Yard)/ 090472 (Old Paint Shop Yard)/ 140472 (Top Yard).
Not listed 211171 and 250472/ 140572/ 100672.

Disposal: St Rollox Works: 0572 (D&ELfS).
Commencing month of cutting-up: 0572 (RO0776).

Notes:
1. D8522/6/7 were recorded in the *Railway Observer* (July 1970) as transferring to Polmadie on 28/04/70. The first sightings of these locomotives after this date was D8522/6 at St Rollox Works on 25/05/70 and D8527 at Polmadie on 24/05/70. Presumably D8522/6 were staged at Polmadie before onward movement to the Works.

D8523   Serial No.: 4365/U 25

D8523, St Rollox Works (Erecting Shop Yard), 1970. (Transport Topics)

Derby St Andrews Goods Yard: 2we020363 (stored, Makers' No.25)/ 030363/ 100363 (stored, Makers' No.25)/ 070463 (primer, Makers' No.25)

New (ScR): 1we130763 (090763)-66A

St Rollox Works: Nil. DLRC: St Rollox: 270264-200364 U

St Rollox Works: Nil. DLRC: St Rollox: 140667-310767 I

Transfer LMR(o/l): 2we200168 (200168/070168)-12A
Transfer LMR(perm): 1we030268 (280168)-12A

Transfer: 1we220668 (170668)-D10

12A Carlisle Kingmoor: 091068

Withdrawn (ex-LMR): 1we121068 (121068) (ex-D10)

12A Carlisle Kingmoor (Steam): 241068/ 071268/ 311268/ 030169/ 110169
Kingmoor Yard: xx0269/ 120569/ 260569/ 060769/ 200869/ 240869/ 040969/ 200969
66A Polmadie: 141269/ 301269
St Rollox Works: 110170/ 270370/ 290370/ 110470 (Works Yard)/ 250570/ 140670/ 110770/ 040870/ 220870/ 120970/ 031070 (old Paint Shop sidings)/ 041070 (beyond old Carriage Shop)/ 101070/ 100171/ 200271 (½mile from Works)/ 140371/ 170471/ 190571 (dumped near old Carriage Shop)/ 310571/ 190671 (¼mile north of Works)/ 070871/ 100871/ 160871/ 210871/ 300871/ 270971/ 161071 (Yard)/ 211171/ 281171/ 090172/ 120272/ 270272/ 010472/ 020472 (Works Yard)/ 090472 (Old Paint Shop Yard)/ 140472 (Top Yard)/ 250472/ 260472/ 140572/ 280572. Not listed 100672.

Disposal: St Rollox Works: 0672 (D&ELfS). Commencing month of cutting-up: 0672 (RO0776).

D8524   Serial No.: 4365/U 26

D8524, 66A Polmadie, 27 July 1963. (Rail-Online)

Derby St Andrews Goods Yard: 2we020363 (stored, Makers' No.26)/ 030363/ 100363 (stored, Makers' No.26) / 070463 (primer, Makers' No.26)

New (ScR): 1we200763 (15 or 160763)-66A

St Rollox Works: Nil. DLRC: St Rollox: 200364-110464 U

St Rollox Works: Nil. DLRC: St Rollox: 030266-300366 I

Transfer LMR(o/l): 3we130168 (250168/070168)-12A
Transfer LMR(perm): 1we 030268 (280168)-12A

Transfers: 1we220668 (170668)-D10

Withdrawn (ex-LMR): 1we121068 (121068) (ex-D10)

12A Carlisle Kingmoor (Steam): 241068/ 071268/ 311268/ 030169/ 110169
Kingmoor Yard: xx0269/120569/ 260569/ 060769/ 200869/ 240869/ 040969/ 200969
66A Polmadie: 141269/301269/ 110170/ 290370/ 040470/ 110470/ 240570/ 130670/ 140670/ 030770/ 300770/ 310770/ 020870. Not listed 220870.
J. MacWilliam, Shettleston: 230870 (intact)

Disposal: J. MacWilliam, Shettleston: 0870 (D&ELfS).

D8525   Serial No.: 4365/U 27

D8525, 66A Polmadie, 29 September 1964.
(Rail-Online)

Derby (near London Road Bridge): 030363
Derby (Sidings opposite Shed): 100363 (stored, Makers' No.27)/ 230363 (stored, Makers' No.27)/ 070463 (primer, Makers' No.27)/ 160663 (primer, Maker's No.27)

New (ScR): 1we170863 (140863)-66A

St Rollox Works: 170564. DLRC: St Rollox: 290464-150564 U

St Rollox Works: Nil. DLRC: St Rollox: 110165-160165 U

St Rollox Works: 070168. DLRC: St Rollox: 011267-060168 I

Transfer LMR(o/l): 2we200168 (180168/070168)-12A
Transfer LMR(perm): 1we 030268 (280168)-12A

Transfer: 1we220668 (170668)-D10

Transfer ScR: 1we121068 (121068/061068)-64B

Transfer: 1we161168 (111168)-66A

St Rollox Works: 160871. Not listed 070871. DLRC: St Rollox: 050771-140871 U

Withdrawn (ex-ScR): 4we091071 (061071) (ex-66A)

66A Polmadie: 191071/271171/ 030172/ 070172/ 270272/ 300372/ 230472/ 140572/ 100672/ 080772/ 170872/ 290872/ 060972/ 160972/ 141072
St Rollox Works: 221072/051172/ 191172/ 100273/ 110273/ 080473/ 210473/ 220473/ 050573/ 120573/ 130573/ 280573/ 290573/ 090673 (Yard)/ 170673/ 050873/ 110873/ 120873/ 140873/ 160873/ 150973/ 131073/ 041173/ 181173/ 130174/ 100374/ 230374/ 130474/ 140474/ 280574/ 150674 (Dump)/ 200674/ 030874/ 240874/ 260974 (N of Traverser)/ 051074 (N of Traverser)/ 221274/ 260175/ 080375/ 200375/ 290375/ 170575 (N of Traverser)/ 270575/ 070675 (Dump)/ mid-0775 (Dump)/ 120875)
Transfer to R.A. King, Norwich: 04-080975 (D8507/8/16/25/9/31/6/52/74, 9X20 dep. Glasgow 03.35 040975, arr March Up Yard 17.55 080975; dep. March 19.45 080975, arr. Norwich Trowse Yard 23.30 080975)
Tyne Yard: 060975
Doncaster: 070975 (with 8507/8/16/29/31/6/52/74)
March: 080975 (hauled by 47360, with D8507/8/16/29/31/6/52/74)
Wensum Yard, Norwich: 270975/241075/ 181275/ 271275
Transfer (Trowse Yard-R.A. King, Norwich): 210176 (with D8529/39)
R.A. King, Norwich: 290276 (partly dismantled)/ 100476 (part dismantled, cab/engines only on baseframe, on bogies)

Disposal: A. King & Son, Norwich: 0176 (D&ELfS).

D8526   Serial No.: 4365/U 28

D8526, 66B Motherwell, 1 September 1963.
(Rail-Online)

Derby (Sidings opposite Shed): 100363 (stored, Makers' No.28)/ 230363 (stored, Makers' No.28)/ 070463 (primer, Maker's No.28)/ 160663 (primer, Maker's No.27)
Ais Gill: 200863 (with D8540)

New (ScR): 1we240863 (19 or 200863)-66A

St Rollox Works: Nil. DLRC: St Rollox: 031165-271165 U

St Rollox Works: 140967 (E.Shop). DLRC: St Rollox: 040967-211067 I

Transfer LMR: 1we300368 (240368)-12A

12A Carlisle Kingmoor (Diesel): 150668
St Rollox Works: 220668. Not listed 200668 and 180768.

Transfer: 1we220668 (170668)-D10

12A Carlisle Kingmoor (Diesel): 091068

Withdrawn (ex-LMR): 1we121068 (121068) (ex-D10)

Note:
1. See Note 1 for D8522.

12A Carlisle Kingmoor (Steam): 071268/ 311268/ 030169/ 110169. Not listed 241068.
Kingmoor Yard: xx0269/120569/ 260569/ 060769/ 200869/ 240869/ 040969/ 200969/ 280370/ 300370/ xx0470
12A Carlisle Kingmoor (Diesel): 030470
Abington: 280470 (8Z42 Carlisle Kingmoor-Polmadie, hauled by D6839, with D8522/7)
66A Polmadie: Nil. Not listed 240570.
St Rollox Works: 250570/140670/ 110770/ 040870/ 220870/ 120970/ 031070 (old Paint Shop sidings)/ 041070 (beyond old Carriage Shop)/ 101070/ 100171/ 200271 (½mile from Works)/ 140371/ 170471/ 190571 (dumped near old Carriage Shop)/ 310571/ 190671 (¼mile north of Works)/ 070871/ 100871/ 160871/ 210871 (Old Paint Shop Yard)/ 300871/ 270971/ 161071 (Yard)/ 211171/ 281171/ 090172/ 120272/ 270272/ 010472/ 020472 (Works Yard)/ 090472 (Old Paint Shop Yard)/ 140472. Not listed 250472/ 140572/ 100672.

Disposal: St Rollox Works: 0572 (D&ELfS). Commencing month of cutting-up: 0572 (RO0776).

D8527   Serial No.: 4365/U 29

Derby St Andrew's Goods Yard: 160663 (primer, Maker's No.29)/ 230663 (stored, Makers' No.29)/ 070763 (stored, Makers' No.29). N.B. Gone by 300763.

New (ScR): 1we310863 (23 or 290863)-66A

66A Polmadie: 260964
St Rollox Works: Nil. DLRC: St Rollox: 240964-161064 U

St Rollox Works: Nil. DLRC: St Rollox: 130265-130265 U

St Rollox Works: 041167 (Works Yard). DLRC: St Rollox: 011167-091267 I

Transfer LMR: 1we300368 (240368)-12A

Transfer: 1we220668 (170668)-D10

Stored ( ): 1we121068 (121068) (ex-D10)

Reinstated: 1we301168 (141068)-D10 (late entry)

Withdrawn (ex-LMR): 2we281268 (281268) (ex-D10)

12A Carlisle Kingmoor (Diesel): 030169/ 110169
Kingmoor Yard: 120569/260569/ 060769/ 200869/ 210869/ 240869/ 040969/ 200969/ 280370/ 300370/ xx0470.
Abington: 280470 (8Z42 Carlisle Kingmoor-Polmadie, hauled by D6839, with D8522/6)
66A Polmadie: 240570/140670/ 030770/ 020870/ 040870. Not listed 300770/ 310770.
St Rollox Works: 220870 (dumped beyond the old carriage shop)/ 031070 (old Paint Shop sidings)/ 041070 (beyond old Carriage Shop)/ 101070/ 100171/ 200271 (½ mile from Works)/ 140371/ 170471/ 190571 (dumped near old Carriage Shop)/ 310571/ 190671 (¼mile north of Works)/ 070871/ 100871/ 160871/ 210871/ 300871/ 270971/ 161071 (Dump)/ 211171/ 281171/ 090172. Not listed 120272/ 270272/ 020472/ 090472/ 140572/ 280572.

Disposal: St Rollox Works: 0672 (D&ELfS). Commencing month of cutting-up: 0672 (RO0776).

Notes:
1. See Section 20.4 for further discussion re. disposal of D8527.
2. See Note 1 for D8522.

D8528   Serial No.: 4365/U 30

Derby (Sidings opposite Shed): 100363 (stored, Makers' No.30)/ 230363 (stored, Makers' No.30)/ 070463 (primer, Maker's No.30)/ 160663 (primer, Maker's No.27)

New (ScR): 1we270763 (230763)-66A

66A Polmadie: 290566
St Rollox Works: Nil. DLRC: St Rollox: 270566-160766 I

St Rollox Works: Nil. DLRC: St Rollox: 061167-021267 U

Transfer to LMR: 1we300368 (240368)-12A

Transfer: 1we220668 (170668)-D10

St Rollox Works: 180768
12A Carlisle Kingmoor (Diesel): 100868

Withdrawn (ex-LMR): 1we121068 (121068) (ex-D10)

12A Carlisle Kingmoor (Steam): 241068/ 071268/ 030169/ 110169. Not listed 311268.
Kingmoor Yard: xx0269/ 120569/ 260569/ 060769/ 200869/ 240869/ 040969/ 200969

64B Haymarket: arrived ex-Carlisle on 101069 (with D8529/31)

Reinstated (ScR): 1we081169 (021169)-64B

St Rollox Works: 251070. DLRC: Glasgow: 300970-141170 C

Withdrawn (ex-ScR): 4we091071 (051071) (ex-64B)

64B Haymarket: 091071
Millerhill Yard: 271171/ 030172/ 080172/ 270272/ 310372/ 230472/ 130572/ 030672/ 230772/ 180872/ 260872/ 020972
St Rollox Works: 141072 (Dump)/ 221072/ 051172/ 191172 (outside)/ 100273/ 110273/ 080473/ 210473/ 220473/ 230473 (Top Yard)/ 050573/ 120573/ 130573/ 280573/ 290573/ 090673 (Yard)/ 170673/ 160873. Not listed 110873/ 120873/ 150973.

Disposal: St Rollox Works: 0873 (D&ELfS).
'...cut up by 16/8/73.' (AHBRD&E5)
'...cut up...about July (*1973*). (RO1273)

D8529 Serial No.: 4365/U 31   N.B. Incomplete DLRC information.

D8529, 66A Polmadie, 28 May 1972. (Rail-Online)

New (ScR): 1we180563 (170563)-66A

66A Polmadie: 200867
St Rollox Works: 140967 (E.Shop). **DLRC: St Rollox:** 180867-071067 C

Transfer LMR: 1we300368 (240368)-12A

Transfer: 1we220668 (170668)-D10

Withdrawn: 1we121068 (121068) (ex-D10)

Reinstated: 1we021168 (021168)-D10 (store)
Reinstated: 1we301168 (041168)-D10 (late entry)

Withdrawn (ex-LMR): 2we281268 (281268) (ex-D10)
12A Carlisle Kingmoor (Diesel): 030169/ 110169
Kingmoor Yard: 120569/ 260569/ 060769/ 200869/ 240869/ 040969/ 200969
64B Haymarket: arrived ex-Carlisle on 101069 (with D8528/31)

Reinstated (ScR): 1we081169 (021169)-64B

Withdrawn: 4we091071 (061071) (ex-64B)

Reinstated: 4we271171 (071171)-64B

Transfer: 5we010172 (061271)-66A

64B Haymarket: 121271
66A Polmadie: 171271
Withdrawn (ex-ScR): 2we010172 (311271) (ex-66A)

66A Polmadie: 030172/ 090172/ 270272 ('working off 66A') (mobile generator?)/ 300372/ 230472/ 140572/ 280572/ 100672/ 080772/ 170872/ 290872/ 060972/ 160972/ 141072
St Rollox Works: 221072/ 051172/ 191172/ 100273/ 110273/ 080473/ 210473/ 220473/ 050573/ 120573/ 130573/ 280573/ 290573/ 090673 (Yard)/ 170673/ 050873/ 110873/ 120873/ 140873/ 160873/ 150973/ 131073/ 041173/ 181173/ 130174/ 100374/ 230374/ 130474/ 140474/ 280574/ 150674/ 200674/ 030874/ 240874/ 260974 (N of Traverser)/ 051074 (N of Traverser)/ 221274/ 260175/ 080375/ 200375/ 290375/ 170575 (N of Traverser)/ 270575/ 070675 (Dump)/ 120675 (N of Traverser)/ mid-0775 (Dump)/ 120875
Transfer to R.A. King, Norwich: 04-080975 (D8507/8/16/25/9/31/6/52/74, 9X20 dep. Glasgow 03.35 040975, arr. March Up Yard 17.55 080975; dep. March 19.45 080975, arr. Norwich Trowse Yard 23.30 080975)
Tyne Yard: 060975
Doncaster: 070975 (with 8507/8/16/25/31/6/52/74)
March: 080975 (hauled by 47360, with D8507/8/16/25/31/6/52/74)
Wensum Yard, Norwich: 270975/ 241075/ 181275/ 271275
Transfer (Trowse Yard-R.A. King, Norwich): 210176 (with D8525/30)
R.A. King, Norwich: 100476 (whole)

Disposal: A. King & Son, Norwich: 0176 (D&ELfS). '. . .cut up by 1/3/76.' (AHBRD&E5)

D8530 Serial No.: 4365/U 32

D8530, 66A Polmadie, 27 July 1963. (Rail-Online)

Derby: 180463 (Derby-Millers Dale test train)

New (ScR): 1we270463 (260463)-66A

St Rollox Works: Nil. DLRC: St Rollox: 051264-051264 U

St Rollox Works: Nil. DLRC: St Rollox: 210665-100765 U
66A Polmadie: 040765

66A Polmadie: 080167
St Rollox Works: Nil. DLRC: St Rollox: 141066-180267 I

Transfer LMR: 1we300368 (240368)-12A

Transfer: 1we220668 (170668)-D10

Withdrawn (ex-LMR): 1we121068 (121068) (ex-D10)

12A Carlisle Kingmoor (Steam): 241068/ 311268/ 030169/ 110169

Kingmoor Yard: xx0269

Reinstated (ScR): 1we240569 (180569)-66A
Withdrawn: 1we070669 (180569) (ex-66A/D10) (late entry)

Kingmoor Yard: 260569/ 060769/ 200869/ 240869/ 040969. Not listed 120569.
12A Carlisle Kingmoor (Diesel): 140969/ 200969 (being prepared for return to service in Scotland)

Reinstated (ScR): 1we111069 (051069)-64B

Withdrawn (ex-ScR): 5we060371 (030371) (ex 64B)

66A Polmadie: 110471 ('dead' road)/ 250471/ 280571 (Works Yard)/ 050871/ 060871/ 070871/ 100871/ 160871/ 210871/ 290871.
St Rollox Works: 270971 (<u>unidentified</u> cut-up remains). Not listed 161071.

Disposal: St Rollox Works: 1071 (D&ELfS). Commencing month of cutting-up: 1071 (RO0776) Disposal not proven.

D8531 Serial No.: 4365/U 33 N.B. Incomplete DLRC information.

D8531, 66A Polmadie, 2 June 1963. (Colour-Rail)

New (ScR): 1we110563 (08 or 090563)-66A

St Rollox Works: Nil. DLRC: St Rollox: 060464-020564 U

XX: 240566 (fire, location unknown) (FDTL)

66A Polmadie: 180966
St Rollox Works: Nil. DLRC: St Rollox: 120966-101166 I

Transfer LMR: 1we300368 (240368)-12A

Transfer: 1we220668 (170668)-D10

Withdrawn (ex-LMR): 1we121068 (121068) (ex-D10)

12A Carlisle Kingmoor (Steam): 241068/ 311268/ 030169/ 110169
Kingmoor Yard: xx0269/120569/ 260569/ 060769/ 200869/ 240869/ 040969/ 200969
64B Haymarket: arrived ex-Carlisle on 101069 (with D8528/9)

Reinstated (ScR): 1we081169 (021169)-64B

Transfer: 1we151169 (141169)-66A

66A Polmadie: 280571 (Repair Bay)

Stored ( ): 5we110971 (250771) (ex-66A) (late entry)

66A Polmadie: 070871/080871/ 160871 210871/ 290871

Withdrawn (ex-ScR): 4we091071 (210971) (ex-66A (store))

66A Polmadie: 260971/271171
Ex-67D Ardrossan: Arrived mid-1271/ 030172/ 090172/ 260272/ 300372/ 250472/ 140572/ 250672/ 170772/ 170872/ 060972
66A Polmadie: 061172/091172/ 191172.
Not listed 141072.
St Rollox Works: 100273/110273/ 080473/ 210473/ 220473/ 050573/ 120573/ 130573/ 280573/ 290573/ 090673 (Yard)/ 170673/ 050873/ 110873/ 120873/ 140873/ 160873/ 150973/ 131073/ 041173/ 181173/ 130174/ 100374/ 230374/ 130474/ 140474/ 280574/ 150674 (Dump)/ 200674/ 030874/ 240874/ 260974 (N of Traverser)/ 051074 (N of Traverser)/ 121074 (N of Traverser)/ 241174 (N of Traverser)/ 221274/ 260175/ 080375/ 200375/ 290375/ 170575 (N of Traverser)/ 270575/ 070675 (Dump) / 120675 (N of Traverser)/ mid-0775 (Dump)/ 100875 (N of Traverser)/ 120875
Transfer to R.A. King, Norwich: 04-080975 (D8507/8/16/25/9/31/6/52/74, 9X20 dep. Glasgow 03.35 040975, arr. March Up Yard 17.55 080975; dep. March 19.45 080975, arr. Norwich Trowse Yard 23.30 080975)
Tyne Yard: 060975
Doncaster: 070975 (with D8507/8/16/25/9/36/52/74)
March: 080975 (hauled by 47360, with D8507/8/16/25/9/36/52/74)
Wensum Yard, Norwich: 270975/241075
R.A. King, Norwich: Nil.

Disposal: A. King & Son, Norwich: 1175 (D&ELfS). Disposal not proven.

D8532   Serial No.: 4365/U 34

D8532, Carlisle (possibly 12A Carlisle Kingmoor (Diesel), 1970. D8532 with D8533/4 awaiting transfer from Carlisle to Polmadie. (Transport Topics)

Derby area: Exact location/dates unknown (stored)

New (ScR): 1we030863 (26 or 290763)-66A

St Rollox Works: Nil. DLRC: St Rollox: 031263-211263 LC

66A Polmadie: 081164
St Rollox Works: Nil. DLRC: St Rollox: 031164-131164 U

66A Polmadie: 100165
St Rollox Works: Nil. DLRC: St Rollox: 060165-290165 U

66A Polmadie: 040566/ 290566
St Rollox Works: Nil. DLRC: St Rollox: 210466-080666 U

St Rollox Works: Nil. DLRC: St Rollox: 110766-140766 Load test

St Rollox Works: 070168. DLRC: St Rollox: 121267-200168 I

Transfer LMR: 1we300368 (240368)-12A (ScR)

Transfer: 1we220668 (170668)-D10

Stored ( ): 1we121068 (121068) (ex-D10)

Reinstated: 1we301168 (141068)-D10 (late entry)

Withdrawn (ex-LMR): 2we281268 (281268) (ex-D10)

12A Carlisle Kingmoor (Diesel): 030169/ 110169
Kingmoor Yard: 120569/260569/ 060769/ 200869/ 240869/ 040969/ 200969/ 280370/ 300370/ xx0470/ 250570
Abington: 280470 (8Z43 Carlisle Kingmoor-Polmadie, with D8533/4)
66A Polmadie: 130670/140670/ 030770/ 300770/ 310770/ 020870/ 040870. Not listed 240570 and 160870.

St Rollox Works: 220870 (dumped beyond the old carriage shop)/ 031070 (Old Paint Shop sidings)/ 041070 (beyond Old Carriage Shop)/ 101070/ 100171/ 200271 (½mile from Works)/ 140371/ 170471/ 190571 (dumped near old Carriage Shop)/ 310571/ 190671 (¼mile north of Works)/ 070871/ 100871/ 160871/ 210871/ 300871/ 270971/ 161071 (Yard)/ 211171/ 281171/ 090172. Not listed 120272/ 270272/ 020472/ 090472/ 140572/ 280572.

Disposal: St Rollox Works: 0672 (D&ELfS). Commencing month of cutting-up: 0672 (RO0776).

Notes:
1. 'Abington. On 27th April (*1970*), 1739 (D05) hauled D8501/9/14 from Kingmoor to Polmadie as 8Z42, and 8518/9/20 followed later as 8Z43. On the 28th 8Z42 consisted of 8522/6/7 hauled by 6839 (66A), and 8Z43 was 8532/3/4' (RO0770).
Comments: Did 8Z43 run on 28 April 1970, or was the *Railway Observer* report actually a booked transfer? D8532-4 were <u>not</u> subsequently seen at Polmadie on 22/05/70, and, sightings exist for D8532-4 at Carlisle (exact location unspecified) on 25/05/70.
It is strange also that the RO report gives the hauling locomotive for 8Z42 but not for 8Z43 (on both days). Therefore, as well as the obvious doubt surrounding D8532-4, it <u>may</u> also be that D8518-20 moved on a different day to that specified.
2. See Section 20.4 for further discussion re. disposal of D8532.

D8533   Serial No.: 4365/U 35

D8533, Carlisle (possibly 12A Carlisle Kingmoor [Diesel]), 1970. Perhaps a career highlight for D8533 (with D8548) was the haulage of the "Night Limited" on 3 April 1968 between Motherwell and Glasgow Central due to a derailment at Shawfield.
(Transport Topics)

Derby area: Exact location/dates unknown (stored)
Ais Gill: 030863 (with D8513 (sic, more likely D8515)

New (ScR): 1we030863 (02 or 030863)-66A

St Rollox Works: Nil. DLRC: St Rollox: 161263-080164 U(HC)

66A Polmadie: 180966/230966/ 021066/ 121166
St Rollox Works: Nil. DLRC: St Rollox: 090966-261166 I

Transfer LMR(o/l): 1we110568 (080568/110568)-12A
Transfer LMR(perm): 1we250568) (190568)-12A

Transfer: 1we220668 (170668)-D10

St Rollox Works: 170868 (E.Shop)
65A Eastfield: 240868/250868

Note:
1. See Note.1 for D8532.

Withdrawn (ex-LMR): 1we121068 (121068) (ex-D10)

12A Carlisle Kingmoor (Steam): 241068/ 311268/ 030169/ 110169. Not listed 071268.
Kingmoor Yard: xx0269/120569/ 260569/ 060769/ 200869/ 240869/ 040969/ 200969/ 280370/ 300370/ xx0470/ 250570
Abington: 280470 (8Z43 Carlisle Kingmoor-Polmadie, with D8532/4)
66A Polmadie: 130670/140670/ 030770/ 300770/ 310770/ 020870/ 040870/ 160870/ 220870/ 041070/ 071170/ 100171/ 140271/ 110471/ 250471. Not listed 240570.
St Rollox Works: 190571 (Yard)/310571/ 190671/ 070871/ 100871/ 160871/ 210871/ 300871/ 270971/ 161071 (Dump)/ 211171/ 281171. Not listed 090172/ 120272.

Disposal: St Rollox Works: 0172 (D&ELfS).
Commencing month of cutting-up: 0172 (RO0776).

D8534 Serial No.: 4365/U 36

D8534, Carlisle (possibly 12A Carlisle Kingmoor [Diesel]), 1970. Twenty-eight withdrawn Claytons were sent to Ardrossan from Polmadie for storage, all arriving by 3 January 1972. Twenty-seven were withdrawn during the last four months of 1971 from Scottish Region books. The exception was D8534 which had been withdrawn as early as October 1968 from Carlisle Kingmoor! At a guess it just got caught up in the mass clearout to make space for the wave of replacement locomotives at Polmadie. (Transport Topics)

New (ScR): 1we150663 (11 or 120663)-66A

St Rollow Works: Nil. DLRC: St Rollox: 261163-201263 LC

66A Polmadie: 050164
St Rollox Works: Nil. DLRC: St Rollox: 301263-110164 U(HC)

66A Polmadie: 020767
St Rollox Works: 170867/210867 DLRC: St Rollox: 190667-190867 I.

St Rollox Works: Nil. DLRC: St Rollox:280867-020967 Rect

Transfer LMR(o/l): 1we110568 (080568/110568)-12A
Transfer LMR(perm): 1we250568 (190568)-12A

Transfer: 1we220668 (170668)-D10

St Rollox Works: 200668. Not listed 250568 and 220668.

12A Carlisle Kingmoor (Diesel): 091068

Note:
1. See Note.1 for D8532.

Withdrawn (ex-LMR): 1we121068 (121068) (ex-D10)

12A Carlisle Kingmoor (Steam): 241068/ 311268/ 030169/ 110169
Kingmoor Yard: xx0269/120569/ 260569/ 060769/ 200869/ 210869/ 240869/ 040969/ 200969/ 280370/ 300370/ xx0470/ 250570
Abington: 280470 (8Z43 Carlisle Kingmoor-Polmadie, with D8533/4)
66A Polmadie: 130670/140670/ 030770/ 300770/ 310770/ 020870/ 040870/ 160870/ 220870/ 041070/ 071170/ 100171/ 140271/ 110471/ 250471/ 280571/ 070871/ 160871/ 210871/ 290871/ 260971/ 271171. Not listed 240570.
Ex-67D Ardrossan: Arrived mid-1271/ 030172/ 090172/ 260272/ 300372/ 250472/ 140572/ 170772/ 170872/ 060972
66A Polmadie: 061172/091172/ 191172. Not listed 141072.
St Rollox Works: 100273 (outside Works)/ 110273. Not listed 080473/ 210473.

Disposal: St Rollox Works: 0273 (D&ELfS).

D8535   Serial No.: 4365/U 21

D8535, 66B Motherwell, undated. After completion of the transfer of D8500-34 to the LMR with D8533/4 in May 1968, the next three numerically, D8535-7, were declared as being surplus to requirements by the Scottish Region at the same time. In the event, however, these three were not withdrawn from service but D8537 did subsequently become the first Clayton withdrawal on 21 July 1968, due to Paxman engine defects. (Transport Topics)

Derby area: Exact location/dates unknown (stored)

New (ScR): 1we200763 (17 or 180763)-66A

St Rollox Works: Nil. DLRC: St Rollox: 091167-161267 I

Withdrawn (ex-ScR): 4we030471 (270371) (ex-66A)

66A Polmadie: 110471 ('dead' road) / 250471 / 280571 / 070871 / 160871 / 210871 / 290871 / 070971
St Rollox Works: 270971. Not listed 161071/ 211171.

Disposal: St Rollox Works: 1171 (D&ELfS). Commencing month of cutting-up: 1171 (RO0776).

D8536 Serial No.: 4365/U 37

D8536, St Rollox Works (Near Traverser), 26 September 1974.
(Anthony Sayer)

Derby (St Andrew's Goods Yard): 160663 (primer, Maker's No.37)/ 230663 (stored, Makers' No.37)/ 070763 (stored, Makers' No.37). Gone by 300763.

New (ScR): 1we310863 (290863)-66A

Transfer (NER): 1we070963 (040963)-NER(52H/56B) (on loan)

Lynemouth-Cambois: 090963 (test train)
56B Ardsley: 140963 (trials (E&W Yorkshire Union Railway route), with D8501)

Transfer (ScR): 1we051063 (300963)-66A

St Rollox Works: Nil. DLRC: St Rollox: 060264-140264 U

XX: 300566 (fire, location unknown) (FDTL)

St Rollox Works: Nil. DLRC: St Rollox: 270167-250367 I

66A Polmadie: 200867
St Rollox Works: Nil. DLRC: St Rollox: 170867-010967 U

St Rollox Works: Nil. DLRC: St Rollox: 310768-170868 U

St Rollox Works: 270370/280370/ 290370/ 110470 (E.Shop). DLRC: Glasgow: 130370-170470 C

Withdrawn (ex-ScR): 2we010172 (311271) (ex-66A)

66A Polmadie: 030172/090172/ 230272/ 300372/ 230472/ 140572/ 100672/ 080772/ 170872/ 290872/ 060972/ 160972/ 141072
St Rollox Works: 221072/051172/ 191172 (outside)/ 100273/ 110273/ 080473/ 210473/ 220473/ 050573/ 120573 (CS)/ 130573/ 280573/ 290573/ 090673 (Yard)/ 170673/ 090773 (CS)/ 050873 (CS)/ 110873/ 120873/ 140873/ 160873/ 150973 (CS)/ 131073/ 231073 (CS))/ 041173/ 181173/ 130174/ 100374/ 230374/ 130474/ 140474/ 010574 (CS)/ 280574/ 150674 (Dump)/ 200674/ 030874/ 240874/ 260974 (N of Traverser)/ 051074 (N of Traverser)/ 221274/ 260175/ 080375/ 200375/ 290375/ 170575 (N of Traverser)/ 270575/ 070675 (Dump)/ mid-0775 (Works Yard)/ 120875
Transfer to R.A. King, Norwich: 04-080975 (D8507/8/16/25/9/31/6/52/74, 9X20 dep. Glasgow 03.35 040975, arr. March Up Yard 17.55 080975; dep. March 19.45 080975, arr. Norwich Trowse Yard 23.30 080975)
Tyne Yard: 060975
Doncaster: 070975 (with D8507/8/16/25/9/31/52/74)
March: 080975 (hauled by 47360, with D8507/8/16/25/9/31/52/74)
Wensum Yard, Norwich: 270975/241075
R.A. King, Norwich: 311075/011175. Not listed 271275.

Disposal: A. King & Son, Norwich: 1075-1275 (D&ELfS).

D8537   Serial No.: 4365/U 40

D8537, St Rollox Works, 6 June 1965. The move of D8537 to 67A Corkerhill following storage is notable; Corkerhill was rarely involved with any form of maintenance work on the Clayton fleet. (Author's Collection)

New (ScR): 1we150663 (11 or 120663)-66A

St Rollox Works: Nil. DLRC: St Rollox: 140164-010264 U(HC)

St Rollox Works: Nil. DLRC: St Rollox: 100364-250364 U

St Rollox Works: Nil. DLRC: St Rollox: 100265-060365 U

St Rollox Works: 060665 (Works Yard). DLRC: St Rollox: 020665-050665 U

St Rollox Works: Nil. DLRC: St Rollox: 220966-231266 I

Stored (u): 1we110568 (050568) (ex-66A)

66A Polmadie: 120568
67A Corkerhill: 200668/180768

Withdrawn (ex-ScR): 1we270768 (210768) (ex-66A (store))

67A Corkerhill: 210768
66A Polmadie: 170868/250868/ 260868/ 290868/ 100968/ 140968/ 061068/ 120169. Not listed 250369.
J. MacWilliam, Shettleston: Nil.

Disposal: J. MacWilliam, Shettleston: 0968-0169 (D&ELfS).
Disposal not proven.

Notes:
1. First Clayton withdrawn. Reason for Withdrawal: Engine defect, surplus.
2. Sold for scrap (January 1969): D8537/53 (to J. MacWilliam, Shettleston, from Polmadie). (RO0469)

D8538   Serial No.: 4365/U 41

D8538, 66A Polmadie, 6 August 1963.
(David Dippie)

New (ScR): 1we220663 (210663)-66A

66A Polmadie: 201264/241264
St Rollox Works: Nil. DLRC: St Rollox: 171264-080165 U

66A Polmadie: 051265
St Rollox Works: Nil. DLRC: St Rollox: 301165-241265 U

XX: 061066 (fire, location unknown) (FDTL)

Beattock North: 140467 (1530 Clydesdale-Carlisle freight (with D8537), fire) (FDTL)

St Rollox Works: 280567 DLRC: St Rollox: 150567-170667 I

Glasgow Central: 250667 (station pilot, ex-Works)

Withdrawn (ex-ScR): 4we091071 (061071) (ex-66A)

66A Polmadie: 061071/271171
Ex-67D Ardrossan: Arrived mid-1271/ 030172/ 090172/ 260272/ 300372/ 250472/ 140572
St Rollox Works: 080672 (CS)/ 100672 (CS)/ 180672/ 060772/ 060872/ 100872/ 130872/ 170872 (CS)/ 280872/ 290872/ 020972/ 141072/ 281072/ 051172. Not listed 221072/ 191172/ 100273.

Disposal: St Rollox Works: 1272 (D&ELfS). Commencing month of cutting-up: 1272 (RO0776).

D8539 Serial No.: 4365/U 38

D8539, 65A Eastfield, Undated. Sometime between September 1974 and December 1975. Used for re-railing exercises. (Colour-Rail)

New (ScR): 1we150663 (140663)-66A

66A Polmadie: 050164
St Rollox Works: Nil. DLRC: St Rollox: 040164-180164 U(HC)

St Rollox Works: Nil. DLRC: St Rollox: 120364-100464 U

66A Polmadie: 020766/ 030766/ 170766/ 310766
St Rollox Works: 080866/ 120866/ 200866 (Test House).
DLRC: St Rollox: 020666-200866 C

66A Polmadie: 011067
St Rollox Works: 241067 (Test House). DLRC: St Rollox: 021067-211067 U

St Rollox Works: Nil. DLRC: Glasgow: 160170-070270 C

Withdrawn (ex-ScR): 4we091071 (061071) (ex-66A)

66A Polmadie: 271171
Ex-67D Ardrossan: Arrived mid-1271/ 030172/ 090172/ 260272/ 300372/ 250472/ 140572/ 170772/ 170872/ 060972

66A Polmadie: 301072/ 061172/ 091172/ 191172/ 100273/ 250273/ 100473/ 120573/ 270573/ 090673/ 170873/ 150973/ 131073/ 181173/ 130174/ 260174/ 170274/ 230374/ 140474/ 280574/ 150674/ 290674/ 030874/ 100874. Not listed 141072.
66B Motherwell: 130874/ 250874
65A Eastfield: 220974/ 280974/ 051074/ 061274/ 110175 / 260175/ 080375/ 160375/ 050475/ 270475/ 300575/ 040775/ 090875/ 120875 (derailed)/ 230875/ 300875/ 041075/ 211075/ 081175/ 151175/ Departed 031275
Tyne Yard: 021275 (9X20 Eastfield-Norwich)
MR March: 071275
Wensum Yard, Norwich: 181275/ 271275 (number painted out)
Transfer (Trowse Yard-R.A. King, Norwich): 210176 (with D8525/29)
R.A. King, Norwich: 100476 (intact)/ 090676 (numbers painted out).

Disposal: A. King & Son, Norwich: 0176-0776 (D&ELfS).

D8540   Serial No.: 4365/U 39

D8540, 66A Polmadie, 25 July 1968.
(RCTS Archive)

Derby St Andrew's Goods Yard: 160663 (primer, Maker's No.39)/ 230663 (stored, Maker's No.39)/ 070763 (stored, Maker's No.39). Gone by 300763.
Ais Gill: 200863 (with D8526)

New (ScR): 1we240863 (19 or 200863)-66A

66A Polmadie: 060964
St Rollox Works: Nil. DLRC: St Rollox: 070964-120964 U (AWS fitment)

St Rollox Works: Nil. DLRC: St Rollox: 271167-301267 I

St Rollox Works: xx0370. Not listed 110170 and 270370/ 280370.
XX: xx0370 (recently ex-shops)

St Rollox Works: 130271. DLRC: Glasgow: 121170-170271 I. P&SP:Repair extended due to waiting Main Generator Armature Bearings.

Withdrawn (ex-ScR): 4we091071 (051071) (ex-66A)

66A Polmadie: 271171
Ex-67D Ardrossan: Arrived mid-1271/ 030172/ 090172/ 260272/ 300372/ 250472/ 140572/ 170772/ 170872/ 060972
66A Polmadie: 061172/091172/ 191172. Not listed 141072.
St Rollox Works: 100273/110273

Disposal: St Rollox Works: 0173 (D&ELfS).

D8541   Serial No.: 4365/U 42

New (ScR): 1we290663 (290663)-66A

66A Polmadie: 120764/ 010864/ 020864
St Rollox Works: Nil. DLRC: St Rollox: 240664-040964 U

St Rollox Works: Nil. DLRC: St Rollox: 210964-250964 U (AWS Fitment)

St Rollox Works: Nil. DLRC: St Rollox: 020265-260265 U
66A Polmadie: 250265

St Rollox Works: Nil. DLRC: St Rollox: 011266-200167 I

St Rollox Works: Nil. DLRC: Glasgow: 281070-281170 C

Withdrawn (ex-ScR): 4we091071 (051071) (ex-66A)

66A Polmadie: 271171
Ex-67D Ardrossan: Arrived mid-1271/ 030172/ 090172/ 260272/ 300372/ 250472/ 140572/ 170772/ 170872/ 060972
66A Polmadie: 061172/ 091172/ 191172. Not listed 141072.
St Rollox Works: 100273 (frames/bogies only) 110273. Not listed 080473.

Disposal: St Rollox Works: 0273 (D&ELfS).
'...by 10/2/73 only the frame and bogies were left... disposed of by 1/3/73.' (AHBRD&E5)

D8542   Serial No.: 4365/U 43

D8542, 66A Polmadie, 26 March 1967. Sunday at Polmadie and typically Claytons everywhere! Miniature snow ploughs fitted and through heating pipe removed.
(Rail-Photoprints)

New (ScR): 1we070963 (060963)-66A

St Rollox Works: Nil. **DLRC:** St Rollox: 061064-091064 U (AWS fitment)

66A Polmadie: 270365/ 110465/ 180465/ 190465
St Rollox Works: Nil. **DLRC:** St Rollox: 030365-010565 U

St Rollox Works: 241067 (E.Shop)/ 041167 (E.Shop)
DLRC: St Rollox: 290967-111167 I

Withdrawn (ex-ScR): 4we091071 (061071) (ex-66A)

66A Polmadie: 271171
Ex-67D Ardrossan: Arrived mid-1271/ 030172/ 090172/ 260272/ 300372/ 250472/ 140572/ 170772/ 170872/ 060972

66A Polmadie: 301072/061172/ 091172/ 191172/ 100273/ 250273/ 100473/ 120573/ 270573/ 090673/ 170873/ 150973/ 131073/ 181173/ 130174/ 260174/ 170274/ 230374/ 140474/ 280574/ 150674/ 290674/ 030874/ 100874/ 250874/ 280974/ 051074/ 260175/ 080375/ 160375/ 290375/ 050475/ 270475. Not listed 141072.
St Rollox Works: 170575 (middle of Carriage Sdgs)/ 270575/ 070675 (Crane Shop)/ mid-0775 (Works Yard)/ 120875
Transfer to J. Cashmore, Great Bridge: 16-xx0975 (D8542/8/50/1, 9X20 via Carlisle/Hellifield)
Chesterfield: 100975 (with D8548/50/1, hauled by 25130)
J. Cashmore Ltd, Great Bridge: Nil.

Disposal: J. Cashmore, Great Bridge: 0975 (D&ELfS). Disposal not proven.

D8543   Serial No.: 4365/U 44

D8543, 66A Polmadie, July 1969. Miniature snow ploughs.
(John Grey Turner)

New (ScR): 1we140963 (09 or 100963)-66A

St Rollox Works: 201064 (E.Shop) DLRC: St Rollox: 191064-231064 U (AWS fitment)

Beattock: 270766 (1510 Mossend-Carlisle freight, fire) (FDTL)

St Rollox Works: 140967 (E.Shop) DLRC: St Rollox: 280867-141067 C

Store ( ): 5we110971 (250771) (ex-66A) (late entry)

66A Polmadie: 070871/ 080871/ 160871/ 210871/ 290871

Withdrawn (ex-ScR): 4we091071 (210971) (ex-66A (store))

66A Polmadie: 260971
St Rollox Works: 211171/ 281171/ 090172. Not listed 120272/ 270272/ 020472/ 090472/ 140572/ 280572.

Disposal: St Rollox Works: 0672 (D&ELfS). Commencing month of cutting-up: 0672 (RO0776).

Note:
1. See Section 20.4 for further discussion re. disposal of D8543.

D8544   Serial No.: 4365/U 45   N.B. No DLRC information.

D8544, Muirhouse Junction, 27 June 1967. (Colour-Rail)

New (ScR): 1we140963 (100963)-66A

65B St Rollox: 100864. Probably visited St Rollox Works before or after this sighting for U repair (inc. AWS fitment).

66A Polmadie: 100466
St Rollox Works: 100466

Stored (s): 1we110169 (050169) (ex-66A)

66A Polmadie: 120169

Reinstated: 1we250169 (190169)-66A

Withdrawn (ex-ScR): 1we070669 (180569) (ex-66A) (late entry)

66A Polmadie: 130769/ 020869/ 160869/ 240869/ 020969/ xx1069/ 141269/ 110170/ 290370/ 040470/ 110470/ 240570/ 130670/ 140670/ 030770/ 300770/ 020870/ 040870/ 160870/ 220870/ 041070/ 071170/ 100171. Not listed 310770 and 140271.
J. MacWilliam, Shettleston: 170471 (intact)

Disposal: J. MacWilliam, Shettleston: 0171-0471 (D&ELfS).
'...broken up by 1/5/71.' (AHBRD&E5)

Notes:
1. Reason for Withdrawal. No.2 engine crankcase damaged.
2. Disposal details required: D8544 (ex-66A). (RO0471).
3. Sold for scrap (January 1971): D8544/54 (to J. MacWilliam, Shettleston, from Polmadie). (RO1071/1271)

D8545   Serial No.: 4365/U 46

D8545, 66A Polmadie, 29 May 1966.
(Author's Collection)

New (ScR): 1we210963 (170963)-66A

Transfer: 1we280963 (230963)-Leith Central
Transfer: 1we161163 (161163)-66A

Falkland Junction: 221063 (crashed through headshunt)/ 231063 (re-railed)
66A Polmadie: 271063
St Rollox Works: 031163 (accident damage). DLRC: St Rollox: 241063-051263 LC

St Rollox Works: Nil. DLRC: St Rollox: 161164-201164 U (AWS fitment)

66A Polmadie: 270365/ 110465/ 180465/ 190465
St Rollox Works: Nil. DLRC: St Rollox: 010365-010565 U

St Rollox Works: Nil. DLRC: St Rollox: 130565-140565 U

St Rollox Works: Nil. DLRC: St Rollox: 250565-280565 U

XX: 240566 (fire, location unknown) (FDTL)

66A Polmadie: 170967
St Rollox Works: 140967 (Works Yard)/ 241067 (Test House). DLRC: St Rollox: 130967-271067 I

St Rollox Works: 250570/ 130670/ 140670. Not listed 100670. DLRC: Glasgow: 040570-130670 C

Withdrawn (ex-ScR): 4we091071 (061071) (ex-66A)

66A Polmadie: 271171/ 030172/ 090172
GEC, Stafford: Arrived 240272, removed 100372
Stoke: 100372 (GEC Stafford-Cockshute depot, with D8512/67)
Stoke Cockshute (Carriage Depot): 060872/ 110872/ 011072
XX: 261072 (Ardrossan Harbour -Cadder Yard freight, hauled by Class 24)
St Rollox Works: 051172/191172 (inside)/ 100273/ 110273/ 080473/ 210473/ 220473/ 050573/ 120573 (Top Yard)/ 130573/ 280573/ 290573/ 090673 (Yard)/ 170673/ 050873/ 110873/ 120873/ 140873/ 160873/ 150973/ 131073/ 041173/ 181173/ 130174. Not listed 141072, 041173 and 130474.

Disposal: St Rollox Works: 1173 (D&ELfS).

Notes:
1. See Section 13.1 for additional details regarding the accident damage sustained by D8545 on 22 October 1963.
2. See D8512 (Note 4) for commentary regarding GEC generator duties in 1972.
3. See Section 20.4 for further discussion re. disposal of D8545.

D8546   Serial No.: 4365/U 47

D8546, 66A Polmadie, undated. GFY livery, D-prefix removed. This is probably late-1970 or 1971. The locomotive will in all likelihood have received a "touch-up and varnish" as part of its Intermediate repair in March 1970. (RCTS Archive)

New (ScR): 1we210963 (170963)-66A

St Rollox Works: Nil. DLRC: St Rollox: 301164-041264 U (AWS fitment)

Ayr: 180566 (excursion)

St Rollox Works: 290566 DLRC: St Rollox: 160566-250666 I

St Rollox Works: Nil. DLRC: Glasgow: 300170-070370 C
65A Eastfield: 070370 (ex-works)

Withdrawn (ex-ScR): 4we091071 (061071) (ex-66A)

66A Polmadie: 271171
Ex-67D Ardrossan: Arrived mid-1271/ 030172/ 090172/ 260272/ 300372/ 250472/ 140572/ 170772/ 170872/ 060972

66A Polmadie: 061172/091172/ 191172/ 100273/ 250273/ 100473/ 120573/ 270573/ 090673/ 170873/ 150973/ 131073/ 181173/ 130174/ 260174/ 170274/ 230374/ 140474/ 280574/ 150674/ 290674/ 030874/ 100874/ 250874/ 280974/ 051074/ 260175/ 080375/ 160375/ 290375/ 050475/ 270475. Not listed 141072 and 161174.
St Rollox Works: 170575/270575/ mid-0775 (Works Yard)/ 120875. Not seen 070675.
Transfer to J. Cashmore, Great Bridge: 17-xx0975 (D8504/46/63, 9X20 via Carlisle/Hellifield)
Hunslet South Sidings: 180975 (with D8504/63). Not listed 260975.
Chesterfield: 110975 (with D8504/63, hauled by 40019)
J. Cashmore, Great Bridge: Nil.

Disposal: J. Cashmore, Great Bridge: 0975 (D&ELfS). '. . .cut up by 10/75.' (AHBRD&E5)
Disposal not proven.

D8547   Serial No.: 4365/U 48

D8547, 64B Haymarket, 20 October 1963. At Haymarket for training purposes prior to the arrival of D8554-87. (Rail-Online)

New (ScR): 1we280963 (270963)-66A

64B Haymarket: 101063 (crew-training)/ 201063

St Rollox Works: Nil. DLRC: St Rollox: xxxxxx-150464 U

St Rollox Works: Nil. DLRC: St Rollox: 141264-181264 U (AWS fitment)

St Rollox Works: Nil. DLRC: St Rollox: 260866-221066 I

Stored (u): 1we180169 (120169) (ex-66A)

66A Polmadie: 120169

Withdrawn (ex-ScR): 1we220269 (160269) (ex-66A (store))

66A Polmadie: 130369/250369/ 290369/ 300369
Bescot: Arrived 240669 (with D8560/76)
Worcester: 240669 (with D8560/76)
Bird's, Long Marston: 280669/290669/ 260769/ 140869/ 190869/ 111069/ 110170/ 250170/ 030370/ 070370 (sidings near site of station). Not listed 050769/ 121069 and 300370.
Bird's, Cardiff: Nil.

Disposal: Bird Group, Long Marston: 0669-0370* (D&ELfS).
* - 'Requiring definitive confirmation'
Disposal not proven.

Notes:
1. Reason for Withdrawal. No.2 engine aluminium crankcase fractured, surplus.
2. Sold for scrap (May 1969): D8547/56/60/4/76/7 (to Birds, Long Marston, from Polmadie). (RO0869)
3. '...On 16th March (1970) D8547/56/64/77 were booked to be moved from Birds, Long Marston to Bird's, Cardiff ...' (RO0670)
   But N.B. D8556/64/77 still at Long Marston on 30/03/70; D8547 not noted.
4. Bird's, Long Marston/Cardiff: 'It remained in the (Bird, Long Marston) yard until removed to Bird, Cardiff Docks for scrap on 16/3/70. Scrapping followed shortly afterwards. Note: D8556/64/77 were also booked to move to Cardiff, but did not go.' (AHBRD&E5)

D8548   Serial No.: 4365/U 49

D8548, 66B Motherwell, undated. D8548 (with D8541) assisted a failed D336 on the 21.00 Perth-Euston from Wishaw to Carstairs on 3 April 1964; further stardom came for D8548 exactly four years later when it was involved with the haulage of the "Night Limited" on 3 April 1968 between Motherwell and Glasgow Central due to a derailment at Shawfield. This time D8548's companion was D8533. It must have been a good locomotive this one, lasting until the final day of the Clayton fleet (31 December 1971). (Transport Topics)

New (ScR): 1we280963 (270963)-66A

St Rollox Works: Nil. DLRC: St Rollox: 050265-060265 U

St Rollox Works: 020965. DLRC: St Rollox: 310865-020965 U (AWS fitment)

Cowlairs: 120366 (LE to 65A Eastfield, fire) (FDTL)

66A Polmadie: 040566
St Rollox Works: 290566. DLRC: St Rollox: 290366-170666 I

St Rollox Works: 070168 DLRC: St Rollox: 221267-130168 U

St Rollox Works: 111069 DLRC: St Rollox: 131069-311069 C/I

Withdrawn (ex-ScR): 2we010172 (311271) (ex-66A)

Ex-67D Ardrossan: Arrived mid-1271/ 030172/ 090172/ 260272/ 300372/ 250472/ 140572/ 250672/ 170772/ 170872/ 060972
66A Polmadie: 091172/191172/ 100273/ 250273/ 100473/ 120573/ 270573/ 090673/ 080773/ 170873/ 150973/ 131073/ 181173/ 130174/ 260174/ 170274/ 230374/ 140474/ 280574/ 150674/ 290674/ 030874/ 100874/ 250874/ 280974/ 051074/ 161174/ 260175/ 080375/ 160375/ 290375/ 050475/ 270475. Not listed 141072.
St Rollox Works: 170575/270575/ mid-0775 (Works Yard)/ 120875. Not listed 070675.
Transfer to J. Cashmore, Great Bridge: 16-xx0975 (D8542/8/50/1, 9X20 via Carlisle/Hellifield)
Chesterfield: 100975 (with D8542/50/1, hauled by 25130)
J. Cashmore, Great Bridge: Nil.

Disposal: J. Cashmore, Great Bridge: 0975 (D&ELfS). Disposal not proven.

D8549   Serial No.: 4365/U 50

D8549, 66B Motherwell, Undated.
(Transport Topics)

New (ScR): 1we121063 (081063)-66A

St Rollox Works: Nil. DLRC: St Rollox: 170264-290264 U

66A Polmadie: 081164
St Rollox Works: Nil. DLRC: St Rollox: 091164-051264 U (AWS fitment)

Hurlford: 230466 (2220 Auchinleck-Bridgeton freight, fire) (FDTL)

St Rollox Works: Nil. DLRC: St Rollox: 061266-280167 I, and, St Rollox; 310167-040267 Rect

66A Polmadie: 071170
St Rollox Works: Nil. DLRC: Glasgow: 061170-051270 C

Withdrawn (ex-ScR): 4we091071 (210971) (ex-66A

66A Polmadie: 260971
St Rollox Works: 211171/281171/ 090172. Not listed 120272/ 270272/ 020472/ 090472/ 140572/ 280572.

Disposal: St Rollox Works: 0672 (D&ELfS). Commencing month of cutting-up: 0672 (RO0776).

Note:
1. See Section 20.4 for further discussion re. disposal of D8549.

D8550   Serial No.: 4365/U 51

D8550, Colchester North, October 1963. The *Railway Observer* (January 1964) reported: 'D8550 on a trial run from Derby to the Davey Paxman works arrived at Colchester North station on 28th October (1963)…..The locomotive returned to Derby on 1st November after exhibition at St Botolph's station.' (Michael Philips)

New (ScR): 1we091163 (2810 or 081163)-66A

Colchester North: 281063
Colchester St Botolphs: xx1063 (on exhibition, returned to Derby on 011163)
16C Derby: 031163 (ex-Works)

St Rollox Works: Nil. DLRC: St Rollox: 040264-150264 U

St Rollox Works: Nil. DLRC: St Rollox: 260165-300165 U (AWS fitment)

St Rollox Works: Nil. DLRC: St Rollox: 100265-050365 U

St Rollox Works: Nil. DLRC: St Rollox: 011167-091267 I

66A Polmadie: 130670/ 140670
St Rollox Works: 110770. DLRC: Glasgow: 080670-110770 U

Withdrawn (ex-ScR): 4we091071 (051071) (ex-66A)

66A Polmadie: 271171
Ex-67D Ardrossan: Arrived mid-1271/ 030172/ 090172/ 260272/ 300372/ 250472/ 140572/ 170772/ 170872/ 060972
66A Polmadie: 061172/ 091172/ 191172/ 100273/ 250273/ 100473/ 120573/ 270573/ 090673/ 170873/ 150973/ 131073/ 181173/ 130174/ 260174/ 170274/ 230374/ 140474/ 280574/ 150674/ 290674/ 030874/ 100874/ 250874/ 280974/ 051074/ 260175/ 080375/ 160375/ 290375/ 050475/ 270475. Not listed 141072 and 161174.
St Rollox Works: 170575 (middle of CS)/ 270575/ 070675 (Works Yard)/ mid-0775 (Works Yard)/ 120875
Transfer to J. Cashmore, Great Bridge: 16-xx0975 (D8542/8/50/1, 9X20 via Carlisle/Hellifield)
Chesterfield: 100975 (with D8542/8/51, hauled by 25130)
J. Cashmore, Great Bridge: Nil.

Disposal: J. Cashmore, Great Bridge: 0975 (D&ELfS). Disposal not proven.

D8551 Serial No.: 4365/U 52

D8551, 66A Polmadie, 28 September 1974. Nearly three years after withdrawal. (Anthony Sayer)

New (ScR): 1we121063 (081063)-66A

66A Polmadie: 231163/241163
St Rollox Works: Nil. DLRC: St Rollox: 251163-141263 LC

St Rollox Works: 201064 (Works Yard). DLRC: St Rollox: 071064-161064 U

St Rollox Works: Nil. DLRC: St Rollox: 180165-200165 U (AWS fitment)

St Rollox Works: 210467 (E.Shop)/ 220467/ 110567. DLRC: St Rollox: 290367-190567 I

Stored (u): 1we240569 (180569) (ex-66A)

66A Polmadie: 130769/020869/ 160869/ 240869/ 020969/ xx1069

Reinstated: 1we111069 (051069)-66A

St Rollox Works: Nil. DLRC: St Rollox: 171069-061169 U (Power unit damage)

St Rollox Works: Nil. DLRC: St Rollox: 041269-061269 U (Power unit).

St Rollox Works: 190671 (E.Shop). Not listed 310571.

Withdrawn (ex-ScR): 4we091071 (051071) (ex-66A)

66A Polmadie: 051171/271171
Ex-67D Ardrossan: Arrived mid-1271/ 030172/ 090172/ 260272/ 300372/ 250472/ 140572/ 170772/ 170872/ 060972
66A Polmadie: 061172/091172/ 191172/ 100273/ 250273/ 100473/ 120573/ 270573/ 090673/ 170873/ 150973/ 131073/ 181173/ 130174/ 260174/ 170274/ 230374/ 140474/ 280574/ 150674/ 290674/ 030874/ 100874/ 250874/ 280974/ 051074/ 161174/ 260175/ 080375/ 290375/ 050475/ 270475.
Not listed 141072 and 160375.
St Rollox Works: 170575 (middle of CS)/ 270575/ 070675 (Crane Shop)/ mid-0775 (Works Yard)/ 120875
Transfer to J. Cashmore, Great Bridge: 16-xx0975 (D8542/8/50/1, 9X20 via Carlisle/Hellifield)
Chesterfield: 100975 (with D8542/8/50, hauled by 25130)
J. Cashmore, Great Bridge: Nil.

Disposal: J. Cashmore, Great Bridge: 0975 (D&ELfS). Disposal not proven.

Note:
1. Reason for Storage (May 1969). No.1 engine/generator coupling adrift.

D8552   Serial No.: 4365/U 53

D8552, St Rollox Works (Near Traverser), 26 September 1974.
(Anthony Sayer)

New (ScR): ScR: 1we191063 (15 or 171063)-66A

St Rollox Works: Nil. DLRC: St Rollox: 030865-060865 U (AWS fitment)

St Rollox Works: Nil. DLRC: St Rollox: 261165-041265 U

66A Polmadie: 080166
St Rollox Works: Nil. DLRC: St Rollox: 110166-280166 U

66A Polmadie: 180966/ 230966/ 021066/ 231066
St Rollox Works: Nil. DLRC: St Rollox: 150966-241266 I

Transfer: 1we170667 (120667)-64B

Transfer: 1we190867 (190867)-66A

St Rollox Works: Nil. DLRC: St Rollox: 110368-230368 U

St Rollox Works: 130271/190271/ 200271 (E.Shop). DLRC: Glasgow: 250171-030371 U

Withdrawn (ex-ScR): 2we010172 (311271) (ex-66A)

66A Polmadie: 030172/090172
66B Motherwell: 270272 (mobile generator?)
66A Polmadie: 300372/140472/ 150472/ 230472/ 140572/ 100672/ 080772/ 170872/ 290872/ 060972/ 160972

St Rollox Works: 141072 (Dump)/221072/ 051172/ 191172 (outside)/ 040273 (CS)/ 100273/ 110273/ 080473/ 210473/ 220473/ 050573/ 120573 (CS)/ 130573/ 280573/ 290573/ 090673 (Yard)/ 170673/ 050873 (CS)/ 110873/ 120873/ 140873/ 160873/ 150973/ 131073/ 041173/ 181173/ 130174/ 100374/ 230374/ 130474/ 140474/ 010574 (CS) / 110574 (CS)/ 280574/ 150674 (Dump)/ 200674/ 030874/ 240874/ 260974 (N of Traverser)/ 051074 (N of Traverser)/ 221274/ 260175/ 080375/ 200375/ 290375/ 170575 (N of Traverser)/ 270575/ 070675 (Dump)/ mid-0775 (Works Yard)/ 120875. Not listed 250273.

Transfer to R.A. King, Norwich: 04-080975 (D8507/8/16/25/9/31/6/52/74, 9X20 dep. Glasgow 03.35 040975, arr. March Up Yard 17.55 080975; dep. March 19.45 080975, arr. Norwich Trowse Yard 23.30 080975)

Tyne Yard: 060975
Doncaster: 070975 (with D8507/8/16/25/9/31/6/74)
March: 080975 (hauled by 47360, with D8507/8/16/25/9/31/6/74)
Norwich mpd: 140975
R.A. King, Norwich (Trowse Upper Junction): 270975 (Entrance)/ xx1075 (whole). Not listed 241075.

Disposal: A. King & Son, Norwich: 0975 (D&ELfS).

D8553   Serial No.: 4365/U 54

New (ScR): 1we191063 (15 or 171063)-66A

St Rollox Works: Nil. DLRC: St Rollox: 060264-190264 U

St Rollox Works: Nil. DLRC: St Rollox: 270265-270265 U

66A Polmadie: 080865
St Rollox Works: 090865/120865 (E.Shop). DLRC: St Rollox: 100865-130865 U (AWS fitment)

St Rollox Works: 100466. DLRC: St Rollox: 230266-300466 L

Transfer: 1we170667 (120667)-64B

Transfer: 1we190867 (190867)-66A

St Rollox Works: 180768. DLRC: St Rollox: 080768-290768 U

Stored (u): 1we051068 (051068) (ex-66A)

66A Polmadie (??): Nil.

Withdrawn (ex-ScR): 1we021168 (271068) (ex-66A (store))

66A Polmadie: 120169. Not listed 250369.
J. MacWilliam, Shettleston: Nil.

Disposal: J. MacWilliam, Shettleston: 0169 (D&ELfS). ' . . . broken up in 2/69.' (AHBRDE5)
Disposal not proven.

Notes:
1. Reason for Withdrawal. Engine defect, surplus.
2. Sold for scrap (January 1969): D8537/53 (to J. MacWilliam, Shettleston, from Polmadie). (RO0469)

D8554 Serial No.: 4365/U 55 N.B. No DLRC information.

D8554, 64B Haymarket, 16 August 1964. First of the batch of 34 Claytons allocated to Haymarket from new (D8554-87). (Rail-Online)

New (ScR): 1we261063 (241063)-64B

64B Haymarket: 251063 (crew-training)/ 271063

Inverurie Works: 091064. Probably for U repair (inc. AWS fitment).

St Rollox Works: 090865/120865 (Yard)/ 190865/ 020965. Not listed 010865.
65B St Rollox: 040965

St Rollox Works: Nil. Probably received I repair during second half of 1966. Not listed 080866/ 120866/ 200866.

Transfer: 1we190867 (190867)-66A

Stored (s): 1we110169 (050169) (ex-66A)

66A Polmadie: 120169

Reinstated: 1we250169 (190169)-66A

Stored (u): 1we150369 (020369) (ex-66A) (late entry)

66A Polmadie: 290369/ 300369

Withdrawn (ex-ScR): 1we070669 (180569) (ex-66A (store)) (late entry)

66A Polmadie: 130769/020869/ 160869/ 240869/ 020969/ xx1069/ 141269/ 110170/ 140270/ 290370/ 040470/ 110470/ 240570/ 130670/ 140670/ 280670 (out of use, partly cannibalised)/ 030770/ 300770/ 310770/ 040870/ 160870/ 220870/ 041070/ 071170. Not listed 020870 and 100171/ 140271.
J. MacWilliam, Shettleston: Nil. Not listed 170471.

Disposal: J. MacWilliam, Shettleston: 0171-0471 (D&ELfS). Disposal not proven.

Notes:
1. Reason for Withdrawal. No.1 engine crankcase corrosion.
2. Disposal details required: D8554 (ex-66A). (RO0471).
3. Sold for scrap (January 1971): D8544/54 (to J. MacWilliam, Shettleston, from Polmadie). (RO1071/1271)

D8555   Serial No.: 4365/U 56

D8555, Strawfrank Junction, Carstairs, 16 June 1969. (Rail-Online)

New (ScR): 1we021163 (291063)-64B

Loanhead branch/Millerhill: 041263 (freight, derailed at Millerhill)
St Rollox Works: Nil. DLRC: St Rollox: 051263-310164 U

Inverurie Works: Nil. DLRC: Inverurie: 091164-121164 U (AWS fitment)

St Rollox Works: 290566. DLRC: St Rollox: 070666-110666 I

Transfer: 1we260867 (200867)-66A

66A Polmadie: 041070

St Rollox Works: Nil. DLRC: Glasgow: 011070-311070 I (?)

St Rollox Works: 190571 (E.Shop). Not listed 310571.

Withdrawn (ex-ScR): 4we091071 (210971) (ex-66A)

66A Polmadie: 260971
St Rollox Works: 211171/281171/ 090172/ 120272/ 270272/ 010472/ 020472 (Works Yard)/ 090472 (Old Paint Shop Yard)/ 140472/ 250472/ 260472/ 140572/ 280572. Not listed 161071 and 100672.

Disposal: St Rollox Works: 0572 (D&ELfS). Commencing month of cutting-up: 0572 (RO0776).

Notes:
1. See Section 13.1 for additional details re. derailment damage sustained by D8555 on 4 December 1963
2. Reason for Withdrawal. Engine/generator coupling adrift.

D8556   Serial No.: 4365/U 57

D8556, Millerhill s.p., 7 June 1965. (Author's Collection)

New (ScR): 1we021163 (291063)-64B

Inverurie Works: Nil. DLRC: Inverurie: 070764-100764 U

St Rollox Works: Nil. DLRC: St Rollox: 240665-260665 U

Inverurie Works: Nil. DLRC: Inverurie: 020865-070865 U (AWS fitment)

63A Perth: 300767. Not listed 020867.
St Rollox Works: 170867/210867. DLRC: St Rollox: 100767-080967 I

Transfer: 1we190867 (190867)-66A
Transfer: 1we260867 (200867)-64B
Transfer: 1we020967 (020967)-66A

Stored (u): 1we080269 (020269) (ex-66A)

66A Polmadie (??): Nil.

Withdrawn: 1we220269 (160269) (ex-66A (store))

66A Polmadie: 130369/250369/ 290369. Not listed 300369.
Bescot: Arrived 270669 (with D8564/72)
Bird's, Long Marston: 290669/050769/ 260769/ 140869/ 190869/ 111069/ 110170/ 250170/ 030370/ 070370 (sidings near site of station)/ 300370. Not listed 280669 and 121069.

Disposal: Bird Group, Long Marston: 0769-0470 (D&ELfS). '...scrapped by 30/4/70.' (AHBRD&E5)

Notes:
1. Reason for Withdrawal. No.2 engine crankcase fractured.
2. Sold for scrap (May 1969): D8547/56/60/4/76/7 (to Birds, Long Marston, from Polmadie). (RO0869)
3. D8556/64/72 arrived Bescot 27/06/69 per RO0969, but D8572 shown as arriving Bescot on 28/06/69 in RO0470.
4. See Note 3 for D8547.

D8557   Serial No.: 4365/U 58

D8557, 66B Motherwell, undated (Slide Processed May 1968). (TOPticl Digital Memories)

New (ScR): 1we091163 (051163)-64B

Inverurie Works: Nil. DLRC: Inverurie: 161164-191164 U (AWS fitment)

St Rollox Works: 090865/120865 (E.Shop). DLRC: St Rollox: 050865-130865 U

St Rollox Works: 170867/210867/ 140967 (Works Yard). DLRC: St Rollox: 050867-160967 I

Transfer: 1we020967 (020967)-66A

Withdrawn (ex-ScR): 4we 091071 (210971) (ex-66A)

66A Polmadie: 260971/271171
Ex-67D Ardrossan: Arrived mid-1271/ 030172/ 090172/ 260272/ 300372/ 250472/ 140572/ 170772/ 170872/ 060972

66A Polmadie: 061172/091172/ 191172/ 100273/ 250273/ 100473/ 120573/ 270573/ 090673/ 080773/ 170873/ 150973/ 131073/ 181173/ 130174/ 260174/ 170274/ 230374/ 140474/ 280574/ 150674/ 290674/ 100874/ 250874/ 280974/ 051074/ 161174/ 260175/ 080375/ 160375/ 290375/ 050475/ 270475. Not listed 141072, 030874.
St Rollox Works: 170575/270575/ / 070675 (Dump)/ 120675 (Carriage Sdgs?)/ 120875. Not listed mid-0775.
Transfer to J. MacWilliam, Shettleston: 260875 (with D8573, D8613)
J. MacWilliam, Shettleston: Nil.

Disposal: J. MacWilliam, Shettleston: 0875 (D&ELfS). '...torn apart during September 1975.' (AHBRD7E5) Disposal not proven.

D8558 Serial No.: 4365/U 59

D8558, 66A Polmadie, 1970. (Transport Topics)

New (ScR): 1we091163 (051163)-64B

Inverurie Works: Nil. DLRC: Inverurie: 190165-220165 U (AWS fitment)

St Rollox Works: Nil. DLRC: St Rollox: 211265-150166 U

St Rollox Works: Nil. DLRC: St Rollox: 281066-091266 I

St Rollox Works: Nil. DLRC: St Rollox: 010267-110267 U
65A Eastfield: 190267

63A Perth: 300767 / 020867
St Rollox Works: 170867. DLRC: St Rollox: 240767-170867 U
65A Eastfield: 200867

Transfer: 1we020967(020967)-66A

St Rollox Works: Nil. DLRC: Glasgow: 201170-311270 I

Withdrawn (ex-ScR): 2we010172 (311271) (ex-66A)

Ex-67D Ardrossan: Arrived mid-1271/ 030172/ 090172/ 260272/ 300372/ 250472/ 140572/ 170772/ 170872/ 060972
66A Polmadie: 061172/ 091172/ 191172. Not listed 141072.
St Rollox Works: 100273 (frames/bogies only)/ 110273. Not listed 080473.

Disposal: St Rollox Works: 0273 (D&ELfS).
'...cab and frames left by 10/2/73...' (AHBRD&E5)

D8559   Serial No.: 4365/U 60

D8559, 66A Polmadie, 28 May 1971. (Colour-Rail)

New (ScR): 1we161163 (121163)-64B

Inverurie Works: Nil. DLRC: Inverurie: 090664-110664 U

Inverurie Works: Nil. DLRC: Inverurie: 031064-061164 U (AWS fitment)

St Rollox Works: 290566. DLRC: St Rollox: 210466-040666 I

Transfer: 1we020967 (020967)-66A

St Rollox Works: Nil. SRSS:St Rollox: 281169-191269 C/I

Withdrawn (ex-ScR): 4we091071 (061071) (ex-66A)

66A Polmadie: 271171

Ex-67D Ardrossan: Arrived mid-1271/ 030172/ 090172/ 260272/ 300372/ 250472/ 140572

St Rollox Works: 180672/060772/ 060872/ 100872/ 120872 (Dump)/ 130872/ 170872/ 280872/ 290872/ 020972/ 141072 (Dump)/ 221072/ 281072 (Top Yard)/ 051172/ 191172/ 100273/ 110273/ 080473/ 210473/ 220473/ 050573/ 120573 (Top Yard)/ 130573/ 280573/ 290573/ 090673 (Yard)/ 170673/ 050873/ 110873/ 120873/ 140873/ 160873/ 150973/ 131073/ 041173/ 181173/ 130174/ 100374/ 230374 (Scrap Road)/ 130474/ 140474/ xx0474 (being-cut)/010574 (unidentified part cut-up remains)/ 150674 (unidentified part cut-up remains)/ 070675 (unidentified part cut-up remains). Not listed 100672.

Disposal: St Rollox Works: 0474 (D&ELfS).

Note:
1. St Rollox Works, 150674: 'frames and bogies of two others were seen on the scrap-road' (RO0974). Assumed to be D8505/59; see Section 20.4.

D8560   Serial No.: 4365/U 61

D8560, 64A St Margarets, 16 August 1964. (Rail-Online)

New (ScR): 1we161163 (121163)-64B

Inverurie Works: Nil. DLRC: Inverurie: 271064-021164 U (AWS fitment)

St Rollox Works: Nil. DLRC: St Rollox: 270267-150467 I

Transfer: 1we021267 (301167)-66A

Stored (u): 1we110169 (050169) (ex-66A)

66A Polmadie: 120169

Withdrawn (ex-ScR): 1we220269 (160269) (ex-66A (store))

66A Polmadie: 130369/ 250369/ 290369/ 300369
Bescot: Arrived 240669 (with D8547/76)
Worcester: 240669 (with D8547/76)
Bird's, Long Marston: 280669/ 290669/ 050769/ 260769/ 140869/ 190869. Not listed 121069/ 030370.

Disposal: Bird Group, Long Marston: 0669-1269 (D&ELfS).

Notes:
1. Reason for Withdrawal. No.1 engine aluminium crankcase fractured.
2. Sold for scrap (May 1969): D8547/56/60/4/76/7 (to Birds, Long Marston, from Polmadie). (RO0869)

D8561   Serial No.: 4365/U 62

D8561, 64A St Margarets, 16 August 1964. Two B1 steam locomotives keep D8561 company and at the same time providing a less than ideal environment for effective diesel operations. (Rail-Online)

New (ScR): 1we231163 (20 or 221163)-64B

Inverurie Works: Nil. DLRC: Inverurie: 260664-290664 U

Inverurie Works: 020965 (E.Shop). DLRC: Inverurie: 300865-030965 U (AWS fitment 040965).

St Rollox Works: 180266. DLRC: St Rollox: 160266-180266 U

St Rollox Works: Nil. DLRC: St Rollox: 241066-161266 C

Transfer: 1we200168 (160168)-66A

Transfer: 1we151169 (141169)-64B

St Rollox Works: 130271/190271/ 200271 (Test House). DLRC: Glasgow: 220171-170271 U

Withdrawn (ex-ScR): 4we271171 (071171) (ex-64B)

64B Haymarket: 271171
Millerhill Yard: 030172/080172/ 270272/ 310372/ 230472/ 130572/ 030672/ 230772/ 180872/ 260872/ 020972
St Rollox Works: 141072 (Dump)/221072/ 281072 (Top Yard)/ 051172/ 191172 (outside)/ 100273/ 110273/ 080473/ 210473/ 220473/ 050573/ 120573/ 130573/ 280573/ 290573/ 090673 (Yard)/ 170673/ 050873/ 110873/ 120873/ 140873/ 160873/ 150973. Not listed 131073/ 041173.

Disposal: St Rollox Works: 0973 (D&ELfS).
St Rollox Works: '. . .noted on 15/9/73 with cutting already started . . .completed by 30/9/73.' (AHBRD&E)
'. . .cut up . . .about July (*1973*). (RO1273)

Note:
1. See Section 20.4 for further discussion re. disposal of D8561.

D8562   Serial No.: 4365/U 63

New (ScR): 1we231163 (20 or 221163)-64B

Inverurie Works: Nil. DLRC: Inverurie: 160664-180664 U

64B Haymarket: 260664/170764/ 020864
St Rollox Works: Nil. DLRC: St Rollox: 240664-140864 U

St Rollox Works: Nil. DLRC: St Rollox: 140964-190964 U (AWS fitment)

XX: 140565 (fire, location unknown) (FDTL)

Gorgie: 190166 (0720 Bathgate-Millerhill freight, fire) (FDTL)

XX: 220666 (fire, location unknown) (FDTL)

XX: 130267 (fire, location unknown) (FDTL)

St Rollox Works: 210467 (E.Shop)/220467. DLRC: St Rollox: 080367-050567 I, and, St Rollox; 060567-080567 Rect

Transfer: 1we160368 (160368)-66A

Stored (u): 1we190469 (130469) (ex-66A)

66A Polmadie (??): Nil.

Reinstated: 1we120769 (060769)-66A

St Rollox Works: Nil. DLRC: Glasgow: 211169-131269 C/I

Withdrawn (ex-ScR): 4we091071 (061071) (ex-66A)

66A Polmadie: 271171/030172/ 090172/ 270272 ('working off 66A', mobile generator?)/ 300372/ 230472/ 140572/ 100672/ 080772/ 170872/ 290872/ 060972/ 160972/ 141072/ 301072/ 061172/ 091172/ 191172
St Rollox Works: 100273/110273/ 210473/ 220473/ 050573/ 120573 (Scrapping Area, one nose-end dismantled)/ 130573. Not listed 280573/ 090673.

Disposal: St Rollox Works: 0673 (D&ELfS).

Note:
1. Reason for Storage (April 1969). No.1 engine/generator coupling adrift.

D8563   Serial No.: 4365/U 64

D8563, Millerhill Yard, July 1970. (Grahame Wareham)

New (ScR): 2we281263 (19 or 201263)-64B

64B Haymarket: 120464
St Rollox Works: Nil. DLRC: St Rollox: 100464-220464 U

Inverurie Works: Nil. DLRC: Inverurie: 221264-281264 U Not marked as such on DLRC but probably AWS fitted during this visit.

Kirkcaldy: 230466 (0345 Millerhill-Thornton freight, fire) (FDTL)

XX: 210766 (fire, location unknown) (FDTL)

63A Perth: 220467/250467
St Rollox Works: 110567/280567/ 010667. DLRC: St Rollox: 190467-270567 I
64B Haymarket: 040667 (ex-Works)

Transfer: 1we230368 (180368)-66A

Transfer: 1we151169 (141169)-64B

St Rollox Works: Nil. DLRC: Glasgow: 010970-160970 U

St Rollox Works: 190571 (E.Shop) DLRC: Glasgow: 040571-220571 C

65A Eastfield: 280571 (ex-Works)

Transfer: 5we010172 (061271)-66A

Withdrawn (ex-ScR): 2we010172 (311271) (ex-66A)

Ex-67D Ardrossan: Arrived mid-1271/ 030172/ 090172/ 260272/ 300372/ 250472/ 140572/ 170772/ 170872/ 060972
66A Polmadie: 061172/091172/ 191172/ 100273/ 250273/ 100473/ 120573/ 270573/ 090673/ 170873/ 150973/ 131073/ 181173/ 130174/ 260174/ 170274/ 230374/ 140474/ 280574/ 150674/ 290674/ 030874/ 100874/ 250874/ 280974/ 260175/ 080375/ 160375/ 290375/ 050475/ 270475. Not listed 141072 and 051074/ 161174.
St Rollox Works: 170575 (West of Old Scrapping Area)/ 270575/ mid-0775 (Works Yard)/ 120875. Not listed 070675.
Transfer to J. Cashmore, Great Bridge: 17-xx0975 (D8504/46/63, 9X20 via Carlisle/Hellifield)
Hunslet South Sidings: 180975 (with D8504/46). Not listed 260975.
Chesterfield: 110975 (with D8504/46, hauled by 40019)
J. Cashmore, Great Bridge: Nil.

Disposal: J. Cashmore, Great Bridge: 0975 (D&ELfS). Disposal not proven.

D8564   Serial No.: 4365/U 65

New (ScR): 2we281263 (19 or 201263)-64B

Inverurie Works: Nil. DLRC: Inverurie: 300664-060764 U

St Rollox Works: 020965. DLRC: St Rollox: 230865-220965 U (AWS fitment)

St Rollox Works: Nil. DLRC: St Rollox: 171167-231267 I

Transfer: 1we040568 (040568)-66A

Stored (u): 1we080269 (020269) (ex-66A)

66A Polmadie (??): Nil.

Withdrawn (ex-ScR): 1we220269 (160269) (ex-66A (store))

66A Polmadie: 130369/250369/ 290369/ 300369
Bescot: Arrived 270669 (with D8556/72)
Bird's, Long Marston: 290669/050769/ 260769/ 140869/ 190869/ 111069/ 110170/ 250170/ 030370 /070370 (sidings near site of station)/ 300370/ xx0470 (cab only on road-trailer)/ 280670 (cab only on road trailer)/ xx0870 (cab only on road-trailer). Not listed 280669, 121069 and 230870.

Disposal: Bird Group, Long Marston: 0669-0470 (D&ELfS).
'…last noted intact on 1/4/70, but had succumbed to the scrap men by 30/4/70.' (AHBRD&E5)

Notes:
1. Reason for Withdrawal. No.2 engine aluminium crankcase corrosion.
2. Sold for scrap (May 1969): D8547/56/60/4/76/7 (to Birds, Long Marston, from Polmadie). (RO0869)
3. D8556/64/72 arrived Bescot 27/06/69 per RO0969, but D8572 shown as arriving Bescot on 28/06/69 in RO0470.
4. See Note 3 for D8547.

D8565   Serial No.: 4365/U 66

D8565, Thornton Yard, 23 May 1970. Signs of minor fire damage on one of the No.2 end access doors. (RCTS Archive)

New (ScR): 2we281263 (20 or 211263)-64B

St Rollox Works: Nil. DLRC: St Rollox: 300164-070264 U

St Rollox Works: 120464. DLRC: St Rollox: 060464-280464 U

St Rollox Works: Nil. DLRC: St Rollox: 191064-301064 L (AWS fitment)

St Rollox Works: Nil. DLRC: St Rollox: 091264-191264 U

XX: 031166 (fire, location unknown) (FDTL)

XX: 060167 (fire, location unknown) (FDTL)
64B Haymarket: 070167

St Rollox Works: 140967 (E.Shop). DLRC: St Rollox: 060967-141067 C

Transfer: 1we040568 (040568)-66A

Transfer: 1we291169 (231169)-64B

St Rollox Works: 190571 (E.Shop)/ 310571. Not listed 190671.

Withdrawn (ex-ScR): 4we271171 (071171) (ex-64B)

Millerhill Yard: 030172/ 080172/ 270272/ 310372/ 230472/ 130572/ 030672/ 230772/ 180872/ 260872/ 020972. Not listed 271171.
St Rollox Works: 141072 (Dump)/221072/ 281072 (Top Yard)/ 051172/ 191172 (outside)/ 040273 (E.Shop Yard)/ 100273/ 110273/ 180373/ 080473/ 210473/ 220473/ 050573/ 120573 (E.Shop Yard)/ 130573/ 160873. Not listed 280573/ 090673/ 170673/ 040873/ 110873.

Disposal: St Rollox Works: 0873 (D&ELfS).
'...the end had come by 16/8/73.' (AHBRD&E5)

Note:
1. See Section 20.4 for further discussion re. disposal of D8565.

D8566   Serial No.: 4365/U 67

D8566, 64B Haymarket, undated.
(Rail-Online)

New (ScR): 2we281263 (20 or 211263)-64B

Inverurie Works: Nil. DLRC: Inverurie: 090165-120165 U Not marked as such on DLRC but probably AWS fitted during this visit.

Whitemyre: 070267 (1800 Millerhill-Alloa freight, fire) (FDTL)
63A Perth: 11-120267/200267

St Rollox Works: 170867/210867. DLRC: St Rollox: 02 or 090867-090967 I

St Rollox Works: 070168. DLRC: St Rollox: 021167-200168 U

Transfer: 1we010668 (260568)-66A

Withdrawn (ex-ScR): 1we071268 (041268) (ex-66A)

66A Polmadie: 120169/130369/ 250369/ 290369/ 300369
J. MacWilliam, Shettleston: Nil.

Disposal: J. MacWilliam, Shettleston: 0469 (D&ELfS). Disposal not proven.

Notes:-
1. Reason for Withdrawal. No.1 engine aluminium crankcase corrosion. No.2 engine cylinder head stud threads stripped in crankcase.
2. Sold for scrap (April 1969): D8566/9/85 (to J. MacWilliam, Shettleston, from Polmadie). (RO0669)

D8567   Serial No.: 4365/U 68

D8567, 66B Motherwell, 25 May 1969. (RCTS Archive)

New (ScR): 1we110164 (060164)-64B

Inverurie Works: Nil. DLRC: Inverurie: 151264-181264 U (AWS fitment)

Cowlairs Works: 050766 (shunting A3 60041)

XX: 280167 (fire, location unknown) (FDTL)

St Rollox Works: 110567/280567. DLRC: St Rollox: 060567-160667 I, and, St Rollox: 200667-220667 Rect

St Rollox Works: 140967 (Works Yard)

Transfer: 1we040568 (040568)-66A

Newton: 040969 (scrap train from Sighthill Yard to Hallside steelworks, derailed and overturned)

St Rollox Works: 111069. DLRC: St Rollox: 220969-031169 U (Collision damage)

64B Haymarket: 040470
St Rollox Works: 110470 (Works Yard). DLRC: Glasgow: 240370-020570 C

St Rollox Works: 100670/130670/ 140670. DLRC: Glasgow: 040670-200670 U

Withdrawn (ex-ScR): 4we091071 (061071) (ex-66A)

66A Polmadie: 271171/030172/ 090172/ xx02-0372
GEC, Stafford: Arrived 240272, removed 100372
Stoke: 100372 (GEC Stafford-Cockshute depot, with D8512/45)
Stoke Cockshute (Carriage Depot): 060872/ 270872/ 011072
St Rollox Works: 051172/191172 (outside). Not listed 100273.
Lowe's scrapyard, Kilnhurst: xxxxxx (cut-up remains)

Disposal: St Rollox Works: 0173 (D&ELfS). Commencing month of cutting-up: 0173 (RO0776).

Note:
1. See D8512 (Note 4) for commentary regarding GEC generator duties in 1972.

D8568   Serial No.: 4365/U 69

D8568, 66A Polmadie, 17 August 1972. D8568 hiding in the depths of Polmadie depot. A large number of the withdrawn Claytons were to be found with 1S83 wound-up in their indicator boxes, 1S83 denoting the 1600hrs Euston-Glasgow Central passenger service. 1S83 was apparently frequently used by Polmadie staff when returning to Scotland after business trips and/or enthusiast excursions! (Anthony Sayer)

New (ScR): 1we110164 (060164)-64B

St Rollox Works: Nil. DLRC: St Rollox: 020364-060364 U

Inverurie Works: Nil. DLRC: Inverurie: 120165-150165 U

Not marked as such on DLRC but probably AWS fitted during this visit.

XX: 170666 (fire, location unknown) (FDTL)
XX: 160766 (fire, location unknown) (FDTL)

St Rollox Works: Nil. DLRC: St Rollox: 151266-090267 I

St Rollox Works: Nil. DLRC: St Rollox: 140267-250267 U

Transfer: 1we070968 (020968)-66A

St Rollox Works: 111069. DLRC: St Rollox: 230969-181069 C

St Rollox Works: Nil. DLRC: St Rollox: 160270-040370 U. P&SP: Engine/generator coupling.

Withdrawn (ex-ScR): 4we091071 (061071) (ex-66A)

66B Motherwell: 271171
66A Polmadie: 030172/ 090172/ 270272/ 300372/ 230472/ 140572/ 210672/ 080772/ 170872/ 290872/ 060972. Not listed 100472.

Rotherham: 110972 (en route to Cupid Green under own power)
Hemel Hempstead Lightweight Concrete Co Ltd, Cupid Green, Hertfordshire: Arrived 110972/ 121072/ 181072/ 220275/ 080476/ 190676/ 070876/ 010177/ Departed 160677
CW Cricklewood: Arrived 160677/ Departed 200677
Transfer to Clitheroe: 20-240677 (via freight trains)
Ribblesdale Cement Ltd, Horrocksford, near Clitheroe: Arrived 240677/ xx0478/ 040481/ 160981/ 141181/ 070682/ 101082/ Departed 090283 (by road)
M62: 090283 (passing Junction 17 on low-loader)
North Yorkshire Moors Railway: Arrived110283 (by road, for preservation)

Disposal: Industry/ Preserved (D&ELfS).

Notes:
1. '8568 has been sold to Hemlite Ltd. at Harpenden. It arrived from Scotland under its own power on 11th September (*1972*) and was noted on 12th and 18th October working six-wagon mineral trains to the Hemlite factory on the Hemel Hempstead branch.' (RO1272)
2. '8568 . . .has been resold to Ribblesdale Cement, Clitheroe, Lancs., and departed from Harpenden to Cricklewood depot for attention during the night of 16th June. It then travelled in freight trains leaving Cricklewood on 20th June . . .arriving at Clitheroe on 24th. ' (RO0877)
3. Sold to North Yorkshire Moors Railway, December 1982 (ExBRiInd2007).
4. 'Rescued for preservation by the Diesel Traction Group . . .D8568 arrived at Pickering on the North Yorkshire Moors Railway on February 11 on a road low-loader.' (RO0483)

D8569, 64A St Margarets, 16 August 1964. (Rail-Online)

D8569 Serial No.: 4365/U 70

New (ScR): 1we180164 (10 or 150164)-64B

64B Haymarket: 290364
St Rollox Works: Nil. DLRC: St Rollox: 310364-110464 U

Inverurie Works: Nil. DLRC: Inverurie: 011264-041264 U (AWS fitment)

St Rollox Works: Nil. DLRC: St Rollox: 260167-240367 I

St Rollox Works: 170867/210867 DLRC: St Rollox: 040867-190867 U

Transfer: 1we070968 (020968)-66A

Withdrawn (ex-ScR): 1we071268 (041268) (ex-66A)

66A Polmadie: 120169/130369/ 250369/ 290369/ 300369
J. MacWilliam, Shettleston: 250569 (being cut-up).

Disposal: J. MacWilliam, Shettleston: 0469 (D&ELfS). '. . .being cut up on 25/5/69, with completion by about 1/6/69.' (AHBRD&E5)

Notes:-
1. Reason for Withdrawal. No.1 engine aluminium crankcase fractured.
2. Sold for scrap (April 1969): D8566/9/85 (to J. MacWilliam, Shettleston, from Polmadie). (RO0669)

D8570   Serial No.: 4365/U 71

D8570, 64A St Margarets, 16 August 1964. (Rail-Online)

New (ScR): 1we180164 (100164)-64B

Inverurie Works: Nil. DLRC: Inverurie: 201064-261064 U (AWS fitment)

Hilton Junction: 300366 (1135 Millerhill-Perth freight, 'serious' fire) (FDTL)
63A Perth: 030466/070466/ 080466/ 100466/ 110466/ 120466
St Rollox Works (??): Nil. Not listed 290566.

St Rollox Works: 070168. DLRC: St Rollox: 081267-130168 I

Transfer: 1we070968 (060968)-66A

Ross Junction: 071168 (scrap metal train from Motherwell, ran through catch-points, derailed)/ 171168

Stored (u): 1we161168 (101168) (ex-66A) (Collision damage)

Withdrawn (ex-ScR): 1we301168 (241168) (ex-66A (store))

66A Polmadie: 120169/250369/ 290369/ 300369. Not listed 130369.
Bromford Bridge: 200569 (in southbound freight at 18.15hrs, hauled by D5233; sheeted over one end)
Bird's, Long Marston: 220569/310569/ 280669/ 290669. Not listed 050769/ 260769/ 190869.

Disposal: Bird Group, Long Marston: 0569 (D&ELfS). '…still intact 20/6/69…broken up completely by 27/6/69.' (AHBRD&E5)

Note:
1. Reason for Withdrawal. Collision damage. See Section 13.1.

D8571 Serial No.: 4365/U 72   N.B. No DLRC information.

D8571 and D8578, 66A Polmadie, 10 July 1969. Less than two months after withdrawal and cannibalisation has already begun. Fading Blue-Star coupling codes are crudely positioned over the previous Red-Diamond codes.
(Alan Walker)

New (ScR): 1we250164 (200164)-64B

Inverurie Works: 091064. Probably for U repair (inc. AWS fitment).

St Rollox Works: 200866 (E.Shop). Not listed 120866.

XX: 230267 (fire, location unknown) (FDTL)

Stored (u): 1we110169 (050169) (ex-64B)

66B Motherwell: 120169

Transfer: 1we180169 (140169)-66A
Reinstated: 1we250169 (190169)-66A

Stored (u): 1we080369 (020369) (ex-66A)

66A Polmadie: 130369/290369. Not listed 300369.

Withdrawn (ex-ScR): 1we070669 (180569) (ex-66A (store)) (late entry)

66A Polmadie: 100769/130769/ 250769/ 020869/ 160869/ 240869/ 020969/ xx1069 (u/s). Not listed 141269.
St Rollox Works: 110170/270370/ 290370/110470 (Works Yard)/ 250570/ 140670/ 110770/ 040870/ 220870/ 120970/ 031070 (Old Paint Shop sidings)/ 041070 (beyond old Carriage Shop)/ 101070/ 100171/ 200271 (½mile from Works)/ 140371/ 170471/ 190571 (dumped near old Carriage Shop)/ 310571/ 190671 (¼mile north of Works).
J. MacWilliam, Shettleston: xx0671

Disposal: J. MacWilliam, Shettleston: 0671 (D&ELfS). '...broken up during July 1971.' (AHBRD&E5)

Notes:
1. Reason for Withdrawal. Main generator defect.
2. Disposal details required: D8571 (ex-66A). (RO0471). No response.

D8572   Serial No.: 4365/U 73   N.B. No DLRC information.

New (ScR): 1we250164 (200164)-64B

XX: 110566 (fire, location unknown) (FDTL)

St Rollox Works: Nil. SRSS:St Rollox: xxxxxx-we110267 U

65A Eastfield: 190267
St Rollox Works: Nil. SRSS:St Rollox: xxxxxx-we180367 Rect

66A Polmadie: 010467
St Rollox Works: Nil. SRSS:St Rollox:xxxxxx-we290467 U. Not listed 220467.

Transfer: 1we180169 (140169)-66A

Store (u): 1we010369 (230269) (ex-66A)

66A Polmadie (??): Nil.

Reinstated: 1we080369 (020369)-66A

Stored (u): 1we290369 (230369) (ex-66A)

66A Polmadie: 290369/300369

Withdrawn (ex-ScR): 1we070669 (180569) (ex-66A (store)) (late entry)

Bescot: Arrived (in error) 270669* (with D8556/64)
Bescot Yard: 060769/ 100769/ 030869/ 150869/ 010969/ 280969 (sidings near Bescot depot)/ 061069/ 091169/ 231169/ 110170/ 220370/ 030570/ 170570/ 310570
J. Cashmore, Great Bridge: 210670/290670

Disposal: J. Cashmore, Great Bridge: 0670-0870 (D&ELfS).

Notes:
1. Probably AWS fitted at Inverurie Works during 1964/65.
2. Reason for Withdrawal. No.2 engine crankcase fractured.
3. D8556/64/72 arrived Bescot 27/06/69. '8572 was sent in error, 8577 should have arrived, but before it could be returned it was damaged by children and considered unfit to travel back to Scotland . . .'. (RO0969)
    N.B. Shown as arriving Bescot on 28/06/69 in RO0470.

D8573   Serial No.: 4365/U 74

D8573, North Berwick, undated. Red Diamond coupling code retained at this point. Recently ex-works so the date maybe July or August 1967. (Colour-Rail)

New (ScR): 1we010264 (300164)-64B

Inverurie Works: Nil. DLRC: Inverurie: 070764-080764 U

St Rollox Works: Nil. DLRC: St Rollox: 100864-150864 U

Inverurie Works: Nil. DLRC: Inverurie: 130965-170965 U (AWS fitment)

XX: 081266 (fire, location unknown) (FDTL)

XX: 090367 (fire, location unknown) (FDTL)

St Rollox Works: Nil. DLRC: St Rollox: 300567-010767 I

Transfer: 1we180169 (140169)-66A

Greskine/Beattock: 020969 (2300 Mossend-Carlisle freight (with D8539), fire) (FDTL)

St Rollox Works: 110170. DLRC: St Rollox: 301269-240170 C

Withdrawn (ex-ScR): 4we091071 (061071) (ex-66A)

66A Polmadie: 271171
Ex-67D Ardrossan: Arrived mid-1271/ 030172/ 090172/ 260272/ 300372/ 250472/ 140572/ 170772/ 170872/ 060972
66A Polmadie: 061172/091172/ 191172/ 100273/ 250273/ 100473/ 120573/ 270573/ 090673/ 080773/ 170873/ 150973/ 131073/ 211073/ 181173/ 130174/ 260174/ 170274/ 230374/ 140474/ 280574/ 150674/ 290674/ 210774/ 030874/ 100874/ 250874/ 280974/ 051074/ 161174/ 260175/ 080375/ 160375/ 290375/ 050475/ 270475. Not listed 141072.
St Rollox Works: 170575/270575/ / 070675 ('Dump')/ mid-0775 (Works Yard)/ 120875
Transfer to J. MacWilliam, Shettleston: 260875 (with D8557, D8613)
J. MacWilliam, Shettleston: xx1075 (intact)

Disposal: J. MacWilliam, Shettleston: 0875 (D&ELfS). '...broken up in early 9/75.' (AHBRD&E5)

D8574   Serial No.: 4365/U 75

New (ScR): 1we010264 (300164)-64B

Inverurie Works: Nil. DLRC: Inverurie: 081264-111264 U (AWS fitment)

Touch South: 030566 (1528 Millerhill-Thornton freight, fire) (FDTL)

St Rollox Works: 170867. DLRC: St Rollox: 170667-120867 C, and, St Rollox: 150867-170867 Rect
65A Eastfield: 200867

Transfer: 1we010269 (270169)-66A

Morningside: 061269 (Shieldmuir/Kingshill Colliery trip freight, fire) (FDTL)

St Rollox Works: 251070

St Rollox Works: 130271/190271/ 200271 (Works Yard)/ 140371. Not listed 170471. P&SP:Repair extended due to waiting Main Generator Armature Bearings, Unclassified Repair.

D8574, 66B Motherwell, 1970.
(Transport Topics)

66A Polmadie: 250471

St Rollox Works: 190571 (Works Yard) /310571. DLRC: Glasgow: xxxxxx-120671 Rect, and, Glasgow: 150671-060771 Rect

Withdrawn (ex-ScR): 4we091071 (061071) (ex-66A)

66A Polmadie (??): Nil.

Reinstated: 4we271171 (071171)-66A

66A Polmadie: 171271

Withdrawn (ex-ScR): 2we010172 (311271) (ex-66A)

66A Polmadie: 030172/090172
66C Hamilton: 270272 (mobile generator?)
66A Polmadie: 300372/230472/ 140572/ 080772/ 170872/ 290872/ 060972/ 160972. Not listed 100672/ 210672.
St Rollox Works: 141072 (Dump)/221072/ 051172/ 191172 (outside)/ 100273/ 110273/ 080473/ 210473/ 220473/ 050573/ 120573/ 130573/ 280573/ 290573/ 090673 (Yard)/ 170673/ 050873/ 110873/ 120873/ 140873/ 160873/ 150973/ 131073/ 231073 (Carriage Sdgs)/ 041173/ 181173/ 120174/ 100374/ 230374/ 130474/ 140474/ 010574 (CS)/ 110574 (CS)/ 280574/ 150674 (Dump)/ 200674/ 030874/ 240874/ 260974 (N of Traverser)/ 051074 (N of Traverser)/ 121074 (N of Traverser)/ 221274/ 260175/ 080375/ 200375/ 290375 (N of Traverser)/ 170575 (N of Traverser) / 270575/ 070675 (Dump)/ mid-0775 (Works Yard)/ 120875
Transfer to R.A. King, Norwich: 04-080975 (D8507/8/16/25/9/31/6/52/74, 9X20 dep. Glasgow 03.35 040975, arr. March Up Yard 17.55 080975; dep. March 19.45 080975, arr. Norwich Trowse Yard 23.30 080975)
Tyne Yard: 060975
Doncaster: 070975 (with D8507/8/16/25/9/31/6/52)
March: 080975 (hauled by 47360, with D8507/8/16/25/9/31/6/52)
Wensum Yard, Norwich: 270975
R.A. King, Norwich: xx1075 (baseframe only, on bogies). Not listed 241075.

Disposal: A. King & Son, Norwich: 0975 (D&ELfS).

D8575   Serial No.: 4365/U 76

D8575, Dunbar,
14 August 1964.
(Rail-Online)

New (ScR): 1we150264 (120264)-64B

Inverurie Works: Nil. DLRC: Inverurie: 300664-020764 U

Inverurie Works: Nil. DLRC: Inverurie: 121165-161065 (*sic 161165?*) U (AWS fitment)

Dalmeny Junction: 160666 (hauling 246 *Morayshire* from Eastfield to Kirkliston)

St Rollox Works: Nil. DLRC: St Rollox: 071166-311266 I

St Rollox Works: Nil. DLRC: St Rollox: 270368-020468 U

64B Haymarket: 121068

Note:
1. Reason for Withdrawal. Fire damage, surplus.

Stored (u): 1we191068 (131068) (ex-64B)

64B Haymarket (??): Nil.

Withdrawn (ex-ScR): 1we021168 (271068) (ex-64B (store))

64B Haymarket: 241168/ 020169/ 120169
J. MacWilliam, Shettleston: 090369 (intact)/ end-0369 (cab only). Not listed 250569.

Disposal: J. MacWilliam, Shettleston: 0369 (D&ELfS). '...still intact on 10/3/69 ... dismantled by 2/4/69.' (AHBRD&E5)

D8576 Serial No.: 4365/U 77

New (ScR): 1we150264 (120264)-64B

Inverurie Works: 030964. **DLRC:** Inverurie: 020964-110964 U (AWS fitment?)
St Rollox Works: 201064 (Works Yard) **DLRC:** St Rollox: 270864-311064 U

St Rollox Works: Nil. **DLRC:** St Rollox: 220866-221066 I

63A Perth: 11-120267/200267
St Rollox Works: Nil. **DLRC:** St Rollox: 140367-200467 U

Transfer: 1we010269 (270169)-66A

Withdrawn (ex-ScR): 1we010369 (280269) (ex-66A)

66A Polmadie: 130369/250369/290369/300369
Bescot: Arrived 240669 (with D8547/60)
Worcester: 240669 (with D8547/60)
Bird's, Long Marston: 280669/290669/260769/140869/190869/111069/110170/250170/030370/070370 (sidings near site of station). Not listed 050769/121069 and 300370.

Disposal: Bird Group, Long Marston: 0669-0370* (D&ELfS).
* - 'Requiring definitive confirmation'
'...last observed intact on 8/3/70... totally dismantled by 28/3/70.' (AHBRD&E5)

Notes:
1. The September/October 1964 Inverurie/St Rollox Works dates are conflicting. Possibly received AWS equipment at Inverurie first, which was the norm at this time, with completion at St Rollox later. Marked as AWS fitted on 31/10/64; however, fitment at Inverurie Works by 11/09/64 was more likely.
2. Reason for Withdrawal. Auxiliary generator damage.
3. Sold for scrap (May 1969): D8547/56/60/4/76/7 (to Birds, Long Marston, from Polmadie). (RO0869)

D8577   Serial No.: 4365/U 78

D8577, 64B Haymarket, 16 August 1964.
(Rail-Online)

New (ScR): 1we220264 (180264)-64B

St Rollox Works: 201064 (Works Yard). DLRC: St Rollox: 170964-231064 U

Inverurie Works: Nil. DLRC: Inverurie: 090265-120265 U (AWS fitment)

St Rollox Works: 140967 (E.Shop). DLRC: St Rollox: 110867-230967 I

Stored (u): 1we080269 (020269) (ex-64B)

Withdrawn (ex-ScR): 1we220269 (160269) (ex-64B (store))

64B Haymarket: 300369
Thornton Junction: 250569
64B Haymarket (??): Nil.
Millerhill s.p.: 090769/130769/ 160869
Carlisle Kingmoor Yard: 200869/190969/ 200969
Bird's, Long Marston: 111069/121069/ 110170/ 250170/ 290270 (sic) (one nose part dismantled)/ 030370/ 070370 (sidings near site of station)/ 300370/ 280670 (partially dismantled)/ 230870

Disposal: Bird Group, Long Marston: 0370-c0770 (D&ELfS). ' . . .still intact on 1/4/70 . . . cut up by 30/4/70.' (AHBRD&E5)

Notes:
1. Reason for Withdrawal. No.1 engine crankcase holed.
2. Sold for scrap (May 1969): D8547/56/60/4/76/7 (to Birds, Long Marston, from Polmadie). (RO0869)
3. See Note 3 for D8547.

D8578 Serial No.: 4365/U 79 N.B. No DLRC information.

D8578, 64B Haymarket, 16 August 1964. Exhaust stack and cowling temporarily removed. (Rail-Online)

New (ScR): 1we070364 (030364)-64B

Perth: 170366 (2230 Millerhill-Inverness freight, fire) (FDTL)

St Rollox Works: Nil. SRSS:St Rollox: xxxxxx-we250267 I
64B Haymarket: 250367
St Rollox Works: Nil. SRSS:St Rollox: xxxxxx-we220467 Rect. Not listed 210467.

St Rollox Works: Nil. SRSS:St Rollox: xxxxxx-we150767 U

Transfer: 1we220269 (190269)-66A

Stored (u): 1we190469 (130469) (ex-66A)

66A Polmadie (??): Nil.

Withdrawn (ex-ScR): 1we070669 (180569) (ex-66A (store)) (late entry)

66A Polmadie: 100769/ 130769/ 020869/ 160869/ 240869/ 020969/ xx1069/ 141269/ 110170/ 290370/ 040470/ 110470/ 240570/ 130670/ 140670/ 280670 (partly cannibalised) / 030770/ 300770/ 310770/ 020870/ 040870/ 160870/ 220870/ 041070/ 071170. Not listed 100171/ 140271.
J. MacWilliam, Shettleston: Nil. Not listed 170471.

Disposal:
Disposal unknown (D&ELfS).
Disposal: 0371 (IHRC)
'…broken up during 8/71.' (AHBRD&E5)
Disposal not proven

Notes:
1. Probably AWS fitted at Inverurie Works during 1964/65.
2. Reason for Withdrawal. No.1 engine crankcase damaged.
3. Disposal details required: D8578 (ex-66A). (RO0471).
4. Sold for scrap (January 1971): D8544/54, plus D8584 (*sic* D8578?) (to J. MacWilliam, Shettleston, from Polmadie). (RO1071/1271)

D8579 Serial No.: 4365/U 80

D8579, 66B Motherwell, undated.
(Transport Topics)

New (ScR): 1we070364 (030364)-64B

Inverurie Works: 120864. DLRC: Inverurie: 110864-130864 U (AWS fitment?)

64B Haymarket: 160864
St Rollox Works: Nil. DLRC: St Rollox: 290764-280864 U

St Rollox Works: Nil. DLRC: St Rollox: 031164-141164 U

St Rollox Works: Nil. DLRC: St Rollox: 191164-271164 U

64B Haymarket: 040165 / 100165
St Rollox Works: Nil. DLRC: St Rollox: 281264-150165 U

XX: 300167 (fire, location unknown) (FDTL)

63A Perth: 250367 / 260367
St Rollox Works: 210467 (E.Shop) / 220467 / 110567.
DLRC: St Rollox: 300367-260567 I
65A Eastfield: 280567

St Rollox Works: Nil. DLRC: St Rollox: 020667-160667 U, and St Rollox: 190667-200667 U

Transfer: 1we220269 (180269)-66A

Transfer: 1we291169 (231169)-64B

St Rollox Works: 250570. Not listed 100670. DLRC: Glasgow: 230470-230570 C
65A Eastfield: xx0670 (ex-Works)

64B Haymarket: 270971 (stored)

Withdrawn (ex-ScR): 4we091071 (041071) (ex-64B)

Millerhill Yard: 271171 / 030172 / 080172 / 270272 / 310372 / 230472 / 130572 / 030672 / 230772 / 180872 / 260872 / 020972
St Rollox Works: 141072 (Dump) / 221072 / 281072 (Top Yard) / 051172 / 191172 / 100273 / 110273 / 080473 / 210473 / 220473 / 050573 / 120573 (Top Yard) / 130573 / 280573 / 290573 / 090673 (Yard) / 050873 / 110873 / 120873 / 140873 / 160873 / 150973 / 131073 / 041173 / 181173 / 130174. Not listed 170673 .

Disposal: St Rollox Works: 1173 (D&ELfS).

Notes:
1. Marked as AWS fitted at St Rollox Works on 28/08/64; however, fitment at Inverurie Works on 13/08/64 is more likely.
2. See Section 20.4 for further discussion regarding the disposal of D8579.

D8580   Serial No.: 4365/U 81

D8580, Millerhill s.p., 7 June 1965.
(Author's Collection)

New (ScR): 1we140364 (10 or 110364)-64B

St Rollox Works: Nil. DLRC: St Rollox: 271064-311264 U Not marked as such on DLRC but probably AWS fitted during this visit.

St Rollox Works (??): Nil. Not listed 311065.
65A Eastfield: 020166

XX: 190466 (fire, location unknown) (FDTL)

St Rollox Works: 241067 (E.Shop)/041167. DLRC: St Rollox: 121067-251167 I

St Rollox Works: Nil. DLRC: Glasgow: 080570-200570 U

64B Haymarket: 270870
St Rollox Works: Nil. DLRC: Glasgow: 120870-190970 U

St Rollox Works: 130271/190271/ 200271 (Works Yard)/ 140371. DLRC: Glasgow: 220171-240371 U

Withdrawn (ex-ScR): 4we091071 (061071) (ex-64B)

64B Haymarket: 271171/121271/ 030172/ 080172
Millerhill Yard: 270272/310372/ 230472/ 130572/ 030672/ 230772/ 180872/ 260872/ 020972
St Rollox Works: 141072 (Dump)/221072/ 281072 (Top Yard)/ 051172/ 191172 (outside)/ 100273/ 110273/ 180373 (Old Paint Shop Yard)/ 080473/ 210473/ 220473/ 050573/ 120573/ 130573/ 280573/ 290573/ 090673 (Yard)/ 170673/ 050873/ 110873/ 120873/ 140873/ 160873/ 150973/ 131073/ 041173/ 181173/ 130174/ 100374/ 230374/ 130474/ 140474/ 280574/ 150674 (Dump)/ 200674/ 030874/ 240874 / 260974 (N of Traverser)/ 051074 (N of Traverser)/ 221274/ 260175/ 080375/ 290375/ 170575 (N of Traverser) / 270575/ 070675 (Dump)/ 120675 (N of Traverser)/ mid-0775 (Dump)/ 120875. Not listed 200375.
Transfer to R.A. King, Norwich: 11-xx0975 (8X79 04.14 Cadder Yard-Millerhill on 110975; then 8X21 03.48 Millerhill-Tyne Yard on 120975)
R.A. King, Norwich: 241075 (partly stripped).

Disposal: A. King & Son, Norwich: 0975 (D&ELfS).
'...disappeared completely by 1/11/75' (AHBRD&E5)

D8581   Serial No.: 4365/U 82

D8581, 64B Haymarket, May 1969. Miniature snow ploughs. Exhibiting minor accident damage to No.2 end. D8581 was stored unserviceable on 11 May 1969 and eventually reinstated on 19 October 1969. The Engine History Card shows a visit to St Rollox Works between 29 October and 15 November 1969 for an Unclassified repair. But for an upturn in traffic during 1969 this locomotive would have been condemned with the likes of D8570/5/85. (Transport Topics)

New (ScR): 1we280364 (23 or 240364)-64B

64B Haymarket: 230465
St Rollox Works: Nil. DLRC: St Rollox: 220465-150565 U Not marked as such on DLRC but probably AWS fitted during this visit.

63A Perth: 150966
St Rollox Works: Nil. DLRC: St Rollox: 220966-191166 I

Bridge of Earn/Glenfarg: 300167 (0855 Perth-Millerhill freight, fire) (FDTL)

St Rollox Works: Nil. DLRC: St Rollox: 250368-060468 U

Stored (u): 1we170569 (110569) (ex-64B) (Collision damage)

64B Haymarket: 200769/ 160869/ 170869/ 280869/ 310869/ 060969

Reinstated: 1we251069 (191069)-64B

St Rollox Works: Nil. DLRC: St Rollox: 291069-151169 C

Withdrawn (ex-ScR): 4we091071 (061071) (ex-64B)

64B Haymarket: 091071/191271
Millerhill Yard: 271171/ 030172/ 080172/ 270272/ 310372/ 230472/ 130572/ 030672/ 230772/ 180872/ 260872/ 020972
St Rollox Works: 141072 (Dump)/221072/ 281072 (Top Yard)/ 051172/ 191172 (outside)/ 100273/ 110273/ 080473/ 210473/ 220473/ 050573/ 120573/ 130573/ 280573/ 290573/ 090673 (Yard)/ 170673/ 050873/ 110873/ 120873/ 140873/ 160873/ 150973/ 131073/ 041173/ 181173. Not listed 130174/ 130474/ 140474.

Disposal: St Rollox Works: 0574 (D&ELfS).

Notes:
1. Reason for Storage (May 1969). Collision damage.
2. See Section 13.1 for additional details re. damage sustained by D8581 during 1969
3. See Section 20.4 for further discussion re. disposal of D8581.

D8582   Serial No.: 4365/U 83

D8582, Millerhill s.p., 30 May 1966.
(Author's Collection)

New (ScR): 1we040464 (310364)-64B

St Rollox Works: 201064 (E.Shop)

Inverurie Works: 300165. DLRC: Inverurie Works:xxxxxx-030265 U (AWS fitment)

St Rollox Works: Nil. DLRC: St Rollox: 070667-140767 I

64B Haymarket: 020169

Withdrawn (ex-ScR): 1we110169 (050169) (ex-64B)

64B Haymarket: 120169/300369
J. MacWilliam, Shettleston: 250569 (intact)

Disposal: J. MacWilliam, Shettleston: 0569 (D&ELfS). '…broken up by 19/6/69.' (AHBRD&E5)

Note:
1. Reason for Withdrawal. No.1 engine aluminium crankcase fractured.

D8583   Serial No.: 4365/U 84

D8583, Millerhill s.p., 1970. (Transport Topics)

New (ScR): 1we180464 (170464)-64B

64B Haymarket: 080767
65A Eastfield: 160767
St Rollox Works: 170867/210867/ 010967. DLRC: St Rollox: 260667-260867 I, and, St Rollox: 280867-020967 Rect

St Rollox Works: Nil. DLRC: St Rollox: 121169-291169 C/I

St Rollox Works: Nil. DLRC: Glasgow: 051269-090170 Rect (suspect new cylinder block)
65A Eastfield: 110170

St Rollox Works: Nil. DLRC: Glasgow: 050370-050370 U

Stored ( ): 5we110971 (250771) (ex-64B) (late entry)

64B Haymarket : 010871/080871/ 190871/ 280871

Note:
1. Probably AWS fitted at Inverurie during 1964/65.

Withdrawn (ex-ScR): 4we091071 (210971) (ex-64B (store))

64B Haymarket: 270971/091071
Millerhill Yard: 030172/080172/ 270272/ 310372/ 230472/ 130572/ 030672/ 230772/ 180872/ 260872/ 270872/ 020972. Not listed 271171.
St Rollox Works: 141072 (Dump)/221072/ 051172/ 191172 (outside)/ 040273 (Top Yard)/ 100273/ 110273/ 080473/ 210473 (Top Yard)/ 230473 (Top Yard)/ 250473 (Old Paint Shop Yard)/ 050573/ 120573 (Top Yard)/ 130573/ 280573/ 290573/ 090673 (Yard)/ 170673/ 050873/ 110873 (being stripped)/ 120873/ 140873/ 160873. Not listed 150973.

Disposal: St Rollox Works: 0873 (D&ELfS).
'...cut up by 16/8/73.' (AHBRD&E5)

D8584 Serial No.: 4365/U 85

D8584, 64A St Margarets, 16 August 1964. (Rail-Online)

New (ScR): 1we020564 (280464)-64B

Inverurie Works: xx0265. Probably for U repair (inc. AWS fitment).

St Rollox Works: Nil. DLRC: St Rollox: 151167-161267 I

Oakley: 301267 (Dunfermline-Thornton freight, fire) (FDTL)
64B Haymarket: 020168

Stored (u): 1we261068 (231068) (ex-64B)

64B Haymarket (??): Nil.

Withdrawn (ex-ScR): 1we021168 (291068) (ex-64B (store))

64B Haymarket: 241168/020169/ 120169/ 300369
J. MacWilliam, Shettleston: 250569 (intact)

Disposal: J. MacWilliam, Shettleston: 0569 (D&ELfS). '. . .still intact on 25/5/69 . . .dismantled by 19/6/69.' (AHBRD&E5)

Notes:
1. Reason for Withdrawal. Engine defect, surplus.
2. Disposal details required: D8584 (ex-64B). (RO0471).
3. Sold for scrap (May 1969): D8584 (to J. MacWilliam, Shettleston, from Haymarket). (RO1071)

D8585   Serial No.: 4365/U 86

New (ScR): 1we160564 (140564)-64B

Inverurie Works: Nil. DLRC: Inverurie; 051265-101265 U (AWS fitment)

King's Road: 281266 (Leith South/Millerhill trip freight, fire) (FDTL)

XX: 030267 (fire, location unknown) (FDTL)

XX: 100267 (fire, location unknown) (FDTL)
66A Polmadie: 190267

Sinclairtown: 100567 (Markinch/Inverkeithing trip, fire) (FDTL)

St Rollox Works: 241067 (Test House)/041167. DLRC: St Rollox: 111067-021267 I

Stored (u): 1we261068 (231068) (ex-64B) (Collision damage)

Withdrawn (ex-ScR): 1we021168 (291068) (ex-64B (store))

62A Thornton Junction: 020169/110169
66A Polmadie: 250369/290369/ 300369. Not listed 130369.
J. MacWilliam, Shettleston: 250569 (intact)

Disposal: J. MacWilliam, Shettleston: 0469-0669 (D&ELfS).
'...Still intact on 25/5/69...scrapped by 19/6/69.' (AHBRD&E5)

Notes:-
1. Reason for Withdrawal. Collision damage, surplus.
2. Sold for scrap (April 1969): D8566/9/85 (to J. MacWilliam, Shettleston, from Polmadie). (RO0669)

D8585, Portobello East Junction, August 1964. Trainload of 13 empty LGW vans en route from Gorgie to Leith Docks for re-loading with grain for the North British Distillery. (Colour-Rail)

D8586   Serial No.: 4365/U 87

D8586, 64B Haymarket, 25 May 1969. The first of the two Class 17/2s fitted with Roll-Royce engines, hence the slightly raised bonnet rooves. During the period August 1968 to March 1969, D8586 was stored temporarily (but not officially) at St Rollox Works for 177 working days, with a further 20 working days in Works for repairs. Unconfirmed rumours abound that one of the Class 17/2s, somewhat perversely, had the Rolls-Royce engines replaced by Paxman equipment; whilst the long period out of traffic for D8586 would have given sufficient time for such a conversion, no firm archive evidence is available to support the rumour. (TOPticl Digital Memories (Roger Hateley))

International Combustion Ltd, Derby: 231064

New (ScR): 1we121264 (07 or 111264)-64B

Inverurie Works: 280566. DLRC: Inverurie: 300566-040666 Mod.
Not marked as such on DLRC but probably AWS fitted during this visit.

Inverurie Works: Nil. DLRC: Inverurie: 130367-210367 U

St Rollox Works: 170868 (E.Shop)/250868/ 260868/ August BH weekend (*310868-020968*)/ 210968 (E.Shop)/ 191068/ 120169/ 130169/ 130369/ 290369/ 310369. DLRC: St Rollox: 080868-290369 C

St Rollox Works: Nil. DLRC: Glasgow: 081269-221269 U (No.1 engine oil seal leak)

Store ( ): 5we110971 (050871) (ex-64B)

64B Haymarket: 190871/ 280871

Withdrawn (ex-ScR): 4we091071 (210971) (ex-64B (store))

64B Haymarket: 270971
Millerhill Yard: 271171/030172/ 080172/ 270272/ 310372/ 230472/ 130572/ 030672/ 230772/ 130872/ 180872/ 260872/ 020972
St Rollox Works: 141072 (Dump)/221072/ 281072 (Top Yard)/ 051172/ 191172 (outside)/ 100273/ 110273/ 210473/ 220473/ 230473 (Top Yard)/ 050573/ 120573/ 130573/ 280573/ 290573/ 090673 (Yard)/ 170673/ 160873. Not listed 080473 and 050873/ 110873.

Disposal: St Rollox Works: 0773 (D&ELfS).
'...cut up by 16/8/73.' (AHBRD&E5)
'...cut up ...about July (*1973*). (RO1273)

D8587   Serial No.: 4365/U 88

D8587, 64A St Margarets, 9 July 1966. Modern traction amongst steam infrastructure. (Colour-Rail [K.Fairey])

International Combustion Ltd, Derby: 231064/ 041264 (E.Shop, primer)/ 020265 (E.Shop, painted)/ 040265 (Works Yard)

New (ScR): 1we060265 (050265)-64B

Inverurie Works: Nil. DLRC: Inverurie: 041165-111165 U Not marked as such on DLRC but probably AWS fitted during this visit.

Inverurie Works: Nil. DLRC: Inverurie: 140366-290366 U

64A St Margarets: 030167
St Rollox Works: Nil. DLRC: St Rollox: 031066-170267 U

Blackford Hill: 070967 (2255 Millerhill-Glasgow High Street freight, fire) (FDTL)

XX: 301267 (fire, location unknown) (FDTL)

Withdrawn (ex-ScR): 4we091071 (061071) (ex-64B)

64B Haymarket: 271171/ 121271/ 191271/ 030172/ 080172/ 220172
Millerhill Yard: 270272/ 310372/ 230472/ 130572/ 030672/ 230772/ 180872/ 260872/ 020972
St Rollox Works: 141072 (Dump)/ 221072/ 281072 (Top Yard)/ 051172/ 191172 (outside)/ 100273/ 110273/ 080473/ 210473/ 220473/ 050573/ 120573 (Top Yard)/ 130573/ 280573/ 290573/ 090673 (Yard)/ 170673/ 090873/ 110873/ 120873/ 140873/ 160873/ 150973/ 131073/ 231073 (CS)/ 041173/ 181173/ 130174. Not listed 130474.

Disposal: St Rollox Works: 0274 (D&ELfS).

D8588   Works No.: 8005

D8588, 51A Darlington, 27 May 1964. First of the Beyer Peacock batch, with Crompton Parkinson electrical equipment, immediately identifiable by the twin works and licence plates on the cab side. No pipework provision for through steam heating. (TOPticl Digital Memories (Roger Hateley))

50A York: 200364 (en route from Gorton to 51L Thornaby)

New (NER): 5we110464 (200364)-51L

Doncaster Works: 120266. DLRC: Doncaster: 110266-230266 U

Transfer: 4we160766 (190666)-52A

St Rollox Works: 210467/220467 DLRC: St Rollox: 040467-290467 C

St Rollox Works: Nil. DLRC: Glasgow: xxxxxx-181170 U. Not listed 251070.

St Rollox Works: 190271/200271 (Works Yard)/ 140371. DLRC: Glasgow: xxxxxx-160371 U

Transfer ScR: 1we010571 (250471)-64B

64B Haymarket: 250471

64B Haymarket: 270971 (stored)

Withdrawn (ex-ScR): 4we091071 (041071) (ex-64B)

64B Haymarket (??): Nil.
Millerhill Yard: 271171/030172/ 080172/ 270272/ 310372/ 230472/ 130572/ 030672/ 230772/ 180872/ 260872/ 020972
St Rollox Works: 141072 (Dump)/221072/ 051172/ 191172 (outside)/ 100273/ 110273/ 080473/ 210473/ 220473/ 050573/ 120573 (Top Yard)/ 130573/ 280573/ 290573/ 090673 (Cutting-up Area). Not listed 170673/ 040873.

Disposal: St Rollox Works: 0773 (D&ELfS).
'...cut up...about July (1973). (RO1273)

D8589   Works No.: 8006   N.B. No DLRC information.

D8589, Blaydon East, 15 February 1967. Banishment of D8589 to the rear of Gateshead depot (presumably the west end) following collision damage in 1970 resulted in much debate regarding the exact time of disposal of this locomotive (see Notes). The minutes of the Performance & Service Problems meetings made continuous references to D8589 ensuring that all spares were removed from the locomotive prior to final disposal, particularly given the apparent scarcity of Crompton Parkinson equipment. (Author's Collection)

New (NER): 4we090564 (040564)-51L

Doncaster Works: 050764 (Works Yard)/ 120764

Transfer: 4we160766 (190666)-52A

St Rollox Works: 240867. Not listed 210867.

St Rollox Works: 110568/250568. Not listed 130468.

St Rollox Works: 201068

Withdrawn (ex-NER): 1we110770 (060770) (ex-52A)

52A Gateshead: 260770/040870 (rear of depot, accident damage, bogies removed)/ 181070. Not listed 130271/ end-0571.

Disposal: Gateshead TMD: 0471 (D&ELfS). (Cut up on site at 52A by W. Willoughby, Choppington).

Notes:
1. 'The body of . . .8589, withdrawn due to accident damage, has been removed from its bogies at the rear of the depot (*52A Gateshead*).' (RO1070)
2. ' . . .8589 (52A), involved in a collision at Derwenthaugh earlier this year and since dumped at Gateshead shed, has had both engines and other fittings cannibalised for use on other members of the Class. Both bogies have now been removed and the body shell now stands on the ground at the west end of the yard, presumably awaiting cutting-up.' (RO1170)
3. Disposal details required: D8589 (ex-52A). (RO0471).
4. 'D8589 is still in process of disposal on site at Gateshead.' (RO0671)
5. 'At the end of May (*1971*) . . .8589 had disappeared at Gateshead, presumably cut up.' (RO0871)
6. ' It has now been established that 8589 . . .was cut up at Gateshead Depot by W. Willoughby of Choppington around the end of that year (*1970*).' (RO0278)
7. ' . . .finally broken up on site by staff from W. Willoughby of Choppington . . . in 4/71. All trace had gone by 31/5/71.' (AHBRD&E5)

D8590, Hexham, 8 September 1967. Repainted yellow panel and buffer beam by the looks of it, with the edge of the yellow panel vertical boundary adjusted adjacent to the tail lights. (Author's Collection)

D8590   Works No.: 8007   N.B. No DLRC information.

New (NER): 5we130664 (120564)-51L

Transfer: 4we160766 (190666)-52A

St Rollox Works: 110367

St Rollox Works: 280567

St Rollox Works: 140371. Not listed 200271. N.B. Not repaired.

Withdrawn (ex-NER): 4we270371 (270371) (ex-52A)

St Rollox Works: 170471/190571 (dumped near old Carriage Shop)/ 310571/ 190671 (¼mile north of Works)/ 070871/ 100871/ 160871/ 210871/ 300871/ 270971/ 161071 (Dump)/ 211171/ 281171. Not listed 090172.

Disposal: St Rollox Works: 1271 (D&ELfS). '...gone by 1/1/72.' (AHBRD&E5)

Note:
1. See Section 20.4 for further discussion re. disposal of D8590.

D8591   Works No.: 8008

New (NER): 5we080864 (080764)-51L

Transfer: 4we160766 (190666)-52A

St Rollox Works: Nil. DLRC: St Rollox: 091166-031266 C

St Rollox Works: Nil. DLRC: St Rollox: 211266-231266 Rect.

St Rollox Works: 070168. DLRC: Glasgow: xxxxxx-030268 U. Not listed 041167.

Withdrawn (ex-NER): 2we281268 (221268) (ex-52A)

St Rollox Works: 170569 (High Bank Sidings). Not listed 310369 and 150669/ 170769/ 030869.

J. MacWilliam, Shettleston: 090869 (intact)

Disposal: J. MacWilliam, Shettleston: 0669-0869* (D&ELfS).
* - 'Requiring definitive confirmation'
'...thought to have been broken up by 30/8/69.' (AHBRD&E5)

Notes:
1. Disposal details required: D8591 (ex-52A). (RO0471).
2. '8591 was sold from Glasgow Works to J. McWilliams, Shettleston during June 1969.' (RO0775)

D8592   Works No.: 8009

D8592, 12A Carlisle Kingmoor (Steam), 2 July 1967. (Colour-Rail)

New (NER): 5we130664 (120564)-52A

Doncaster Works: 031065/101065 (Stripping Shop)/ 161065/ xxxx65 (Test House). DLRC: Doncaster: 021065-041265 U
36A Doncaster: 051265

Doncaster Works: 050666/100766. DLRC: Doncaster: 120566-300766 U

Doncaster Works: Nil. DLRC: Doncaster: 150667-230667 U
65A Eastfield: 250667

64B Haymarket: xx0469
St Rollox Works: Nil. DLRC: Glasgow: 160469-090569 U (Generator cables)

St Rollox Works: 170769/020869/ 030869. DLRC: Glasgow: 040869-130869 U (No.2 engine defect)

St Rollox Works: 040870 (E.Shop)/220870. DLRC: Glasgow: xxxxxx-290870 C. Not listed 110770.

Transfer ScR: 1we120671 (060671)-64B

Withdrawn (ex-ScR): 4we091071 (210971) (ex-64B)

64B Haymarket: 270971/091071
St Rollox Works: 211171/281171/ 090172/ 120272/ 270272/ 020472 (Scrapping Area, cab remains). Not listed 090472/ 140572.
Lowe's scrapyard, Kilnhurst: xxxxxx (cut-up remains)

Disposal: St Rollox Works: 0372 (D&ELfS). Commencing month of cutting-up: 0372 (RO0776).

Notes:
1. The June 1967 Doncaster Works DLRC entry is believed to be an error, and should be St Rollox. Note the sighting at Eastfield on 25/06/67.
2. Some reports show the June 1971 transfer to 64B Haymarket as 06/06/71 on-loan, 20/06/71 permanent.
3. The P&SP Meeting minutes for 8 September 1971 indicate that D8592 was stored by that date.

D8593   Works No.: 8010

D8593, Location and date unknown. The location is possibly Thornton Junction Yard s.p., and, if so, the date would be sometime between June and early-October 1971. Typical Scottish Region single digit reporting code. Shocking condition.
(Rail-Online)

New (NER): 5we130664 (280564)-52A

50A York: 010566 (en route to Doncaster, hauled by D257)
Doncaster Works: 070566. DLRC: Doncaster: 050566-030666 C

65B St Rollox: 170766
St Rollox Works: 080866/120866/230866. DLRC: St Rollox: 010866-190866 U
65B St Rollox: 200866/210866

St Rollox Works: Nil. DLRC: St Rollox: 280966-011066 U

St Rollox Works: 180768/010868/ xx0868. DLRC: Glasgow: xxxxxx- 100868 C. Not listed 220668.
65A Eastfield: 170868

Doncaster Works(?): Nil. P&SP:Exhauster circuit fire damage.

Note:
1. Some reports show the June 1971 transfer to 64B Haymarket as 06/06/71 on-loan, 20/06/71 permanent.

St Rollox Works: 250570/100670/ 130670/ 140670/ 280670/ 110770/ 040870 (Works Yard)/ 220870. DLRC: Glasgow: xxxxxx-070970 C

Transfer ScR: 1we120671 (060671)-64B

Withdrawn (ex-ScR): 4we091071 (041071) (ex-64B)

64B Haymarket: 091071
Millerhill Yard: 271171/030172/ 080172/ 270272/ 310372/ 230472/ 130572/ 030672/ 230772/ 180872/ 260872/ 020972
St Rollox Works: 141072 (Dump)/221072/ 051172/ 191172 (outside)/ 100273/ 110273/ 080473/ 210473/ 220473/ 230473 (Top Yard)/ 050573/ 120573 (Top Yard)/ 130573/ 280573/ 290573/ 090673 (Yard)/ 170673/ 090873/ 110873/ 120873/ 140873/ 160873/ 150973. Not listed 210573 and 131073.

Disposal: St Rollox Works: 0973 (D&ELfS).

D8594 Works No.: 8011

D8594, Newcastle, undated.
(Rail-Online)

New (NER): 3we040764 (030764)-52A

St Rollox Works: Nil. DLRC: St Rollox: 300966-051166 C

St Rollox Works: Nil. DLRC: St Rollox: 211166-231166 U

XX: 021170 (fire, location unknown) (FDTL)
St Rollox Works: Nil. DLRC: Glasgow: xxxxxx-191270 C

Transfer ScR: 3we290571 (230571)-64B

Withdrawn (ex-ScR): 4we091071 (210971) (ex-64B)

64B Haymarket: 270971/091071
St Rollox Works: 211171/281171/ 090172/ 120272/ 270272. Not listed 090472.

Disposal: St Rollox Works: 0372 (D&ELfS). Commencing month of cutting-up: 0372 (RO0776).

D8595 Works No.: 8012

New (NER): 5we080864 (210764)-52A

Doncaster Works: 060266/120266/ 030466. DLRC: Doncaster: 120165(*sic 66*)-130466 C

St Rollox Works: 110468/120468/ 130468. DLRC: Glasgow: xxxxxx-040568 C. Not listed 090368.
65A Eastfield: 050568

St Rollox Works: 010868/xx0868. Not listed 180768 and 170868.

Withdrawn (ex-NER): 2we281268 (221268) (ex-52A)

Heaton s.p.: 050169
52A Gateshead: 290369/130469
64B Haymarket: 230669 (with D8596)
J. MacWilliam, Shettleston: 150869 (cab only).

Disposal: J. MacWilliam, Shettleston: 0569-0869 (D&ELfS).

Notes:
1. Sold for scrap (May 1969): D8595/6 (to J. MacWilliam, Shettleston, from Gateshead). (RO0869)
2. Disposal details required: D8595 (ex-52A). (RO0471)

D8596   Works No.: 8013

D8596, Tyne Yard, 16 May 1965. Note the Diesel Brake Tender attached to D8596. These DBTs were used fairly extensively with the North-Eastern Region allocated Claytons to ensure adequate braking capability when working the heavy unfitted mineral trains. The more expensive alternative was to provide a second locomotive. Later deployment of heavier Co-Co Class 37s on trains increasingly with "fitted heads" ultimately negated the need for the DBTs. (Author's Collection)

New (NER): 5we080864 (280764)-52A

St Rollox Works: Nil. DLRC: St Rollox: xxxxxx-251167 ?. Not listed 241067/ 041167.

52A Gateshead: Nil. P&SP: Stopped 230268-110368 (awaiting replacement turbo-blower).

Withdrawn (ex-NER): 2we281268 (221268) (ex-52A)

Heaton s.p.: 050169/180169
52A Gateshead: 290369/130469
64B Haymarket: 230669 (with D8595)
J. MacWilliam, Shettleston: 150869 (cab only).

Disposal: J. MacWilliam, Shettleston: 0569 (D&ELfS).

Notes:
1. Sold for scrap (May 1969): D8595/6 (to J. MacWilliam, Shettleston, from Gateshead). (RO0869)
2. Disposal details required: D8596 (ex-52A). (RO0471)

D8597   Works No.: 8014

D8597, St Rollox Works (Old Paint Shop Yard), 12 May 1973. In shocking condition and awaiting the inevitable. (Rail-Online)

New (NER): 5we080864 (210764)-52A

Doncaster Works: 041064 (Works Entrance)/ 111064 / 011164 (Paint Shop) / 031164. Not listed 031064.

St Rollox Works: Nil. DLRC: Glasgow: 171069-221069 U (Main generator)

St Rollox Works: 110170 DLRC: Glasgow: 171269-140270 C/I

Transfer ScR: 3we290571 (230571)-64B

Withdrawn (ex-ScR): 4we091071 (061071) (ex-64B)

64B Haymarket (??): Nil.
Millerhill Yard: 271171/030172/ 080172/ 270272/ 310372/ 230472/ 130572/ 030672/ 230772/ 180872/ 260872/ 020972
St Rollox Works: 141072 (Dump)/221072/ 281072 (Top Yard) / 051172/ 191172 (outside)/ 100273/ 110273/ 080473/ 210473/ 220473/ 050573/ 120573 (Top Yard)/ 130573/ 280573/ 290573/ 090673 (Cutting Area)/ 160873. Not listed 170673/ 050873/ 110873.

Disposal: St Rollox Works: 0773 (D&ELfS).
'...scrapped by 16/8/73.' (AHBRD&E5)
'...cut up...about July (1973). (RO1273)

D8598   Works No.: 8015

D8598, Millerhill s.p.,
8 September 1971.
(Rail-Online)

New (NER): 5we080864 (060864)-52A

Doncaster Works: 161065. DLRC: Doncaster: 151065-181065 U

?? Works: Nil. DLRC: ??: 090267-xxxxxx ?

St Rollox Works: 041070 (E.Shop)/251070. DLRC: Glasgow: xxxxxx-101070 C. Not listed 220870.

64B Haymarket: 271070

Transfer ScR: 3we290571 (230571)-64B

Withdrawn (ex-ScR): 4we091071 (061071) (ex-64B)

64B Haymarket (??): Nil.

Reinstated: 4we271171 (071171)-64B

Transfer: 5we010172 (061271)-66A

Withdrawn (ex-ScR): 2we010172 (311271) (ex-66A).

66A Polmadie: 030172/090172/ xx02-0372
9A Longsight: 120372
Transfer Longsight-Derby RCD: 280372
Derby RCD: 310372/010472 (newly out-shopped in blue)
9A Longsight: xx0472
Wilmslow: 230472/ xx0572 (stabled with *Hermes*)
Transfer Longsight-Mickleover Research Sidings: 050572 (dropped E26048 off at Crewe Works en route)
Test trains (Styal line): 22-250572 and w/c 050672 (with *Hermes*)
9A Longsight: 030672/040672/ 050772
Derby RTC: 090872
Swindon Works: 160872 (collecting *Hermes*)
Transfer Derby RCD-Mickleover test site: 170872 (D8598 hauling S18521 and D5901)
Transfer Derby RCD-Wilmslow: 180872 (hauling *Hermes*)
9A Longsight: 210872/220872
Wilmslow: 300872 (test train)
9A Longsight: 100972
Willesden Yard: 200972
WN Willesden: 290972/011072/ 041072
9A Longsight: 221072
Crewe: 301072 (with *Hermes*)
9A Longsight: 110273
Test trains (Styal line): w/e 130173 and w/e 090273 (with *Hermes*)
Carlisle: 130273 (Kingmoor-Slochd, with *Hermes*, transponder testing in snowy conditions)
9A Longsight: 170273

Derby Works: 180373/ 150473 (Klondyke Sdgs)/ 190573/ 100673 (Klondyke Sdgs)/ 050873 (Klondyke Sdgs)/ 250873 (Klondyke Sdgs)/ 211073/ 111173 (Klondyke Sdgs)/ 250274 (E.Shop)/ 100374 (E.Shop)/ 210474 (Yard)/ 120574 (Test House)/ 230674 (Test House Yard)/ 040874 (E.Shop)/ 310874/ 220974/ 201074 (Test House Yard)

LO Longsight: 141174/171174 (immaculate external condition, with *Hermes*)/ 020275
Longsight (passing): 140275 (with *Hermes*)
Derby RCD/Derby area: 290375/120475
LO Longsight: 260475/300475/ 150675
Wilmslow: 140775 (with *Hermes*)/160775
LO Longsight: 200775
Wilmslow: 230775 (with Test Coach)/ 300775 (with Test Coach)
LO Longsight: 110875/240875
Derby RCD/Derby area: 140176/160676/ 280676/ 140776/ 280776/ 130876/ 040976/ 260976/ 101076
LO Longsight: 061176(with *Hermes*)
Derby RCD (Mickleover): 081176
Derby RCD/Derby area: 271176/150677/ 090777/ 100877/ 100977/ 261177/ 301277/ 180278/ 050378/ 010478/ 070578/ 100678/ 220778/ 050878/ 180878/ 090978/ 220978
Booked transfer: 9Z10 11.15 Derby RCD-Carlisle 071078, then Carlisle-St Rollox Works 091078 (with S18521)
Chesterfield: 071078 (hauled by 46027, with S18521)
Rotherham: 091078 (with S18521)
Carlisle: xx1078
St Rollox Works: 141078/151078/ 191178/ 251178 (Scrapping Area)/ 161278/ 130179/ 210479 (cut-up remains)/ 190579 (cut-up remains). Not listed 250279.

Disposal: Departmental 8598; St Rollox Works: 0379 (D&ELfS).
'…broken up by 21/11/71.' (AHBRD&E)

Note:
1. RTC Allocations. Reinstated: 28/03/72, Withdrawn: 07/10/78.

D8599   Works No.: 8016

D8599, Newcastle, undated. (Rail-Online)

New (NER): 5we120964 (280864)-52A

64B Haymarket: 141266
St Rollox Works: 161266. DLRC: St Rollox: 191266-210167 C+Collision Damage

Transfer: 4we270468 (310368)-51L

Transfer: 4we150369 (020369)-52A

St Rollox Works: Nil. DLRC: Glasgow: xxxxxx-300171 C, and, Glasgow: xxxxxx-120271 U. Not listed 100171.

Transfer ScR: 1we010571 (250471)-64B

64B Haymarket: 250471

Withdrawn (ex-ScR): 4we091071 (051071) (ex-64B)

64B Haymarket: 091071
St Rollox Works: 161071 (Works Yard)/211171 (being cut-up)/ 281171. Not listed 090172/ 120272.

Disposal: St Rollox Works: 1171 (D&ELfS). Commencing month of cutting-up: 1171 (RO0776).

D8600   Works No.: 8017

D8600, Newcastle, Undated. (Rail-Online)

New (NER): 5we080864 (060864)-52A

St Rollox Works: 150267. DLRC: St Rollox: 150267-180367 C
65A Eastfield: 250367/ 260367

Transfer: 4we270468 (310368)-51L
Transfer: 4we150369 (020369)-52A

XX: 040570 ('serious' fire, location unknown) (FDTL)
St Rollox Works: 250570/ 100670/ 130670/ 140670/ 280670/ 110770/ 040870 (Works Yard)/ 220870/ 251070. Not listed 041070. DLRC: Glasgow: xxxxxx-031270 U, and, Glasgow: xxxxxx-171270 U

Transfer ScR: 1we010571 (250471)-64B

64B Haymarket: 250471

Withdrawn (ex-ScR): 4we091071 (061071) (ex-64B)

Millerhill Yard: 271171/ 030172/ 080172/ 270272/ 310372/ 230472/ 130572/ 030672/ 230772/ 180872/ 260872/ 020972
St Rollox Works: 141072 (Dump)/ 221072/ 281072 (Top Yard)/ 191172 (outside)/ 100273/ 110273/ 220473/ xx0473 (Cutting-up Area, cabside panels)/ 120573 (cab). Not listed 051172/ 080473/ 210473.

Disposal: St Rollox Works: 0473 (D&ELfS).

D8601   Works No.: 8018

D8601, 51L Thornaby, 1969. (Transport Topics)

New (NER): 5we120964 (020964)-52A

Transfer: 5we240268 (280168)-51L

St Rollox Works: Nil. DLRC: Glasgow: xxxxxx-060668 U. Not listed 250568.

Transfer: 7we040470 (150370)-52A

St Rollox Works: 251070. DLRC: Glasgow: xxxxxx-141170 C. Not listed 041070.

Transfer ScR: 1we120671 (060671)-64B

Withdrawn (ex-ScR): 4we091071 (061071) (ex-64B)

64B Haymarket: 271171/ 121271/ 030172/ 080172/ 220172
Millerhill Yard: 270272/ 310372/ 230472/ 130572/ 030672/ 230772/ 180872/ 260872/ 020972
St Rollox Works: 141072 (Dump)/ 221072/ 281072 (Top Yard)/ 191172 (outside)/ 100273/ 110273/ 080473/ 210473/ 220473/ 050573/ 120573 (Top Yard)/ 130573/ 280573/ 290573/ 090673 (Yard)/ 170673/ 050873/ 110873/ 120873/ 140873/ 160873/ 150973/ 131073/ 231073 (CS)/ 041173/ 181173. Not listed 051172 and 130174.

Disposal: St Rollox Works: 0274 (D&ELfS).

Notes:
1. See Section 20.4 for further discussion re. disposal of D8601.
2. Some reports show the June 1971 transfer to 64B Haymarket as 06/06/71 on-loan, 20/06/71 permanent.

D8602 Works No.: 8019

New (NER): 5we120964 (090964)-52A

St Rollox Works: Nil. DLRC: St Rollox: 060766-190766 U
64B Haymarket: 210766

Transfer: 5we240268 (280168)-51L

51L Thornaby: 140468/280468. P&SP: Stopped 280368-250468 awaiting replacement turbo-blower.

64B Haymarket : 171069
Doncaster Works: Nil. DLRC: Doncaster: 171069-131169 U. SRSS/P&SP:Exhauster circuit fire damage.
64B Haymarket: 2we191169 ('recent visitor')

64B Haymarket: 160370
St Rollox Works: 190370/270370/ 290370/ 110470 (E.Shop). DLRC: Glasgow: xxxxxx-250470 C

Transfer: 6we160570 (030570)-52A

XX: 040171 ('serious' fire, location unknown) (FDTL)
64B Haymarket: 100171
St Rollox Works: 130271/190271/ 200271 (In works).
DLRC: Glasgow: xxxxxx-030371 U

Transfer ScR: 1we010571 (250471)-64B

64B Haymarket: 250471

Withdrawn (ex-ScR): 4we091071 (061071) (ex-64B)

64B Haymarket: 271171
St Rollox Works: 090172/120272/ 270272/ 020472 (Scrap Road, being cut-up)/ 090472 (Cutting-up Area). Not listed 250472/ 140572.

Disposal: St Rollox Works: 0472 (D&ELfS).
'. . .being cut up on 3/4/72 . . .completion by 20/4/72.' (AHBRD&E5)
Commencing month of cutting-up: 0472 (RO0776).

Note:
1. The October/November 1969 Doncaster Works DLRC entry is believed to be an error, and should be St Rollox. Note sighting at Haymarket on 17/10/69.

D8603   Works No.: 8020

D8603, 51L Thornaby, 1969. (Transport Topics)

New (NER): 5we120964 (070964)-52A

Doncaster Works: 030466/280466/ 010566/ 020566/ 070566. DLRC: Doncaster: 230366-140566 C

St Rollox Works: 090368. DLRC: Glasgow: xxxxxx-090368 C. Not listed 070168.

Transfer: 4we270468 (310368)-51L

St Rollox Works: Nil. DLRC: Glasgow: 140769-150769 U (Main generator change)
64B Haymarket: 200769

Transfer: 4we300570 (030570)-52A

St Rollox Works: Nil. DLRC: Glasgow: xxxxxx-141170 U. Not listed 251070.

Transfer ScR: 1we010571 (250471)-64B

64B Haymarket: 250471

Withdrawn (ex-ScR): 4we091071 (061071) (ex-64B)

64B Haymarket: 271171
St Rollox Works: 090172/120272/ 270272/ 020472 (Scrap Road, being cut-up)/ 090472 (Cutting-up Area). Not listed 250472/ 140572.

Disposal: St Rollox Works: 0472 (D&ELfS).
'. . .noted on 3/4/72 being cut up . . .completed by 20/4/72.' (AHBRD&E5)
Commencing month of cutting-up: 0472 (RO0776).

D8604   Works No.: 8021

New (ER): 3we031064 (150964)-41A

Transfer: 4we010565 (050465)-41E

Transfer ScR: 1we280566 (220566)-64B

Transfer NER (o/l): 1we170667 (120667)-52A (ScR)/ 4we150767-52A (ER)
Transfer NER (perm): 1we091267 (031267)-52A

St Rollox Works: Nil. DLRC: St Rollox: xxxxxx-081267 U. Not listed 041167.

Transfer: 5we240268 (280168)-51L

St Rollox Works: August BH weekend (310868-020968). DLRC: Glasgow: xxxxxx-140968 C. Not listed 260868.
65A Eastfield: 180968

Doncaster Works: Nil. DLRC: Doncaster: xxxxxx-041068 U
St Rollox Works: Nil. DLRC: Glasgow: xxxxxx-091068 U
64B Haymarket: 121068

Transfer: 4we300570 (030570)-52A

Transfer ScR: 1we120671 (060671)-64B

Withdrawn (ex-ScR): 4we091071 (041071) (ex-64B)

64B Haymarket: 091071
St Rollox Works: 211171/ 281171/ 090172/ 120272/ 270272. Not listed 090472/ 140572.

Disposal: St Rollox Works: 0472 (D&ELfS). Commencing month of cutting-up: 0472 (RO0776).

Notes:
1. The October 1968 Doncaster Works DLRC entry is believed to be an error, and should be St Rollox. Note Haymarket sighting on 12 October 1968.
2. Some reports show the June 1971 transfer to 64B Haymarket as 06/06/71 on-loan, 20/06/71 permanent.

D8605 Works No.: 8022

D8605, 41B Sheffield Darnall, 1 August 1965. (Author's Collection)

New (ER): 3we031064 (011064)-41A

Transfer: 4we010565 (050465)-41E

Transfer ScR: 1we280566 (220566)-64B

Thornton North: 310567 (Markinch trip, fire) (FDTL)
Dysart: 010667 (2335 Thornton-Millerhill freight, fire) (FDTL)

Transfer NER (o/l): 1we170667 (120667)-52A (ScR)/ 4we150767-52A (ER)
Transfer NER (perm): 1we091267 (031267)-52A

Transfer: 5we240268 (280168)-51L

Store ( ): xwexxxxxx (041068) (ex 51L)

Withdrawn (ex-NER): 4we161168 (281068) (ex-51L (store))

51L Thornaby: 040169/150369/ 290369/ 100569
Draper's, Neptune Street Yard, Hull: Arrived 050569/ Arrived 140569/ Arrived Neptune Street yard 180569/ 210569/ 040669/ 290669/ 300869/ 011169/ 291269/ 090370/ 180470/ 230670 (intact)/ 120870

Disposal: A.Draper, Neptune Street, Hull: 0569-0670 (D&ELfS).

Notes:
1. It appears that D8605 never visited a Main Works.
2. 'D8605 . . .arrived (*at Draper's, Hull*) from Thornaby on 5th May (*1969*)' (RO0869)
3. 'On May 14 . . .D8605 arrived at Hull from Thornaby destined for Draper's yard.' (RW0769)
4. Ian Scotney and Brian Egan ('Draper's Scrapyard, Hull', 2015) record D8605 arriving at Neptune Street yard on 18 May 1969 and cut-up on 22 March 1971.
5. The arrival date discrepancies might be explained by the first date being the arrival date in the sidings outside Neptune Street as opposed to arrival in Draper's scrapyard proper.

D8606   Works No.: 8023

D8606, 64B Haymarket, 1970. Miniature snow ploughs. The only Beyer Peacock built locomotive to carry blue livery. (Transport Topics)

New (ER): 2we171064 (051064)-41A

Transfer: 1we010565 (050465)-41E

Transfer ScR: 1we280566 (220566)-64B

St Rollox Works: Nil. DLRC: St Rollox: 190167-250267 U, and, St Rollox: 070367-230367 Rect
65A Eastfield: 250367/ 260367

St Rollox Works: 070168 DLRC: St Rollox: 181267-200168 I

St Rollox Works: 210869. DLRC: St Rollox: 110869-190869 U (No.2 power unit change)

64B Haymarket: 200371

Withdrawn (ex-ScR): 4we030471 (270371) (ex-64B)

64B Haymarket: 250471/ 160571/ 280571/ 200671/ 010871/ 080871/ 150871/ 190871/ 280871/ 091071
St Rollox Works: 211171/ 281171/ 090172/ 120272/ 270272/ 280572. Not listed 090472/ 140572.

Disposal: St Rollox Works: 0472 (D&ELfS). Commencing month of cutting-up: 0472 (RO0776).

D8607   Works No.: 8024

D8607, 64B Haymarket, 1970. Miniature snow ploughs. The adjacent Class 26 carries tablet catching equipment unlike the Clayton; as far is known no Clayton ever carried tablet catching equipment despite all being fitted the necessary cab side recesses. On 29 May 1969 D8607 found itself assisting Class 29 D6100 on the 07.10 Glasgow Queen Street-Dundee passenger service from Perth to Dundee. (Transport Topics)

New (ER): 2we171064 (161064)-41A

16C Derby: 181064

Doncaster Works: 070265 (Works Yard)

Transfer: 4we010565 (050465)-41E

Doncaster Works: 070166/090166 (Dismantling Shop Yard)/ 100166

Transfer ScR: 1we280566 (220566)-64B

St Rollox Works: 070168. DLRC: St Rollox: 221267-030268 I

St Rollox Works: Nil. DLRC: St Rollox: 220469-020569 U (Main generator defect)

St Rollox Works: Nil. DLRC: Glasgow: 200270-210370 U
65A Eastfield: 260370/290370

St Rollox Works: Nil. DLRC: Glasgow: 301170-301270 U

Transfer: 4we290571 (050571)-66A

Withdrawn (ex-ScR): 4we 091071 (051071) (ex-66A)

66A Polmadie: 271171
Ex-67D Ardrossan: Arrived mid-1271/ 030172/ 090172/ 260272/ 300372/ 250472/ 140572/ 170772/ 170872/ 060972
66A Polmadie: 301072/061172/ 091172/ 191172/ 100273/ 250273/ 100473/ 120573/ 270573/ 090673/ 170873/ 150973/ 131073/ 181173/ 130174/ 260174/ 170274/ 230374/ 140474/ 280574/ 150674/ 290674/ 030874/ 100874/ 250874/ 280974/ 051074/ 260175/ 080375/ 160375/ 290375/ 050475/ 270475. Not listed 141072 and 161174.
St Rollox Works: 170575/270575/ 070675 ('Dump')/ 120675 (Carriage Sidings?)/ mid-0775 (Works Yard)/ 120875
Transfer to J. MacWilliam, Shettleston: 260875 (with D8608/12)
J. MacWilliam, Shettleston: Nil.

Disposal: J. MacWilliam, Shettleston: 0875 (D&ELfS). '. . .to Shettleston, Glasgow on 26/8/75, with disposal by 25/9/75.' (AHBRD&E5)
Disposal not proven.

D8608 Works No.: 8025

D8608, Millerhill s.p., July 1970. Miniature snow ploughs.
(Grahame Wareham)

New (ER): 2we311064 (271064)-41A

Doncaster Works: 100165 (Works Yard). DLRC: Doncaster: xxxxxx-160165 U

Transfer: 4we010565 (050465)-41E

Doncaster Works: 030466/ 010566

Transfer ScR: 1we280566 (220566)-64B

St Rollox Works: 090368. DLRC: St Rollox: 060268-160368 I

Transfer: 4we290571 (050571)-66A

Withdrawn (ex-ScR): 4we 091071 (051071) (ex-66A)

66A Polmadie: 271171

Ex-67D Ardrossan: Arrived mid-1271/ 030172/ 090172/ 260272/ 300372/ 250472/ 140572/ 170772/ 170872/ 060972

66A Polmadie: 061172/ 091172/ 191172/ 100273/ 250273/ 100473/ 120573/ 270573/ 090673/ 170873/ 150973/ 131073/ 181173/ 130174/ 260174/ 170274/ 230374/ 140474/ 280574/ 150674/ 290674/ 030874/ 100874/ 250874/ 280974/ 051074/ 161174/ 260175/ 080375/ 160375/ 290375/ 050475/ 270475. Not listed 141072.

St Rollox Works: 170575 (middle of CS)/ 270575/ 070675 (Works Yard)/ mid-0775 (Works Yard)/ 120875

Transfer to J. MacWilliam, Shettleston: 260875 (with D8607/12)

J. MacWilliam, Shettleston: Nil.

Disposal: J. MacWilliam, Shettleston: 0875 (D&ELfS). Disposal not proven.

D8609 Works No.: 8026

D8609, 41E Barrow Hill, 1 August 1965. Inside the famous roundhouse which is still open for business today. Note the number painted in the buffer beam, a typical feature of the Barrow Hill allocated Claytons.

New (ER): 4we281164 (231164)-41A

Derby: 291164 (Derby-Sheffield L.E. with D5690)

Transfer: 4we010565 (050465)-41E

Transfer ScR: 1we280566 (220566)-64B

63A Perth: 11-120267 / 200267
St Rollox Works: Nil. DLRC: St Rollox: 140267-010467 U

St Rollox Works: 280567. DLRC: St Rollox: 100567-030667 U

Withdrawn (ex-ScR): 1we191068 (131068) (ex-64B)

62A Thornton Junction: 020169 / 110169 / 130369 / 020469
J. MacWilliam, Shettleston: xx0569 (intact). N.B. Not listed 240869.

Disposal: J. MacWilliam, Shettleston: 0669* (D&ELfS).
* - 'Requiring definitive confirmation'

Notes:
1. Reason for Withdrawal. Requiring Intermediate Repair/Surplus.
2. Disposal details required: D8609 (ex-64B). (RO0471)
3. Sold for scrap (May 1969): D8609 (to J. MacWilliam, Shettleston, from Haymarket). (RO1071)

D8610   Works No.: 8027

D8610, Millerhill s.p., 1970. D8610 is believed to have been the only Clayton to have reached Ballater, with this being achieved on the final freight service to this location on 15 July 1966. (Transport Topics)

New (ER): 3we191264 (071264)-41A

Transfer: 4we010565 (050465)-41E

Transfer ScR: 1we280566 (220566)-64B

St Rollox Works: 090368. DLRC: St Rollox: 070268-160368 I

Transfer: 4we290571 (050571)-66A

Withdrawn (ex-ScR): 4we 091071 (051071) (ex-66A)

66A Polmadie: 271171
Ex-67D Ardrossan: Arrived mid-1271/ 030172/ 090172/ 260272/ 300372/ 250472/ 140572/ 170772/ 170872/ 060972
66A Polmadie: 061172/091172/ 191172. Not listed 141072.
St Rollox Works: 100273/110273/ 080473/ 210473/ 220473/ 050573/ 120573 (Top Yard)/ 130573/ 280573/ 290573/ 090673 (Yard)/ 170673/ 110873 (being stripped)/ 160873. Not listed 120873 and 150973.

Disposal: St Rollox Works: 0873 (D&ELfS).

D8611   Works No.: 8028

D8611, 64B Haymarket, 25 May 1969. Less than four and a half years old and already awaiting scrapping; surplus to requirements, its first Classified works repair was denied. (TOPticl Digital Memories (Roger Hateley))

New (ER): 3we191264 (171264)-41A

Transfer: 4we010565 (050465)-41E

Darlington Works: 060665 (accident damage, collision with lorry at level crossing)

Transfer ScR: 1we280566 (220566)-64B

St Rollox Works: Nil. DLRC: St Rollox: 050167-280167 U

Withdrawn (ex-ScR): 1we191068 (131068) (ex-64B)

62A Thornton Junction: 020169/110169. Not listed 130369.
64B Haymarket: 300369/250569
J. MacWilliam, Shettleston: xx0669 (intact)/ 240869

Disposal: J. MacWilliam, Shettleston: 0569-0869* (D&ELfS).
* - 'Requiring definitive confirmation'
'...Cut up by 2/9/69.' (AHBRD&E5)

Notes:
1. Reason for Withdrawal. Requiring Intermediate Repair/Surplus.
2. Disposal details required: D8611 (ex-64B). (RO0471).
3. Sold for scrap (June 1969): D8611 (to J. MacWilliam, Shettleston, from Haymarket). (RO1071)

D8612   Works No.: 8029

D8612, 66B Motherwell, 11 September 1971.
(Rail-Online)

New (ER): 2we060265 (270165)-41A

Transfer: 4we010565 (050465)-41E

Doncaster Works: 101265 (E.Shop)

Transfer ScR: 1we280566 (220566)-64B

63A Perth: 220467/250467/ 270467
St Rollox Works: 110567/280567 DLRC: St Rollox: 040567-010667 U

St Rollox Works: 210968 (E.Shop). DLRC: St Rollox: 050968-280968 C

St Rollox Works: 020969 DLRC: St Rollox: 030969-050969 U (Main generator defect)

66A Polmadie: 220870/041070/ 071170
St Rollox Works: 100171/130271. DLRC: Glasgow: 040870-120371 U

Transfer: 4we290571 (050571)-66A

Withdrawn (ex-ScR): 4we 091071 (051071) (ex-66A)

66A Polmadie: 271171
Ex-67D Ardrossan: Arrived mid-1271/ 030172/ 090172/ 260272/ 300372/ 250472/ 140572/ 170772/ 170872/ 060972
66A Polmadie: 301072/061172/ 091172/ 191172/ 100273/ 100473/ 120573/ 270573/ 090673/ 170873/ 150973/ 131073/ 181173/ 130174/ 260174/ 170274/ 230374/ 140474/ 280574/ 150674/ 290674/ 030874/ 100874/ 250874/ 280974/ 051074/ 260175/ 080375/ 160375/ 290375/ 050475/ 270475. Not listed 141072 and 250273/ 161174.
St Rollox Works: 170575/270575/ 070675 (Dump)/ mid-0775 (Works Yard)/ 120875
Transfer to J. MacWilliam, Shettleston: 260875 (with D8607/8)
J. MacWilliam, Shettleston: xx1075 (intact)

Disposal: J. MacWilliam, Shettleston: 0875 (D&ELfS)

D8613 Works No.: 8030

D8613, 66A Polmadie, 28 September 1974.
(Anthony Sayer)

New (ER): 1we230165 (220165)-41A

Transfer: 4we010565 (050465)-41E

Seymour Junction/Gladwell Colliery: 161065 (RCTS 'Midlands Locomotive Requiem Rail Tour' assisting 43953 on Seymour Jct-Glapwell Colliery Sidings section)

Transfer ScR: 1we280566 (220566)-64B

Inverurie Works: 220866

64B Haymarket: 020967
St Rollox Works: 140967 (E.Shop). DLRC: St Rollox: 250867-150967 U

St Rollox Works: Nil. DLRC: St Rollox: 290967-131067 U

64B Haymarket: 130468/140468
St Rollox Works: 110568/190568. DLRC: St Rollox: 160468-180568 C
65A Eastfield: 200568 (ex-Works)

St Rollox Works: 020969. DLRC: Glasgow: 270869-100969 U (Collision damage)

St Rollox Works: 270370/280370/290370/110470 (Works Yard)/250570. Not listed 100670. DLRC: Glasgow: 180270-230570 U

64B Haymarket: 030870/080870

66A Polmadie: 220870/041070
St Rollox Works: 251070. DLRC: Glasgow: 060770-311070 U. Not listed 041070.

St Rollox Works: Nil. DLRC: Glasgow: 231170-261170 U

St Rollox Works: Nil. DLRC: Glasgow: 161270-090171 U
65A Eastfield: 100171

St Rollox Works: Nil. DLRC: Glasgow: 130171-030271 U

St Rollox Works: 140371. DLRC: Glasgow: 080371-200371 U

Transfer: 4we290571 (050571)-66A

Withdrawn (ex-ScR): 4we091071 (061071) (ex-66A)

66A Polmadie: 271171

Ex-67D Ardrossan: Arrived mid-1271/ 030172/ 090172/ 260272/ 300372/ 250472/ 140572/ 170772/ 170872/ 060972
66A Polmadie: 061172/091172/ 191172/ 100273/ 250273/ 100473/ 120573/ 270573/ 090673/ 080773/ 170873/ 150973/ 131073/ 211073/ 181173/ 130174/ 260174/ 170274/ 230374/ 140474/ 280574/ 150674/ 290674/ 030874/ 100874/ 250874/ 280974/ 051074/ 161174/ 260175/ 080375/ 160375/ 290375/ 050475/ 270475. Not listed 141072.
St Rollox Works: 170575 (middle of Carriage Sds)/ 270575/ 070675 (Crane Shop)/ mid-0775 (Works Yard)/ 120875
Transfer to J. MacWilliam, Shettleston: 260875 (with D8557/73)
J. MacWilliam, Shettleston: Nil.

Disposal: J. MacWilliam, Shettleston: 0875 (D&ELfS).
'...broken up by 5/9/75.' (AHBRD&E5)
Disposal not proven.

D8614 Works No.: 8031

New (ER): 2we200265 (110265)-41A

Transfer: 4we010565 (050465)-41E

Transfer ScR: 1we280566 (220566)-64B

64B Haymarket: 020168
St Rollox Works: 070168. DLRC: St Rollox: 040168-100268 I

64B Haymarket: 200769/160869/ 170869/ 060969
St Rollox Works: Nil. DLRC: St Rollox: 150969-041069 U (Main generator)

Transfer: 4we290571 (050571)-66A

Withdrawn (ex-ScR): 4we 091071 (051071) (ex-66A)

66A Polmadie: 271171
Ex-67D Ardrossan: Arrived mid-1271/ 030172/ 090172/ 260272/ 300372/ 250472/ 140572
St Rollox Works: 080672/100672 (CS)/ 180672/ 060772/ 060872/ 100872/ 130872/ 170872 (CS)/ 280872/ 290872/ 020972 (CS)/ 141072/ 221072/ 191172. Not listed 051172 and 100273.

Disposal: St Rollox Works: 1272 (D&ELfS).
Commencing month of cutting-up: 1272 (RO0776).

D8615   Works No.: 8032

D8614 and D8615, St Rollox Works (Carriage Sidings), 17 August 1972. On 7 February 1967 D8615 was the station pilot at Hawick, when it was called upon to assist 'Peak' D15 on the 21.15 London St Pancras-Edinburgh Waverley passenger from Hawick to Edinburgh Waverley. (Anthony Sayer)

New (ER): 2we030465 (100365)-41A

16C Derby: 140365

Transfer: 4we010565 (050465)-41E

Transfer ScR: 1we280566 (220566)-64B

Thornton Yard: 260667 (0710 Thornton-Markinch trip, fire, rescued by D8566) (FDTL)

64B Haymarket: 180668
St Rollox Works: 180768. DLRC: St Rollox: 240668-130768 I

St Rollox Works: Nil. DLRC: St Rollox: 200669-270669 U (power unit defect)

St Rollox Works: Nil. DLRC: Glasgow: 020270-210270 U

Transfer: 4we290571 (050571)-66A

Withdrawn (ex-ScR): 4we 091071 (051071) (ex-66A)

66A Polmadie: 271171
Ex-67D Ardrossan: Arrived mid-1271/ 030172/ 090172/ 260272/ 300372/ 250472/ 140572
St Rollox Works: 080672/ 100672 (CS)/ 180672/ 060772/ 060872/ 100872/ 130872/ 170872 (CS)/ 280872/ 290872/ 020972 (CS)/ 141072/ 221072/ 191172/ 100273/ 110273/ 080473/ 210473/ 220473/ 050573/ 120573/ 130573/ 280573/ 290573/ 090673 (Yard)/ 170673/ 050873/ 110873/ 120873/ 140873/ 160873/ 150973/ 131073/ 041173/ 181173/ 130174. Not listed 051172 and 130474.

Disposal: St Rollox Works: 1273 (D&ELfS).

D8616 Works No.: 8033

D8616, Marylebone, 30 April 1965. Following the 7th International Congress on Combustion Engines (CIMAC), held in London during 25-29 April 1965, BR, in conjunction with the Locomotive and Allied Manufacturers' Association of Great Britain (LAMA), arranged a "Diesel Power on British Railways" exhibition at the Marylebone Goods Yard on 30 April and 1 May. D8616 was the last Clayton built, but that was not the original plan! (Author's Collection)

New (ER): 1we010565 (230465)-41E

16C Derby: 250465
Marylebone Goods Yard: 300465/010565 (both BR/LAMA exhibition)

Transfer ScR: 1we280566 (220566)-64B

St Rollox Works: 090368. **DLRC: St Rollox:** 160268-300368 I

64B Haymarket: 080870/ 270870/ 300870 (Accident damage to No.2 end)
St Rollox Works: 041070 (E.Shop). **DLRC: Glasgow:** 070870-241070 U

Transfer: 4we290571 (050571)-66A

Withdrawn (ex-ScR): 4we 091071 (210971) (ex-66A)

66A Polmadie: 051171/ 271171

Ex-67D Ardrossan: Arrived mid-1271/ 030172/ 090172/ 260272/ 300372/ 250472/ 140572/ 170772/ 170872/ 060972
66A Polmadie: 061172/ 091172/ 191172/ 100273/ 250273/ 100473/ 120573/ 270573/ 090673/ 170873/ 150973/ 131073/ 181173/ 130174/ 260174/ 170274/ 230374/ 140474/ 280574/ 150674/ 290674/ 100874/ 250874/ 280974/ 051074/ 260175. Not listed 141072, 161174 and 080375.
Dundee (freight yard near diesel depot): 170375/ 250375/ 300375/ 260575/ 040775/ 080775/ 040875
Transfer to J. MacWilliam, Shettleston: 20-220875 (8X08 23.30 (190875) Aberdeen-Thornton Yard (from Dundee) on 200875, then 8X35 05.40 Thornton Yard-Grangemouth on 210875, then 8X12 20.00 Grangemouth-Cadder on 210875, then tripped to Shettleston)

J. MacWilliam, Shettleston: Nil.

Disposal: J. MacWilliam, Shettleston: 0875 (D&ELfS). Disposal not proven.

Chapter 11

PERFORMANCE AND SERVICE PROBLEMS: ENGINES & ASSOCIATED EQUIPMENT

11.1 Paxman Engine Issues.

11.1.1 Background.
Richard Carr's 'Paxman History Pages' website indicates that 243 engines were supplied for use in the Class 17s i.e. 234 for the locomotives themselves, plus 9 spares. With D8586/7 being given Rolls Royce engines these figures became 230 and 13 respectively. The 'Performance & Service Problems of Clayton Type 1 Locomotives' Meeting minutes of 19 September 1967 recorded that a further engine had been supplied to St Rollox Works, without turbocharger, with the comment that 'This brings the total of spare engines to 14.'

11.1.2 Early Problems.
Problems with the Claytons manifested themselves soon after the introduction of D8500 in September 1962. The Paxman 6ZHXL engine and GEC main generator combination resulted in significant early issues which ultimately resulted in BR placing an embargo on delivery of further locomotives on 8 December 1962 pending resolution of the problems. At that date fourteen locomotives (D8500-13) had been delivered.

Crankshaft failures began to appear after between 750 and 800 engine hours. The problem was diagnosed as severe torsional and longitudinal vibration, each occurring at a different engine speed. A test bed engine subsequently showed that, at a rotational speed of about 1300rpm, crankshafts could fail after less than twenty hours use. A report by the Chief Mechanical Engineer entitled *Diesel Train Locomotive Performance* (dated 20 August 1963) described the problem as 'whirling [main generator] armature shafts causing engine crankshaft failures'.

It was to take four months for the problem to be resolved and was eventually overcome by the fitting of vibration dampers to the front end of the engine crankshaft.

11.1.3 Consequences of the 1962/63 Delivery Embargo.
From December 1962, the 14 Claytons delivered to Scotland were placed in store and steam locomotives were reinstated to take up the slack while the problems were resolved. Minute 63/95 of the Scottish Railway Board Meeting held in Glasgow on 17 April 1963 commented:

'The Board noted that the manufacturers are now carrying out modifications to the locomotives but expressed deep concern that the non-availability of these locomotives, plus the retarded delivery of the remainder has resulted in a serious shortage of locomotives, necessitating a delay in the condemnation of steam locomotives and a loan of locomotives from other Regions.'

Most, if not all, of the Claytons already in Scotland were initially stored at 66A Polmadie. D8507/8 subsequently migrated south to Derby, whilst D8500-6/9-13 were moved to 65C Parkhead. Of the Parkhead batch, at least D8500/1/3-6/9-12 were moved to 63A Perth for remedial work prior to re-introduction to traffic; D8502/13 may also have received attention at Perth but available sighting information does not confirm or deny this. Relevant sightings are listed below:

16/12/62	66A: D8500-5/7/9/11
17/02/63	66A: D8500-6/9-12
21/03/63	65C: D8500-6/9-13
04/04/63	65C: D8500-6/9-13
12-14/04/63	63A: D8503/4/11 65C: D8500-2/5/6/9/10/2/3
19/04/63	63A: D8503/4/11
20/04/63	65C: D8500-2/5/6/9/10/2/3
12/05/63	63A: D8500/1/3-6/9/10/2
01/06/63	63A: D8500/1/3/4/10
22/06/63	63A: D8510
23/07/63	63A: D8503 (returned for rectification?)

It is assumed that D8507/8 were repaired by Clayton Equipment at Derby before returning north. Photographs of Claytons stored at Parkhead can be found in *The Railway Magazine* (May 1963, p369) and *Traction* (September 2000, pp42/3).

In the meantime, however, Clayton Equipment continued to produce locomotives, even though these could not be accepted by BR. This was a sensible decision

D8503, 63A Perth, 1963. Replacement engines installed for all to see and undergoing testing in the depot yard. Paxman and/or BR fitters riding 'shot-gun'. (Transport Treasury)

in that it kept the Clayton workforce fully employed and subsequently allowed for a more rapid introduction of locomotives into traffic once a solution of the vibration problem had been found. As a consequence, the numbers of semi-complete Clayton locomotives, both fully painted or in works primer, began to build up, some spilling out of the International Combustion Works into the wider Derby area.

Sightings of Claytons around the Derby area are listed below:-

24/02/63	St Andrews Goods Yard: D8508/14-9/21 and three not numbered.
01/03/63	St Andrews Goods Yard: D8508/14-9/21 and three in primer (Makers' Nos.24-6).
10/03/63	Sidings opposite Shed: D8507 painted and three in primer (Makers' Nos.27/8/30).
	St Andrews Goods Yard: D8508/14-9/21 and three in primer (Makers' Nos.24-6).
23/03/63	Sidings opposite Shed: D8507 painted and three in primer (Makers' Nos.27/8/30).
07/04/63	Sidings opposite Shed: D8507 painted and three in primer (Makers' Nos.27/8/30).
	St Andrews Goods Yard: D8515-9/21 painted and three in primer (Maker's Nos.24-6).
xx/05/63	St Andrews Goods Yard: 'During May it was noted that the batch in St Andrews Goods Yard decreased, first to nine in number, then to seven, then for a short period back to eleven, but all movement was in the un-numbered ones!'
26/05/63	'Several painted and unpainted Claytons were lying about in various sidings.'
16/06/63	Sidings opposite shed: Three Claytons in primer (Makers' Nos.27/8/30).
	St Andrew's Goods Yard: D8515/21 painted and three in primer (Makers' Nos.29/37/9).
07/07/63	St Andrews Goods Yard: D8515 and three in primer (Makers' Nos.29/37/9).
30/07/63	St Andrews Goods Yard: Nil.

At least seven Claytons stored in Derby St Andrews Goods Yard, Summer 1963. The Goods Yard was located immediately to the south-west of Derby Midland station; some of the station buildings can be seen in the distance to the right of the photograph. (Author's Collection)

Derby St Andrews Goods Yard, circa March 1963. Given the work-worn condition, the nearest locomotive is probably D8508. (Rail-Online)

Three stored and presumably engineless Claytons at Derby, 19 May 1963. To the extreme left of the picture is the Way & Works signal box and in the background the construction of the new Railway Technical Centre Test Hall seems to be well advanced. 17A Derby depot is to the left of the photographer. The three Claytons, all still in Works primer, are probably D8525/6/8 (Maker's Nos. 27/8/30), although the number wound-up in the headcode box of the nearest locomotive seems to be trying to confuse matters. (Colour-Rail)

D8514, Derby St Andrews Goods Yard, 1963. (Colour-Rail)

The August 1963 edition of the *Railway Observer* stated that Claytons with Makers' Nos.21/4-30/4/5/7/9 '...have - at some time since February last – been stored at Derby, either in St Andrews Goods Yard or opposite the shed'. These Works. Nos. relate to D8535/22-28/32/33/36/40 respectively.

Deliveries of complete 'new' locomotives resumed in April 1963 with D8516 and D8518, with locomotives appearing in random order. It took until D8542 in September 1963 for deliveries of locomotives in sequential order to be achieved.

The impact that the engine/generator vibration problems had on the availability and usage of the Clayton fleet is illustrated in the following two tables covering 1962 and 1963:

Performance and Service Problems: Engines & Associated Equipment

'Availability by Classes of Locomotives and Power Cars' (Source: BR.1712/454).

<u>Scottish Region.</u>
Not Available Works   In Works & Awaiting Works
Not Available Depots   Under or Awaiting Repairs and Under or Awaiting Exams

<u>Clayton Type 1 - Average Weekday Position (1962/63).</u>
Number of Locomotives and Percentage of Net Operating Stock

Week Ending	Net Operating Stock	Available		Not Available							
				Works		Depots		Awaiting CM&EE Decision		Awaiting Materials	
	No.	No.	%	No.	%	No.	%	No.	%	No.	%
15/09/1962	1	1	100.0	0	0.0	0	0.0	0	0.0	0	0.0
22/09/1962	1	1	100.0	0	0.0	0	0.0	0	0.0	0	0.0
29/09/1962	1	1	100.0	0	0.0	0	0.0	0	0.0	0	0.0
06/10/1962	2	2	100.0	0	0.0	0	0.0	0	0.0	0	0.0
13/10/1962	4	4	100.0	0	0.0	0	0.0	0	0.0	0	0.0
20/10/1962	6	4	66.7	2	33.3	0	0.0	0	0.0	0	0.0
27/10/1962	7	4	57.1	3	42.9	0	0.0	0	0.0	0	0.0
03/11/1962	9	6	66.7	3	33.3	0	0.0	0	0.0	0	0.0
10/11/1962	9	6	66.7	3	33.3	0	0.0	0	0.0	0	0.0
17/11/1962	11	10	90.9	1	9.1	0	0.0	0	0.0	0	0.0
24/11/1962	12	11	92.7	1	8.3	0	0.0	0	0.0	0	0.0
01/12/1962	12	11	92.7	1	8.3	0	0.0	0	0.0	0	0.0
08/12/1962						No data					
15/12/1962						No data					
22/12/1962	14	12	85.7	1	7.1	0	0.0	0	0.0	1	7.1
26/01/1963	14	12	85.7	0	0.0	2	14.3	0	0.0	0	0.0
23/02/1963	14	4	28.6	2	14.3	0	0.0	8	57.1	0	0.0
23/03/1963	14	0	0.0	3	21.4	0	0.0	11	78.6	0	0.0
06/04/1963	14	0	0.0	2	14.3	0	0.0	12	85.7	0	0.0
18/05/1963	20	12	60.0	0	0.0	0	0.0	8	40.0	0	0.0
15/06/1963	23	20	87.0	0	0.0	0	0.0	3	13.0	0	0.0
13/07/1963	29	26	89.7	0	0.0	3	10.3	0	0.0	0	0.0
03/08/1963	37	36	97.3	0	0.0	1	2.7	0	0.0	0	0.0
07/09/1963	41	35	85.3	2	4.9	4	9.8	0	0.0	0	0.0
05/10/1963	49	39	79.6	2	4.1	8	16.3	0	0.0	0	0.0
02/11/1963	56	46	82.1	6	10.7	3	5.4	1	1.8	0	0.0
30/11/1963	63	53	84.1	7	11.1	3	4.8	0	0.0	0	0.0
21/12/1963	67	58	86.5	5	7.5	3	4.5	1	1.5	0	0.0

'Miles of Traction Units' (Source: BR.1712/452).
Scottish Region.

Week Ending	Net Operating Stock	Average No. Used per Day	Total Miles Worked (Weekdays)	Mile per Weekday per Unit	
				Locomotives In Stock	Locomotives In Use
15/09/1962	1	1	228	38	38
22/09/1962	1	0	0	0	0
29/09/1962	1	0	0	0	0
06/10/1962	2	1	634	63	127
13/10/1962	4	1	1862	78	310
20/10/1962	6	1	1871	52	312
27/10/1962	7	4	2257	54	94
03/11/1962	9	6	2806	52	78
10/11/1962	9	6	2043	38	57
17/11/1962	11	10	3814	58	64
24/11/1962	12	11	4092	57	62
01/12/1962	12	11	4665	65	71
08/12/1962			No data		
15/12/1962			No data		
22/12/1962	14	12	6758	80	94
26/01/1963	14	12	6474	77	90
23/02/1963	14	4	1666	20	69
23/03/1963	14	0	0	0	0
06/04/1963	14	0	0	0	0
18/05/1963	20	12	5342	45	74
15/06/1963	23	20	9017	56	75
13/07/1963	29	26	18676	92	104
03/08/1963	37	36	17633	68	84
07/09/1963	41	34	22883	93	112
05/10/1963	49	38	21347	73	94
02/11/1963	56	45	27710	82	103
30/11/1963	63	53	33198	88	104
21/12/1963	67	55	33996	85	103

11.1.4 Post-Embargo Problems.

A number of sources have been used to fully ascertain the various issues encountered by the Clayton fleet and the actions taken to minimise or eliminate the problems; these are credited in the text. There are, however, two major primary sources of information which illustrate in some depth the key difficulties experienced:

- The minutes of the BRB/Davey Paxman (BRB/DP) Liaison Meetings which were held between at least 7 April 1966 and 31 August 1967. Minutes of six of these meetings have been examined.
- The minutes of the *Performance and Service Problems of Clayton Type 1 Locomotives* (P&SP) meetings. This was a multi-functional group set up in 1967 with the remit to identify and tackle the various issues afflicting the Clayton fleet. Meetings were held roughly every two to three months with the first on 4 July 1967 and the last on 8 September 1971; in all twenty meetings were held. Mr V. Atkinson (CM&EE ScR) chaired the meeting with Mr W. McRobbie acting as Secretary. The group included representatives from CM&EE (ScR, LMR, ER); CE(T&RS) BRB; CE(T&RS) Director of Design, Derby; BR Workshops (Headquarters and St Rollox Works); Davey Paxman & Co Ltd (subsequently English Electric Diesels Ltd (Paxman Engine Division), and, later Ruston Paxman Diesels Ltd).

Various availability and reliability statistics were reported at these meetings; these have been collated and presented in the following four tables. There are some issues regarding the consistency of the data provided (as explained in the notes attached to each table) but the information provides a very useful insight into the issues being addressed.

Availability of Paxman Locomotives (Source: BRB/Davey Paxman Liaison Meetings).

Meeting Date Reported	Region	No. in Stock	No. Out of Service (OOS)	Availability %	Engine Repairs	
					No.	% of Total OOS
18/01/1967	NER	16	4	75.0	2	50.0
	ScR	99	37	62.6	28	75.7
29/03/1967	NER	16	6	62.5	4	66.7
	ScR	99	32	67.7	25	78.1
07/06/1967	NER	16	1	93.7	1	100.0
	ScR	99	35	64.6	25	71.4
23/08/1967	NER	18	0	100.0	0	0.0
	ScR	93	30	67.7	14	46.7

Scottish Region Clayton Type 1 Locomotives: Miles per Casualty (Source: P&SP).

Year	Period (4 week ended)	No. of Locos	Miles per Casualty	No. of Casualties	Calculated Mileage per Loco per Period	Comments
1967	Period 09 (4we 09/09/67)	79	19824	?		See Note (1).
	Period 10 (4we 07/10/67)	79	16606	?		See Note (1).
	Period 11 (4we 04/11/67)	71	140695	0	1982	See Notes (1) and (2).
	Period 12 (4we 02/12/67)	66	116476	0	1764	See Notes (1) and (2).
	Period 13 (4we 30/12/67)	65	113568	0	1747	See Notes (1) and (2).

Year	Period (4 week ended)	No. of Locos	Miles per Casualty	No. of Casualties	Calculated Mileage per Loco per Period	Comments
1968	Period 01 (4we 27/01/68)	60	117648	0	1961	See Notes (1) and (2).
	Period 02 (4we 24/02/68)	60	21546	?		See Note (1).
	Period 03 (4we 23/03/68)	60	128762	0	2146	See Notes (1) and (2).
	Period 04 (4we 20/04/68)	53	21876	?		
	Period 05 (4we 18/05/68)	51	101235	0	1985	See Note (2).
	Period 06 (4we 15/06/68)	51	35456	?		
	Period 07 (4we 13/07/68)	51	102079	??	2002	Presumably 0 or 1 casualty (max).
	Period 08 (4we 10/08/68)	50	12645	?		
	Period 09 (4we 07/09/68)	53	24548	?		
	Period 10 (4we 05/10/68)	53	26194	?		
	Period 11 (4we 02/11/68)	49	8993	?		
	Period 12 (4we 30/11/68)	48	20333	?		
	Period 13 (4we 28/12/68)	48	31871	?		
1969	Period 01 (4we 25/01/69)	45	22255	?		
	Period 02 (4we 22/02/69)	40	20562	?		
	Period 03 (4we 22/03/69)	39	85389	0	2189	See Note (2).
	Period 04 (4we 19/04/69)	39	82884	??	2125	Presumably 0 or 1 casualty (max).
	Period 05 (4we 17/05/69)	39	81039	??	2078	Presumably 0 or 1 casualty (max).
	Period 06 (4we 14/06/69)	39	20363	?		
	Period 07 (4we 12/07/69)	39	23010	?		
	Period 08 (4we 09/08/69)	39	62250	?		
	Period 09 (4we 06/09/69)	39	10532	?		
	Period 10 (4we 04/10/69)	40	30940	?		
	Period 11 (4we 01/11/69)	48	21824	?		
	Period 12 (4we 29/11/69)	48	17785	?		
	Period 13 (4we 27/12/69)	48	24281	?		
1970	Period 01 (5we 31/01/70)	48	93754	?	1953	Presumably 0 or 1 casualty (max).
	Period 02 (4we 28/02/70)	48	42386	6		
	Period 03 (4we 28/03/70)	48	85238		1776	Presumably 0 or 1 casualty (max).
	Period 04 (4we 25/04/70)	48	26820			
	Period 05 (4we 23/05/70)	48	19357	28		
	Period 06 (4we 20/06/70)	48	30707			
	Period 07 (4we 18/07/70)	48	17510			
	Period 08 (4we 15/08/70)	48	39788			
	Period 09 (4we 12/09/70)	48	13228	15		
	Period 10 (4we 10/10/70)	48	18578			

Year	Period (4 week ended)	No. of Locos	Miles per Casualty	No. of Casualties	Calculated Mileage per Loco per Period	Comments
	Period 11 (4we 07/11/70)	48	45847			
	Period 12 (4we 05/12/70)	48	17267	13		
	Period 13 (4we 02/01/71)	48	32118			
1971	Period 01 (4we 30/01/71)	48	23013			
	Period 02 (4we 27/02/71)	48	12160	8	2027	
	Period 03 (4we 27/03/71)	44	13657	4	1242	
	Period 04 (4we 24/04/71)	44	41560	2	1889	
	Period 05 (4we 22/05/71)	44	9096	8	1653	
	Period 06 (4we 19/06/71)	44	16372	5	1860	
	Period 07 (4we 17/07/71)	44	16738	4	1522	

Explanatory notes:
1. For the periods up to and including 03/1968, statistics were specified as covering Scottish Region locomotives only and excluding Beyer Peacock built (11) and Rolls-Royce engined locomotives (2). This explains the difference between the 'No. of locos' listed above as compared with the number of Scottish Region allocated locomotives in Section 8.
   From Period April to December 1968 (inclusive) the information provided was stated as 'the position for the Scottish Region'. However, the 'No. of Locos' quoted for Periods 4/1968 to 6/1968 are consistent with Scottish Region Class 17/1s only.
   From Period 7/1968 the P&SP reports did not include the numbers of Clayton in operating stock. Therefore, from this point known Period-end Class 17/1 stocks are used (see blue numbers in 'No. of Locos' column); where numbers of casualties are known, calculations indicate a relatively constant 2000 miles per locomotive per 4 weekly period. This suggests that the Miles per Casualty data used by the P&SP group was restricted to the Scottish Class 17/1s only.
2. No casualties reported; therefore, mileage quoted is the total mileage operated in the specified period.
3. Results highlighted in red look suspect with mileage too high relative to number of casualties or vice versa.
4. The 1970 Miles per Casualty target set for 1970 was 20,000.

Clayton Type 1 Locomotives: Availability (Source: P&SP).

Early data recorded in the P&SP Meeting minutes appeared to be very crude, as follows:

Meeting Date	Time Span	Availability (%)			
		Region Unspecified	Scottish Region	Eastern Region	London Midland Region
04/07/1967	?		68%	85%	N/A
19/09/1967	?	~70%			
10/01/1968	?	~80%			
13/03/1968	?		80%	78%	78%
08/05/1968	Last 2 months		78%	75-77%	78%
14/08/1968	Since previous meeting		83.9%	89.9%	76.7%
31/10/1968	Since previous meeting		85%	88.9%	?
21/01/1969	Since previous meeting		87%	?	N/A
26/03/1969	Since previous meeting		85.5%	?	N/A
11/06/1969	Since previous meeting		86.1%	86%	N/A

More detailed reports followed from Period 5 1969:

Year	Period (4 weeks ended)	ScR Average Daily Availability	ER Average Daily Availability	Comments
1969	Period 05 (4we 17/05/69)	87.8%		
	Period 06 (4we 14/06/69)	81.1%		
	Period 07 (4we 12/07/69)	87.5%		
	Period 08 (4we 09/08/69)	90.7%		
	Period 09 (4we 06/09/69)	80.6%		
	Period 10 (4we 04/10/69)	83.3%		
	Period 11 (4we 01/11/69)	80.2%		
	Period 12 (4we 29/11/69)	78.1%		
	Period 13 (4we 27/12/69)	72.2%		
1970	Period 01 (5we 31/01/70)	77.6%		1970 ScR Target: 80%.
	Period 02 (4we 28/02/70)	75.6%	78.4%	
	Period 03 (4we 28/03/70)	80.2%	69.2%	
	Period 04 (4we 25/04/70)	80.5%	61.5%	
	Period 05 (4we 23/05/70)	78.5%		
	Period 06 (4we 20/06/70)	77.5%		
	Period 07 (4we 18/07/70)	81.9%		
	Period 08 (4we 15/08/70)	81.4%		
	Period 09 (4we 12/09/70)	77.7%		
	Period 10 (4we 10/10/70)	78.1%		
	Period 11 (4we 07/11/70)	73.1%		
	Period 12 (4we 05/12/70)	75.4%		
	Period 13 (4we 02/01/71)	72.6%		
1971	Period 01 (4we 30/01/71)	74.0%		
	Period 02 (4we 27/02/71)	77.8%		
	Period 03 (4we 27/03/71)	73.7%		
	Period 04 (4we 24/04/71)	83.2%		
	Period 05 (4we 22/05/71)	86.1%		
	Period 06 (4we 19/06/71)	85.1%		
	Period 07 (4we 17/07/71)	82.6%		

Explanatory notes:
1. It is unclear as to the extent of Class coverage but in all probability it included only the Scottish Region Class 17/1 locomotives (as per the Miles per Casualty statistics).
2. By way of comparison a National Traction Plan Progress Report dated 18 April 1966 gave actual availability for the full fleet of 117 locomotives as 67 per cent (contrasting with 91 per cent for the 138 English Electric Type 1s then in traffic).

Clayton Type 1 Locomotives: Reported Defects (Source: P&SP).

Year	Reporting Period	Cooling System	Fuel System	Lub Oil System	Engine Parts	Hydrostatic System	Electrical	Other
	3 Months ending 01/09/67 *	40	17	25	113	?	?	
	3 Months ending 01/12/67 *	37	9	13	103	?	?	
	3 Months ending 02/03/68 *	35	9	1	41	?	?	
	03/03/68 - 27/04/68 *	15	8	5	68	?	?	
	09/05/68 - 30/07/68 *	8	8	8	54	10	?	
	01/08/68 - 23/10/68 *	7	11	8	40	2	12	3 (inc. 1 Col.Dam.)
	29/10/68 - 13/01/69	8	6	7	61	0	22	2 (inc. 1 Col.Dam.)
	13/01/69 - 12/03/69	4	4	6	52	1	13	2
	13/03/69 - 04/06/69	3	4	2	61	7	7	6 (inc. 1 Fire, 1 Col.Dam., 3 Derailment)
	08/06/69 - 07/09/69	2	6	1	76	4	10	7 (inc. 2 Fire, 1 Col.Dam., 4 Derailment)
	07/09/69 - 18/10/69	0	1	4	35	2	10	1
	Period 11 (4we 01/11/69)	1	2	1	30	0	7	1
	Period 12 (4we 29/11/69)	2	6	5	38	1	12	3
	Period 13 (4we 27/12/69)	1	1	2	26	0	9	1
1970	Period 01 (5we 31/01/70)	2	4	4	23	4	10	3
	Period 02 (4we 28/02/70)	2	5	3	19	1	9	6
	Period 03 (4we 28/03/70)	2	2	2	24	1	9	1
	Period 04 (4we 25/04/70)	3	2	1	30	3	7	2
	Period 05 (4we 23/05/70)	2	5	0	26	2	7	3
	Period 06 (4we 20/06/70)	1	7	4	23	3	14	5
	Period 07 (4we 18/07/70)	2	3	5	23	4	10	2
	Period 08 (4we 15/08/70)	2	2	0	18	1	4	2
	Period 09 (4we 12/09/70)	2	2	0	13	0	5	5 (inc. 2 Fire)
	Period 10 (4we 10/10/70)	0	4	2	19	0	4	2
	Period 11 (4we 07/11/70)	4	3	1	21	0	4	0
	Period 12 (4we 05/12/70)	3	7	2	18	5	4	2 (both Col. Dam.)
	Period 13 (4we 02/01/71)	2	5	2	10	1	13	2 (inc. 1 Col.Dam.)
1971	Period 01 (4we 30/01/71)	4	9	1	19	5	11	0
	Period 02 (4we 27/02/71)	1	5	3	9	1	13	4
	Period 03 (4we 27/03/71)	2	5	1	11	1	23	4 (inc. 1 Col.Dam.)
	Period 04 (4we 24/04/71)	1	8	1	19	1	23	6
	Period 05 (4we 22/05/71)	3	6	6	25	2	31	5

Year	Reporting Period	Cooling System	Fuel System	Lub Oil System	Engine Parts	Hydrostatic System	Electrical	Other	
	Period 06 (4we 19/06/71)	6	3	8	22	5	21	2	
	Period 07 (4we 17/07/71)	5	5	7	18	4	21	2	
	Total (period 11/1969 to 07/1971)	53	92	61	484	45	271	63	Total - 1069
	% of Total	5.0	8.6	5.7	45.3	4.2	25.4	5.8	

Explanatory notes:
1. The P&SP 'Defects' statistics are fraught with numerous issues:-
    a) The time period of analysis varied considerably, only settling down to 4-weekly periods from October 1969.
    b) The Regional coverage of the class is unclear. The statistics appear to relate to Scottish Region allocated locomotives only, although only the first six sets of data (marked *) specifically confirm this.
    c) Locomotive coverage is equally unclear.
    Certainly the Class 17/1 locomotives are included throughout. However, the extent to which the Beyer Peacock Class 17/3 locomotives are included seems to vary over time. Where specific locomotives experiencing defects are mentioned (during reporting period November 1969 to November 1970 and May to June 1971, some Class 17/3s are specifically mentioned, and, indeed, Beyer Peacock is explicitly mention in the Report Title for Periods 05/1970 to 08/1970. However, as has been seen with the Miles per Casualty statistics, the Class 17/2s and 17/3s were explicitly excluded up to Period 3/1968; did this also apply to Defects reporting?
    The Roll-Royce engined pair appear to be totally excluded from all reports, presumably being regarded as outside the remit of the P&SP group.
2. Note the variable lengths of time periods prior to Period 11 1969, necessitating care when attempting to determine trends.
3. Defect issues recorded were expanded over time, settling down in late 1968. A full breakdown of issues within the high-level Defect categories were provided with the Meeting minutes from March 1968 although some reasons for failure were open to considerable interpretation.

The table summaries above clearly illustrate the fact that the Paxman engines and associated equipment suffered from a whole range problems, including:

- Low lubricating oil pressure.
- Engine water level.
- Piston seizures.
- Fuel supply difficulties.
- Cylinder head fractures.
- Crankcase failures (due to cracking and corrosion).
- Leaking cylinder head joints.
- Turbocharger failures.
- Exhaust system problems.
- Radiator shutters and hydrostatic system problems.
- Engine/generator coupling issues.

Each of these are dealt with in subsequent sections.
    In addition to the above, a Technical Report to the BRB, produced by the Chief Engineer (T&RS) dated 17 March 1967 presented comparative information to put the Clayton situation in perspective (see table below).
    The following supporting comment was provided:

'…..it is considered that inherent features of the (*Clayton*) design will not permit the National Traction Plan level of 90% availability to be achieved, the forecast level showing a shortfall of some 10% (equivalent to 12 locomotives).'

Type	No. of Locos	Average Miles per Casualty 52 we 31/12/1966	Anticipated Miles per Casualty in Dec. 1970 (CE(T&RS) forecast)	% Increase by 1970
Type 1 900h.p. Paxman*	115	8,449	20,000	136
Type 1 1000h.p. English Electric	228	62,831	100,000	59

* Two Rolls-Royce engined locomotives excluded.

11.1.5 Low Lubricating Oil Pressure.

Lubricating oil low pressure problems were identified early on and dealt with via Modification MB39/29; the modification involved fitting larger lubricating oil pumps and was applied on a 'campaign' basis.

Minutes of the joint BRB/Davey Paxman meeting held on 7 June 1966 recorded that three engines had been fitted with the larger pumps and by a further meeting on 15 June 1967 it was reported that 106 engines had been fitted by both Works and depots. By the 19 September P&SP meeting, 144 engines had been fitted, increasing to 222 by 10 January 1968, with the requisite number of pumps available to complete the modification.

11.1.6 Engine Water Level.

Difficulty was experienced with the Beyham Water Level Switches early in the life of the Claytons due to their inability to 'fail safe', i.e. when they failed they tended to stick in the full position irrespective of water level. Given the propensity for the Paxman engine to leak water, there was an increased risk of overheating with associated potential damage to engine parts, notably the cylinder heads.

A modified type of switch was developed and fitted under ModificationMB36/50 with work undertaken at depots. The BRB/DP meeting on 15 June 1967 reported that 50 modified switches had been fitted.

11.1.7 Piston Seizures.

A Memorandum from the BRB CME (J.F. Harrison) to the BR Management Committee dated 21 February 1964 entitled 'Diesel Availability' made reference to the spasmodic occurrence of piston scuffing with consequential seizure in some cases. At that point in time it had not been established what was the root cause of the trouble.

In December 1964, Harrison submitted a 'Report on the Working of the CME Department' to the BRB; Appendix C to this report gave some details of the problems connected with the Clayton fleet, including:

> 'There have been a series of difficulties with this engine. From the start there were piston seizures, resulting from poor running in. A quick bedding piston ring was introduced, so far successfully.'

11.1.8 Fuel Supply Difficulties.

Fuel starvation was an issue throughout the life of the Clayton fleet although it was rarely mentioned in the P&SP Meeting minutes. The minutes of the BRB/DP Meetings make cursory references to modifications to the CAV pumps to improve fuel lift from the fuel tanks to the injectors; by the 19 January 1967 meeting these modifications were recorded as having been completed.

Visitors to the North Yorkshire Moors Railway in the 1980s will recall the 'Heath Robinson' header tank attached to the bonnet roof of D8568. This tank provided additional gravitational pressure in the fuel system to significantly improve the supply problem. Further subsequent modifications provided the solution to finally resolve the 'fuel lift' problem.

11.1.9 Cylinder Head Fractures.

The aluminium cylinder heads gave trouble though cracking in the exhaust port area after about 1,000hrs; this problem was considered to be one of the 'top-four' issues suffered by the Clayton fleet.

Agreement was reached on 7 March 1966 between BRB and Paxman to replace the aluminium cylinder heads by cast iron heads on the D8500-85 batch; this involved the replacement of 1,110 cylinder heads (185 engines including the 13 spares). The policy was to fit these at depots as aluminium heads were removed for any cause and also at works during any engine repair under Modification MB39/24. Although not stated as such, the replacement of the cylinder heads was undertaken as close to a 'campaign' basis as could be achieved, restricted only by Paxman's ability to supply.

Minutes of a BR (ScR)/DP meeting held on 7 June 1966 recorded that orders for 200 cast iron cylinder heads had been received by Davey Paxman. At that point 18 had been supplied, and 57 were due for delivery to Glasgow and Doncaster Works during June/July; further batches were due in August (25), September (25), October (50) and November (25). However, the minutes recorded that:

> 'This rate of supply of cylinder heads was hardly adequate to meet the rate at which they were required, and in fact the 45 heads being supplied this week would barely clear outstanding demands by works and depots. Today 7 locomotives

at the maintenance depots were standing waiting cylinder heads, while Glasgow Workshops were also in the position of having engines under overhaul but without heads . . .

'The Chairman stressed very strongly the serious position we appeared to be approaching if the rate of supply of cylinder heads was not improved.'

As an interim measure, aluminium heads were being refurbished to offset the supply shortfall, although experience was showing that virtually 50 per cent of all the aluminium heads sent in for refurbishing were only fit for scrap.

The meeting on 7 June was actually a pre-meeting in anticipation of the BRB/DP Liaison Meeting three days later, chaired by CME J.F. Harrison. At this point, the aim delivery rate was increased to sixty per month from September, with the placement of additional orders; however:

'The Chairman stressed the importance of the supply of heads as this was the main course for improving the availability of locomotives in Scotland which would enable the withdrawal of steam power, bearing in mind that Workshop repair facilities are being withdrawn in October 1966 [with the closure of Cowlairs] and that any [steam] locomotives requiring repair would have to be sent to Crewe.'

At the 19 August 1966 BR/DP Liaison Meeting:

'Davey Paxman reiterated that there was no difficulty with the production of heads and at the continued rate all engines should be fitted with cast iron heads by the end of April 1967.'

However by the same meeting on 2 November:

'Messrs. Paxman agreed that they were behind in deliveries but assured BR that they would catch up by the end of December 1966, after which they would deliver cylinder heads at the rate of 120 per month. This would ensure full delivery by September 1967.'

This statement implied that 1,080 cylinder heads had still to be supplied in the January-September 1967 period, with approximately 384 delivered by the end of December 1966.

The minutes of the BRB/DP Meeting on 6 April 1967 reported that 120 engines had been fitted with (720) cast iron heads; by 15 June 1967 780 cast iron heads had been fitted to *Scottish Region* locos (i.e. 130 engines) and it was confirmed that orders had been placed on Davey Paxman for cast iron heads to cover the remainder of the fleet plus spares.

By the 19 September 1967 P&SP meeting a total of 1,080 cast iron cylinder heads had been fitted across the full fleet (180 engines), leaving 384 (64 engines) to be fitted. Davey Paxman stated that '40 were outstanding [for delivery] under the concessionary arrangements' (i.e. within the D8500-85 sub-fleet), although 30 should have been quoted. By the 10 January 1968 meeting 1200 cast iron cylinder heads had been fitted and 'Messrs. Davey Paxman confirmed that all C.I. cylinder heads to be supplied under concessionary arrangements had, in fact, been supplied'.

On 24 October 1967, the Works and Equipment Committee (W&EC) (Min.3004, Item 7) authorised the replacement of aluminium cylinder heads on Paxman engines by 1,458 cast iron cylinder heads (i.e. all 243 engines); also authorised at the same time were 132 cast iron crankcases, giving an overall estimated cost of £179,896. Clearly this authorisation was largely retrospective! See Section 11.3 for additional details regarding the concessionary pricing arrangements for the replacement heads.

By the P&SP Meeting of 25 September 1969, it was somewhat belatedly reported that the conversion to cast iron cylinder heads had been completed. Based on the rate of conversion between the 15 June 1967 BRB/DP Liaison Meeting and the 10 January 1968 P&SP meeting, this conversion process is likely to have been completed by early/mid-1968.

The figures above suggest that about ten engines were being converted per month between June 1967 and January 1968.

The cast iron heads proved considerably more successful than the aluminium versions although were not immune to failure mainly due to overheating caused by low water (i.e. defective level switches) rather than any inherent defect in the heads themselves. As early as June 1967 BR were reporting that:

> '...the CI heads together with cast iron crankcases, increased torque loading and more careful fitting had reduced the incidence of water leaks and gas blows considerably, to the point where these are no longer a problem.'

11.1.10 Crankcase Failures.

Aluminium crankcase failures, the second of the 'top-four' major issues which afflicted the Claytons, resulted from both fractures/cracking (variously associated with thermal issues, piston seizures, longitudinal crankshaft movement, etc) and corrosion (attributed to coolant water, and engine vibration).

Harrison's previously mentioned 'Report on the Working of the CME Department' (Appendix C) dated December 1964 mentioned that:

> '... recently there has been a spate of cracked aluminium crankcases. These are claimed by the engine builder to be due to poor annealing and the suspect castings are being replaced. Our Research Department confirm that the defects are basically material and not design.'

The minutes of a joint BR (ScR)/Davey Paxman meeting held on 2 November 1966 further explained that the early crankcase fractures were associated with those crankcases which were inadequately stress-relieved by the manufacturer, Stones and that thirty-seven crankcases had been incorrectly manufactured.

At a meeting between BRB, Clayton Equipment and Davey Paxman on 7 March 1966, it was agreed that with respect to the D8500-85 batch:

> 'Any crankcase of the non stress-relieved batch which becomes defective without external cause within four years of the dates of the acceptance of the locomotives will be replaced by Paxman free of charge. Any other crankcase of the balance of 185 engines which becomes defective without external cause in the same period will be replaced by Paxman at 50% of the normal price to the Board. The work of replacement will be carried out by the Board.'

The guarantee period for the 172 engines was operative from the acceptance date of the eighty-six locomotives (D8500-85) and for the thirteen spare engines from the date of despatch from Paxman's Works in Colchester. The 4-year warranty period will have expired progressively up to May 1968, four years after D8585 was accepted. Whether the same guarantees were ever applied with respect to D8588-D8615 is unknown.

The Works & Equipment Committee on 24 October 1967 authorised the purchase of 132 cast iron crankcases to replace aluminium crankcases; this, with the 1,458 cylinder heads, amounted to an estimated gross cost of £179,896.

As with the cylinder heads, the W&EC authorisation of 24 October 1967 was partially retrospective given that by the BRB/DP Liaison meeting on 19 January 1967 twenty-six cast iron crankcases had been fitted and fifty-eight by 4 July 1967. The order for the crankcases was by a 'call-off' arrangement, rather than bulk purchase. Unlike the cylinder heads, a 'campaign' changeover to cast iron crankcases was considered to be unnecessary.

Modification MB39/39 covered the replacement of aluminium crankcases by cast iron.

The whole saga of the cast iron crankcases as it unfolded was both fascinating and revealing in the way that BR attempted to manage their 'marginal' locomotive fleet against the background of fluctuating market conditions. The Clayton's 'bread-and-butter' duties were trip freights and local yard shunting duties, business which was associated with high operating costs and generally low returns. In the 1960s and 1970s, this 'marginal' traffic was constantly the focus for 'rationalization' (i.e. reduction) to improve overall profitability, particularly so in 1968 and then again in 1971 when the freight market conditions in which BR operated was particularly brutal.

Key comments from the Minutes of the P&SP Meetings illustrate how BR attempted to grapple with the investment decisions regarding the purchase of cast iron crankcases, how they attempted to control stocks to limit the amount of capital tied up, how the various BR Departments worked with different and frequently conflicting objectives, and how ultimately a significant quantity of money was wasted. The whole subject would make a perfect Business Case Study!

Meeting 4 July 1967.
- Fifty-eight cast iron (CI) crankcases fitted to date, including thirty-three fitted free of charge (i.e. non stress-relieved).
- 127 remain to be fitted including four non-stress relieved crankcases.
- Paxman indicated that no bulk orders had yet been received for crankcases and if production stopped, delivery would be protracted to nine months.

Meeting 19 September 1967.
- Sixty-five CI crankcases fitted to date. One non stress-relieved crankcase outstanding to fit.
- Paxman stated four crankcases were outstanding against existing orders.
- It was agreed that Workshops place an order now for twenty-four crankcases and then re-order every time stock falls to eighteen, in batches of six to be delivered at two per month, which Workshops indicated would cover the failure rate. *N.B. There is no subsequent minuted indication that this ordering mechanism was introduced.*

Meeting 10 January 1968
- Sixty-four CI crankcases fitted to date (*N.B. Not consistent with previous report*).
- Paxman stated they had no orders outstanding for delivery.

Meeting 13 March 1968
- No new order for CI crankcases received by Paxman.

Meeting 8 May 1968.
- Confirmed that an order for twenty-three CI crankcases had been placed on Paxman.

Meeting 14 August 1968.
- Paxman stated that 7 of the 23 CI crankcases had been delivered.

Meeting 31 October 1968.
- Seventy engines fitted with CI crankcases to date.
- Twenty-three crankcases on order (7 delivered, 16 outstanding). Final delivery expected early-December.
- The Chairman referred to the run-down of these locomotives in accordance with the National Traction Plan (NTP) and in view of locos now out of service surplus to requirements, spares were available to meet any requirements for locomotives remaining in service. No spares to be ordered without reference to CM&EE (ScR).
- Workshops taking steps to dispose of all major Clayton spares (except five power units, and turbo-chargers).

Meeting 22 January 1969.
- Final seven crankcases of twenty-three ordered about to be delivered.
- Glasgow Works representative stated that ten spare crankcases were on hand.
- No spare engines available in built-up condition in Workshops and should any be required, any engines assembled should incorporate the CI crankcase.

Meeting 26 March 1969.
- Glasgow Works representative stated that seventeen spare crankcases were on hand and these presented a storage problem. Workshops would like to get rid of these. The Chairman instructed that these should be retained.

Meeting 11 June 1969.
- The Chairman stated that since a number of these locomotives would now be required in service for some time, the policy must be to hold adequate spares.
- Paxman reminded the Chairman that owing to no spares now being ordered, none were being made. Ordering of new spares would inevitably be subject to extended delivery times.

Meeting 25 September 1969.
- The Chairman referred to NTP and Freight Plan changes i.e. significantly increased traffic levels and a need to improve the level of movement resources.
- The consequential life extension of the Claytons meant that it was essential that spares be made available to cover this. Specific reference was made to CI crankcases, of which 16 spares were on hand at Glasgow Works.
- The Scottish Region requested Glasgow Works to fit these on engine overhauls to displace any aluminium crankcases, which meantime should be retained if serviceable for eventual further use if this proved necessary.
- Agreed that no further order for additional CI crankcases would be placed subject to review at the next meeting.
- Paxman pointed out that a twelve month lead time prevailed for new crankcases.

Meeting 6 November 1969.
- Glasgow Works confirmed that CI crankcases were being fitted to engines undergoing overhaul and the sixteen spares on hand had now reduced to eight.

- It was again stressed that any serviceable aluminium crankcases should be retained for possible further use, although to date of the displaced crankcases 75 per cent were found to be unfit for further use.
- The supply of new CI crankcases was discussed. The latest scheduled engine repairs planned for 1969/70 amounted to fifty-two (forty ScR, twelve ER). Twelve of the ScR engines already had cast iron crankcases, leaving twenty-eight to be fitted. Assuming 50 per cent of the ER locomotives were already fitted, six remained to be done and therefore the total number of crankcases required was thirty-four. Deducting the sixteen already supplied (eight fitted, eight in stock), a balance of eighteen was required.
- Paxman indicated that the lead time would be up to twelve months.
- Following the meeting, Paxman advised cost and delivery details as follows:-

Cost: 1-5 (£1760 each), 6-15 (£1700), 16-29 (£1640), 30+ (£1580).

Delivery: To commence five to six months from date of order.

- It was noted that some cast iron crankcases might be available from LMR locomotives stored at Carlisle. It was agreed that the CE(T&RS) BRB representative would look into the possibility of making these locomotives available for spares prior to disposal. Subsequently agreed post-meeting.
- Reference was also made to D8512/21 being utilised by Derby Research which could also yield useful parts. The CE(T&RS), Derby, representative agreed to check their potential return from the LMR and whether any of the four installed engines included CI crankcases.

Meeting 21 January 1970
- Glasgow Works indicated that only two CI spare crankcases were available and in the next week or so, recovered aluminium crankcases would have to be used for engine overhaul. It was pointed out that only 30 per cent of aluminium crankcases passing through shops were fit for further use, hence at some stage engine repairs were likely to cease for lack of crankcases.
- The Chairman said that, taking into account all existing CI crankcases and bearing in mind the proposed scale of scheduled repairs, a case had been submitted to BRB for authority to purchase fifty-nine CI crankcases.
- Regarding D8512 and D8521 at Derby Research, the Derby representative indicated that Research wished to keep both locomotives.

Meeting 23 March 1970.
- The Chairman indicated that an order for fourteen cast iron crankcases had been placed on Paxman and that a tentative order had been submitted to bring that total up to sixty (subsequently agreed at the Investment Committee on 24 March 1970). Delivery was expected to start in September at the rate of one per week.

N.B. The Supply Committee Meeting of 2 April 1970 (Minute 678(ii)) approved a contract with Paxman for sixty CI crankcases '…at a firm price of £1,738 each net, ex-works, at a total cost of £104,280, in accordance with the Supplies Manager's memorandum dated 25th March 1970.' Delivery was expected to commence in November 1970 at a rate of four per month, with completion in January 1972.

Meeting: 17 June 1970.
- Paxman confirmed receipt of order to manufacture CI crankcases.
- No spare crankcases were now available in Workshops but with the curtailment of the Clayton shopping programme, it was felt that the expected delivery rate would not cause any embarrassment.

Meeting 7 October 1970
- Paxman confirmed that delivery of the sixty CI crankcases would commence mid-October.

Meeting 13 January 1971
- Given availability of new CI crankcases, all engines now passing through shops were being released with CI crankcases. Serviceable aluminium crankcases were put to one side for emergency use.
- Discussions addressed the subject of further crankcase requirements. The sixty crankcases already ordered were considered to be sufficient to cover 1971 Classified Repair requirements together with failures in service. It was considered that no further new crankcases would be required even allowing for continued Clayton operation up to end-1974.

- St Rollox Works confirmed that of the 60 crankcases on order, five had been delivered and five were en route. Paxman's revised delivery plan was for ten crankcases per four-weekly period with completion in August 1971.

Meeting 21 April 1971
- The Chairman informed Paxman that no further crankcase orders would be placed.
- Paxman had despatched thirty-eight out of the sixty ordered leaving twenty-two still to be supplied. Twenty-seven had been fitted or were in course of erection and Works had eleven in stock.

Meeting 7 July 1971
- Paxman stated thirteen crankcases remained to be delivered, with twenty in stock at Works and twenty-seven fitted.

Meeting 8 September 1971
- St Rollox Works stated that all sixty crankcases had been delivered with thirty fitted, leaving thirty in stock surplus to requirements!

Comments:-
1. Sixty-four CI crankcases had been fitted by the P&SP Meeting on 10 January 1968. An order for twenty-three was placed during early-1968 followed by 60 in April 1970. This suggests that a total of around 147 CI crankcases were manufactured. 132 crankcases were authorised by the Works & Equipment Committee in October 1967 and may have retrospectively covered the earlier orders.
2. Fifteen CI crankcases were damaged beyond repair during the period up to October 1970. Two were reported at the 21 January 1970 meeting, three at the 23 March 1970 meeting (nine in total up to that date), a further three by 17 June 1970 (12) and three more by 07 October 1970 (15). Subsequent meetings reported no further failures. Is it coincidence that the number of CI crankcases manufactured, less those damaged beyond repair, was 132 (the number authorised in October 1967), with the damaged fifteen re-manufactured within the authorised number?
3. Of the sixty CI crankcases ordered in April 1970, thirty never saw use in the Clayton fleet and the other thirty less than one year's use. The thirty unused crankcases can be valued at £52,140. Whether any were sold on for further use in industry is unknown.
4. Given that 30 CI crankcases were never installed it would appear that only about 117 engines were fitted. The 15 damaged crankcases would reduce this figure to 100 engines (net). N.B. This does not equate to fifty locomotives as several ran in traffic with one aluminium and one CI crankcase.
5. Despite investigation by the CM&EE Inspectorate and St Rollox Works staff, no explanation for the CI crankcase failures was published via the P&SP Minutes, although blockage of the oil coolers was suspected.
6. Locomotives withdrawn during 1968 may have been removed from traffic with CI crankcases in situ, particularly some of the recently shopped LMR examples. Although one would like to think that these were subsequently recovered for further use, this cannot be stated with any certainty. Similarly, one might assume that the locomotives reinstated in May and October 1969 from Carlisle were examples selected on the basis of engine condition (including modification with CI crankcases), but again this cannot be confirmed. On the basis of the comments made in the P&SP Meeting of 6 November 1969, the Engineers and Operations Managers were quite clearly unaware of critical information about the assets in their care.

As a footnote to the crankcase story, the following from 'Richard Carr's Paxman History Pages' is pertinent:

'In service, major problems arose with cracking of the cast aluminium engine frame and the engines had to be rebuilt … with cast iron frames, at great cost to the Company (*Paxman*). John Cove, who worked for Paxman at the time, has told me that British Railways had previously tested the first pair of engines with aluminium crankcases before placing an order for a quantity. By that time Paxman had had some experiences with aluminium castings on the YHA and possibly also the YGA air-cooled engine. Both types had experienced troubles with threads and failures in cast aluminium. John went on to say: "Consequently we suggested to

BR that we supply the engines with cast iron crankcases from the start but BR were quite adamant that they wanted the engines to be exactly the same as the ones they had tested and which had given no trouble. We were so keen to get the order that we failed to stand up for what we believed was necessary and so supplied them in aluminium. But before long those engines in service had run longer hours than the test engines and troubles began to become apparent and we had to change all the crankcases to iron. The troubles then ceased but this would have been unnecessary if we had taken a stronger line before the order was placed."'

Clearly BR should have listened to the experts!

11.1.11 Leaking Cylinder Head Joints.

Leaking cylinder head joints represented the third major issue. The P&SP 13 March 1968 Meeting minutes recorded that the incidence of leaks on the Scottish Region Claytons over the previous three monthly periods were forty-three, thirty-five and twenty. These figures included both coolant leaks and gas blows.

Discussion in ensuing minutes centred on the procedure for tightening heads (including torque loading), and the best type of cylinder head/crankcase joint material to be used (i.e. soft copper, copper clad steel, soft iron).

The 31 October 1968 meeting recorded that trouble with cylinder head/crankcase joints constituted about one quarter of all engine defects. At the same meeting 'Paxman offered to supply 6 [soft-iron] engine sets for the same price as the standard copper joints, delivery being 4/6 weeks and it was agreed that these would be fitted to engines as opportunity presents itself.'

By the 22 January 1969 meeting, joint failures had increased to 37 per cent of the total. Subsequent to the meeting, an order was placed on Paxman for the supply of six sets of soft-iron joints. These had been supplied by the 11 June 1969 meeting and were to be deployed under Experiment DL51, although by 6 November 1969 no joints had been fitted.

Yet again, at the Meeting on 21 January 1970, the Chairman stated that 'this type of failure was still the biggest single source of failure'. The St Rollox Works representative referred to difficulties experienced in fitting the soft-iron joints due to their 'being 0.020in. proud of bore' which required removal by filing at extra cost. However, the 'Chairman agreed that an additional charge would have to be authorised for this' and he asked the Works representative 'to fit these rings without further delay'.

Eventually, the Minutes of the Meeting on 23 March 1970 recorded the fitment of soft-iron joints to D8573 (No.1 engine), D8539 (No.2 engine) and D8546 (both engines), a total of four. The Chairman asked that the remainder of the joints be fitted at the earliest possible time so that performance could be assessed and a decision reached regarding their suitability.

Seventeen months had now passed since the original suggestion to fit soft-iron joints, despite the fact that virtually every month, the joint leakage issue had been the biggest single engine failure item and that members of the meeting group were confident that the soft-iron joints were the solution.

The 17 June 1970 Meeting minutes reported that D8597 (No.2 engine) had also been fitted with soft-iron joints with a further two complete units on the shop floor at St Rollox Works awaiting to be fitted. This, of course, indicated seven engines, although only six joints had been supplied! However, of the engines in service, no adverse reports had been received. At this meeting the suggestion was made to increase the trial to include twenty engines.

By the 17 October 1970 meeting, the situation had deteriorated into something of a farce. It transpired that the previously specified engines installed in D8539/46/73 had not been fitted with soft-iron joints at all. Thus only D8597 No.2 engine was <u>confirmed</u> as having been fitted with such with the St Rollox Works representatives saying that they 'did not recollect any advice being given relative to D8539 and D8573 and . . .that D8550 and D8592 were the two locomotives involved'. The Chairman must have been incandescent at this point and 'instructed that this matter must be cleared and put on a correct basis and he would arrange for the HQ representative dealing with experiments to make contact with Glasgow Works staff in order to ensure that this was done'. However, the Minutes recorded that

'The point was made that during the time soft-iron joints were fitted no leaks had been experienced'! How could they say this when it was largely unknown which locomotives had been modified?

The minutes of the Meeting on 13 January 1971, reported that the cylinder head joint problem again accounted for the highest proportion of engine defects, amounting to ten out of fifty-three i.e. about twenty per cent; and yet again it was 'confirmed that these defects should greatly diminish with the introduction of soft-iron joints'. By this time, fifteen engines with cast iron joints were recorded as installed (i.e. D8541/6/9/50/8/94 (both engines), D8607 (No.1) and D8592/7 (No.2)) and that they were trouble-free. Note that D8546 had reappeared in the list. Once again the 'Scottish Region again suggested that the soft-iron joints should be adopted forthwith since the position could not be worsened and no additional cost was involved'.

The minutes of the Meeting on 21 April 1971 included an entry stating that 'Experiment DL51 . . . had been concluded successfully and the Railway Technical Centre had tabled an engineering instruction which had been issued showing the fitting of these joints whenever cylinder heads were removed. The Scottish Region pointed out that additional details (torque loadings and method of tightening) should be incorporated in this engineering instruction; RTC agreed and asked the Scottish Region CM&EE to supply the necessary details.

By this date, St Rollox Works had no soft-iron joints in stock although Polmadie had 144. Polmadie were requested to share their stocks with the Works. It transpired that the Works held *kits* of joints rather than individual joints, so whilst joints were technically in stock, there was an unwillingness to order full kits of parts due to the extra cost involved! To obviate the problem, the Works ordered 500 of the critical joints from Paxman. These had been delivered by the next meeting on 7 July 1971. Regarding the joints in service at this point, no adverse reports had been received; however, the RTC indicated that the full information required to compile the engineering instruction for fitting of soft-iron joints had not yet been received from the Scottish Region CM&EE!

By the 9 September 1971, details of the final run down of the Class was conveyed to the team and the comment was made that 'the soft-iron joints continue to give satisfactory results and it was agreed that had these been fitted to the majority of the fleet, the major part of engine trouble would have been eliminated'.

How can it be that a solution to what was perceived to be one of the biggest reasons for engine failure took virtually three years to commence full-scale implementation? In the overall scheme of things a soft-iron joint was a minor consumable part of minimal cost, but which evidently had a significant positive impact on locomotive performance. It has to be asked how effective this meeting group really was in driving improvement; the group was, of course, a form of matrix management, an arrangement always fraught with difficulty in any large bureaucratic organisation.

11.1.12 Napier MS90 Turbocharger Failures.

The Napier turbochargers (also termed turbo-blowers) in the Clayton fleet gave problems throughout their career on BR and represented the fourth major issue for the Clayton fleet. The normal operating temperature of the MS90 blower was about 600°C and the turbine blades were considered to be incapable of coping with temperatures above 650°C. However, tests showed that turbine inlet temperatures were about 750°C. The use of Nimonic steel blades was regarded as a solution to the problem and a proposal to fit a small number of Holset blowers was also considered following the satisfactory use of these on the Rolls-Royce engines fitted to D8586 and D8587; however, this was subsequently rejected on cost grounds (£1160 per locomotive).

At a detailed level, the key problems with the Napier turbochargers were:

- turbine blades becoming detached.
- bearing seizures due to contamination of the lubricating oil by exhaust gases, caused by the sealing at the turbine end being ineffective against the exhaust gas pressure at low engine speeds. Such low speeds were commonplace with the Claytons given their allotted duties. A heavy build up of carbon in the turbine end of the turbochargers, led to the contamination of the bearing lubricating oil sump and the collapse of the bearing ball race.

- failure caused by unbalanced rotor assemblies
- delays with the repair and balancing of rotor assemblies by Napier.

At the first meeting of the P&SP group on 4 July 1967 it was reported that:

'…..the spares position relative to the supply of repaired turbo-blowers is still extremely serious so far as the Scottish Region is concerned. At the present time 11 locos are standing out of service awaiting replacement turbo-blowers'.

At this meeting the ScR Clayton availability was quoted as 68 per cent with the caveat that:

'This is of course greatly affected by the turbo-blower position which, when put right, should lift availability to something over 80%.'

Scheduled turbocharger changes required the supply of approximately seventy-eight repaired items for Scottish Region locomotives in 1967; unscheduled failures obviously added significantly to the repair workload. By 1971, the annual requirement to cover scheduled changes had reduced to 67, reflecting the reduced fleet size.

By mid-1967, five turbochargers (fitted to D8562 (1), D8588 (2) and D8604 (2)) had been equipped with Nimonic 80A steel turbine wheels in an attempt to cure the turbine blade problem. Reference to the Nimonic equipment did not re-occur until the P&SP meeting on 17 October 1970, when the comment was made that 'Further replacement turbine wheels in Nimonic steel would not be available until March 1971', suggesting that installation was expanded beyond the initial experimental five.

In addition, six turbochargers had been fitted with an external pressure-fed oil feed arrangement from the engine lubricating oil system to the turbocharger bearings (under Experiment M/D/L/1733) to address the bearing seizure problem; the trial locomotives involved were D8512 (fitted 30/09/66), D8561 (16/12/66) and D8575 (30/12/66) (both engines in all three cases).

Following failures, the P&SP group insisted on the balancing of all rotor assemblies prior to reinstallation after repair. Rotor assembly balancing could not be undertaken at St Rollox Works, hence a total reliance on Napier. A turn round time of ten weeks for shaft assemblies was required by Napier. However, by 1970/71 Crewe Works were assisting with balancing work.

By the 19 September 1967 meeting, matters had slightly improved with the number of locomotives stopped waiting for repaired blowers reduced to five, with sixty-four rotor shaft assemblies under repair at Napier. The 10 January 1968 Meeting indicated a further improvement:

'The position is satisfactory so far as the supply of replacement blowers to Depots is concerned. Workshops confirmed that they were now satisfied with spares position and they should soon be in a position to supply all Depots with a float i.e. 6 each to Haymarket and Polmadie Depots and also meet any requisitions from other Regions.'

At the 8 May 1968 Meeting it was confirmed that 'all pressure-fed turbochargers were working satisfactorily'. However, the Chairman referred to the incidence of turbocharger failures during the first three months of 1968 i.e. fourteen at Carlisle, thirteen at Haymarket and four at Polmadie. The failure rate, mainly bearing related, impacted on the number of days lost waiting repaired blowers from St Rollox Works with twenty days being quoted by the Scottish Region over the previous 3 months'. The Eastern Region also described the position as 'not acceptable' quoting D8596 and D8602 as stopped awaiting material for sixteen and twenty-nine days respectively. The St Rollox Works representative at the meeting said there was no spares shortage ex-Napier with logistical issues between the Works and the depots being blamed for locomotive standing time. By the 14 August 1968 meeting 'All were agreed that turbochargers were now freely available from Workshops.'

Such was the confidence with the experimental blowers that the Director of Design, Derby, at the 31 October 1968 meeting when referring to the withdrawal of locomotives from traffic 'expressed the hope that the . . .pressure-fed blowers would be kept in service as long as possible'.

During late 1968 and early 1969, turbochargers recovered from stored and withdrawn Claytons were used to cover for an increased number of incidences of blower failures (listed as fifteen on the Scottish Region during the ten-week period prior to the 22 January 1969 P&SP Meeting and a further six during the seven-week period prior to the 26 March 1969 meeting). At this point, conflicting Departmental objectives started to surface with the Works Supplies Assistant, Glasgow, suggesting that the existing stock of seventeen spare blowers be reduced as part of the campaign to reduce stock holding'!

The minutes of the Meeting held on 11 June 1969 reported that no more engines would be converted to external pressure-fed oil supply presumably as a result of the wind down of the fleet applicable at that time. At the 25 September 1969 meeting, the comment was made that the position relative to the turbocharger failures was well known and that 'all concerned appreciate the design problems have to be lived with in this case. Nevertheless, maintenance at Depots had been tightened up to improve performance i.e. cleaning of restrictors, flushing of lubricating oil sumps, etc.

During May 1969, five LMR Claytons were reinstated to replace five withdrawn on the Scottish Region and during October and November nine further LMR locomotives were reinstated (reflecting an upturn in freight traffic). Turbine-end fractures were found during the preparation of eight of these locomotive for service, consistent with frost damage (including both blowers on D8513).

With the news that the Scottish Region Clayton fleet of fifty-nine locomotives was to be retained, the P&SP minutes of the 6 November 1969 meeting recorded that:

'...the demand for [turbocharger] spares was likely to continue in view of the known design defects which had to be lived with and Glasgow Works were asked to keep an eye especially on the position so that no delay would arise in supply.'

From late 1969, turbocharger failures were increasing, exacerbated by the problems with the reinstated LMR locomotives, such that at the 21 January 1970 P&SP Meeting:

'Reference was made to the critical shortage of turbo-blowers for depot and works use, in consequence of which scheduled turbocharger changes at depots were overdue. These amount to 26 at Polmadie and 18 at Haymarket, giving a total of 44 which are all potential failures... To this figure should be added the normal quota of depot spares (8) which brings the total outstanding to 52.'

The depot 'float' had presumably been reduced to four each for Polmadie and Haymarket by this stage. Five months later it was reported at the 17 June 1970 meeting that:

'The failure rate of (*Scottish Region*) turbochargers has increased during the periods under notice, this increase being due to the shortfall of spares at depots, which means that turbo-blowers are running beyond the change period...

'ER representative pointed that the incidence of failures on their locomotives had risen also and that some difficulty was being experienced in obtaining supplies but as far as change period was concerned, they were keeping abreast.'

By 7 October 1970, it was noted that:

'...turbocharger failures remained at a high level. Again this was due to the unsatisfactory position regarding replacements... Scottish Region indicated that 15 turbochargers in service were overdue for scheduled change.'

Matters deteriorated further such that by 13 January 1971:

'Scottish Region indicated that 9 failures had occurred over the three periods, this being far too high and was symptomatic of the highly unsatisfactory position which still exists so far as the supply of overhauled turbo-blowers was concerned.

'The current position is that 34 blowers overdue scheduled change were still in service, all being potential failures. In addition, 13 defective turbo-blowers have been left in position on locomotives remaining in service such that the total immediate Regional requirement is 47.

'The Chairman asked Glasgow Works representatives the position regarding overhauled turbo-blowers. Glasgow Works representatives stated that owing to the highly unsatisfactory position so far as component parts from manufacturers was concerned, it is highly unlikely that any great improvement could be made in the foreseeable future.'

Similarly the minutes of the 21 April 1971 meeting stated:

'The rate of failures of turbo-blowers continues at a very high rate [nine arising in the first three four-weekly periods of 1971] and the situation is still most unsatisfactory. At present there are 30 overdue scheduled changes, all being potential failures, with 14 in service known to be defective.

'The delivery of spare parts from Napier...has not been adhered to, Napier not being able to meet demand. The general opinion at the meeting was that the situation would worsen because of the present difficulties currently being experienced by Napier.

'Scottish Region at present require 44 turbo-blowers [immediately]...'

At the 07 July 1971 Meeting the situation was described as:

'...still very serious. 32 overdue change with a further 12 engines in service with defective turbo-blowers.

'It was pointed out to Workshops that there was a very high rate of failure of the turbo-blowers...and, due to the acute shortage, these blowers had to remain on the locomotives and consequently are being subject to hot air gases which may hide the primary cause of failure.'

D8503, Hamilton, 11 August 1971. Exhaust emission such as this will probably already have caused the failure of the turbo-charger at No.2 end and at this late stage it is unlikely that D8503 was rectified before final withdrawal less than two months later. (Malcolm Best)

During this meeting, the BRB, London, representative stated that, in view of the envisaged reduction of the fleet, purchase of new spares would be completely blocked. However, 'So far as turbo-blowers are concerned, these will still require to be maintained . . .'

By the last P&SP meeting on 8 September 1971, the final rundown of the Class was well known. At this point, nine locomotives were working with engines without operational turbo-blowers and a further twenty-nine were running with turbo-blowers overdue for scheduled change. The St Rollox Works representatives at the meeting indicated that two blowers were available and that a further four sets of parts were available which could be built up into blowers. It was agreed that these four machines should be assembled making six available to be held by Polmadie against the ten locomotives planned to be kept in traffic until 31 January 1972 (subsequently revised to end of 1971). All other work was stopped.

Over the years, and in the absence of 'cost-effective' solutions to the problem, the most effective way of protecting against turbocharger failure was the frequent cleaning and strict scheduled replacement of air intake filters, including both the generator compartment and blower intake filters, to ensure adequate air for combustion within the engine. Ignition of unburnt fuel in the exhaust system resulted in increased exhaust temperatures and associated damage to the turbo-chargers. Photographs of Clayton in traffic during 1970/71 frequently illustrated locomotives emitting copious amounts of black exhaust thereby significantly contributing to turbocharger failures.

11.1.13 Exhaust System Problems.

Difficulties were experienced with locomotives passing excessive quantities of lubricating oil despite care in fitting components and detailed running-in procedures. Due to the exhaust system pipe joints leaking, this oil was being discharged into the generator compartment causing damage to the insulation of the generator armature. Such problems were evident early on.

As already mentioned in Section 11.1.12, Napier turbocharger problems had led to the suggestion of deploying Holset blowers instead which would have necessitated modifications to the engine exhaust system. Drawings were developed incorporating the Holset blower together with the re-routing and re-design of the exhaust manifold to avoid the generator compartment altogether. However, by September, the Holset blower idea had been ditched on cost grounds. The cost of re-routing the exhaust manifold on its own was calculated as £600/£800 per engine, or £1,200/£1,600 per locomotive, involving stiffening of the bulkhead, as well as additional elbow and body work. This was also considered too expensive.

As an alternative arrangement, the Scottish Region thought that a better plan would be to eliminate joints by welding, where possible, and to use expansion joints instead of bellows, thereby making the exhaust pipework through the generator area a solid single-piece assembly, which could be lifted out by crane after removal of the bonnet roof.

At the 10 January 1968 P&SP Meeting, the Director of Design, Derby, indicated that such a solid assembly 'was practical and a figure of £16 for so doing had been obtained from Workshops'. This would have to be justified economically before the matter proceeded further and Scottish Region and Workshops were asked to estimate their annual savings which could be used to offset the cost.

By the next meeting on 13 March 1968 it had been calculated that the Scottish Region depot maintenance estimated cost saving would amount to £1,226 17s. 6d. per annum and the Workshop saving would be £9 1s. 8d. per locomotive, based on twenty-six classified repairs and an estimated twenty-three unclassified repairs involving dismantling of the exhaust system during 1968. The avoidance of contamination of the main generator armature windings would increase its life, although the financial benefit of this was not determined.

With the identified benefits, Modification MB39/38 was launched on 21 May 1968. By the P&SP Meeting on 14 August, D8589 had been suitably modified.

How many locomotives were modified is unknown, although it is believed that only locomotives receiving classified repairs after May 1968 were so treated. In his book *British Rail Standard Diesels* of the 1960s (2009), David Clough mentioned:

'There was a strong reluctance on the part of the LMR when it was asked to take an allocation of surplus locomotives from Scotland. Eventually agreement was reached, provided that the first one was ex-works.

'The LMR insisted on double-skin exhaust manifolds being fitted to curb leakage and fracturing.'

11.1.14 Radiator Shutters and Hydrostatic System Problems.

Occasional comments have suggested that part of the engine cooling problem associated with the Claytons was as a result of a design inadequacy i.e. the limitation resulting from having the radiator for each engine on one side of the locomotive only (at each bonnet end).

Persistent Paxman engine cooling problems resulted in the temporary expedient of fixing the radiator shutters in the open position during the summer months of 1967; an instruction to this effect was issued on 12 June. By the first P&SP Meeting on 04 July, about half Scottish Region fleet has been dealt with.

Problems experienced with the cooling system included:

- Hydrostatic pump drive failures.
- Breakage of hydrostatic ram linkage which controlled the shutter opening and closing.
- Oil leakages from hydrostatic ram pipe joints.
- Difficulty in cleaning radiator elements.

With the onset of winter, hydrostatic operation was reinstated by 31 October. Ultimately, the hydrostatic pump drive problems were addressed by Modification MB39/37 which involved the fitting of improved parts. Despite the dirty nature of the job, cleaning of radiator elements was given a high priority.

11.1.15 Engine/Generator Coupling Problems.

Nine incidences of loose engine/generator couplings were reported during the P&SP Meetings held on 21 January 1970 (D8551/68/81), 24 March 1970 (D8568 (again), D8607), 7 October 1970 (D8555), 13 January 1971 (D8607 (again)), 21 April 1971 (D8606) and 08 September 1971 (D8555 [again]).

It was pointed out that each power unit concerned had received scheduled repairs by Glasgow Works just prior to failure. During 1970, the CM&EE Inspectorate and St Rollox Works staff were tasked to carry out an investigation, with assistance from the Regional Metallurgist and Paxman, and to report to CM&EE Scottish Region. The conclusion was that the failure was due to badly fitting dowels in the flange coupling. It was generally agreed that a repair procedure was necessary, involving the use of one or two stages of oversize dowel. Paxman agreed to supply the Scottish Region with the relevant drawing indicating the tolerances involved, material details, etc., appertaining to the 'as new' product.

As a consequence, during subsequent scheduled repairs, a check was required of the dowel holes and the dowels themselves for ovality and wear respectively ensuring consistency with the Paxman drawing tolerances. Should the results found to be not in accordance with this drawing, oversize dowels would have to be provided to overcome the problem.

11.2 Rolls-Royce Engine Issues.

The original Clayton order was for eighty-eight with Paxman engines; in the event, the final two (D8586/7) were built with Roll-Royce DV8T engines presumably to test an alternative engine type in the light of experience with the Paxman engines. No archive material has been found regarding the deployment of the Rolls-Royce equipment; it was presumably a Clayton Equipment Co. Ltd. proposal but installation was clearly sanctioned by BR as indicated by J.F. Harrison's comment in his 'Report on the Working of the C.M.E. Department' (Appendix C) dated December 1964:

'Two locomotives will be fitted . . .with a new 'R' type Rolls Royce engine, which is very lowly rated, and has been extensively tested under static conditions with a view to adopting such an engine as 'standard' if the modified Paxman flat engines give any further unsatisfactory performance, since the locomotive design with its centre cab is ideal for shunting and trip working, and much liked by Enginemen.'

To all intents and purposes, there was no mention of D8586 and D8587 in the P&SP Meeting minutes. Mark Alden in his article 'Clayton Countdown' in Issue 3 of the *Classic Diesels & Electrics* magazine suggests that 'because there were only two, they unfortunately tended to get tarred with the same brush' as the Paxman engine examples, claiming that the Rolls-Royce DV8T engines 'performed impeccably'. D. Bromley commented on the 'End of the Line' website forum that 'I remember the foreman at Dalry Road taking us round [the depot] and commenting that these were much more reliable than the rest.'

The only archive material found which provides comparative information between Scottish Region locomotives with Paxman and Rolls-Royce engines was in the minutes of the Scottish Railway Board meeting of 31 January and 1 February 1967. Availability (percentage) details are given in the table below together with those for the Scottish English Electric Type 1 fleet.

R.M. Tufnell, in his book *The Diesel Impact on British Rail* (1970), indicated that two of the four Rolls-Royce engines suffered from broken crankcases and were replaced by Paxman engines, but otherwise these engines gave satisfactory service. An official BR Works report lists D8586 as spending 177 working days stored and 20 working days under repair (233 calendar days in total) in St Rollox Works between 8 August 1968 and 29 March 1969; this might support the Tufnell comment regarding the substitution of Roll-Royce engines with Paxman engines. If this was indeed the case, then some form of bed-plate modification would have been required.

11.3 Engine Modifications: Attribution of Costs (D8500-85).

Issues with the Paxman engines were progressively resolved by engineering modifications, most commonly by the replacement of 'defective' components with improved alternatives. This inevitably incurred considerable cost and through 1965/66 BR negotiated long and hard with Clayton Equipment, Paxman and Napier to establish a correct allocation of responsibility for the various issues and an appropriate attribution of the costs incurred.

A basis of an overall settlement had been suggested by BR in a letter to Clayton Equipment dated 29 October 1965; unfortunately this letter has not been seen but the Minutes of a Meeting held on 7 March 1966 involving BRB, Clayton Equipment and Paxman gives a strong indication of its content. One key element was the proposal that:

'…..Paxman agreed to all Clayton's rights under the sub-contract with Paxman being transferred to the Board (*BRB*) so that the Board could deal direct with Paxman to resolve outstanding troubles on the engine.'

Clayton Equipment were happy to accept this, but Paxman required resolution on certain contractual issues, most notably the subject of the start date for engine guarantees i.e. the so-called 'interim' guarantee based on the acceptance of the engine by Clayton versus the 'final' guarantee based on the acceptance of the complete locomotive by BR. This distinction was critical for Paxman, given that the use of the completed locomotive guarantee effectively extended the guarantee period.

It is not absolutely clear from the Minutes of the Meeting on 7 March 1966 how Paxman's reservations were addressed, although it appears that the acceptance of the completed locomotive became the guarantee reference point, with, for example, the guarantee period for the crankcases being set to an agreed four years. Thus:

'The guarantee for 172 engines would be operative from acceptance of the 86 locomotives and for the 13 spare engines from the date of dispatch from Paxman's works at Colchester.'

Availability (%)

Type	No. of Locos.	Availability			
		National Traction Plan (%)	Budget Forecast for 1967 (%)	Performance over 12 months (%)	Anticipated Performance during 1967 (%)
EE Type 1	75	90	90	88	87
Clayton/Paxman Type 1	99	90	70	66	68
Clayton/Rolls-Royce Type 1	2	90	70	79	75

A chart (No.6ZHXL.99048 Issue No.24, again unseen), produced by Paxman and issued on 9 February 1966, listed twenty-seven modifications. At the meeting in March 1966, it was agreed that no further action would be required from Paxman on seventeen of the modifications; in addition, modifications involving the Napier turbocharger would be dealt with by BR direct with the manufacturer. Agreement was reached with Paxman with respect to the remaining nine items; seven of these were relatively minor in nature and were resolved by Paxman paying either 50 or 100 per cent of the material costs, with BR covering fitment costs.

The two 'big-ticket' items were the replacement of aluminium cylinder heads and crankcases with cast iron. Concessionary arrangements for these two items were ultimately agreed as follows:

Cast Iron Cylinder Heads.

'Paxman will supply a total of 1,110 cast iron cylinder heads on all 185 engines at 50% of the normal price to the Board or allow like credit for cast iron cylinder heads already supplied. The work of replacement will be carried out by the Board.

Cast Iron Crankcases.

'Any crankcase of the non-stress relieved batch which becomes defective without external cause within 4 years from the dates of acceptance of the locomotives will be replaced by Paxman free of charge. Any other crankcase of the balance of 185 engines which becomes defective without external cause in the same period will be replaced by Paxman at 50% of the normal price to the Board. The work of replacement will be carried out by the Board. All future replacement crankcases will be in cast iron.'

A Memorandum to the Supply Committee from R.B. Hoff (Supplies Manager) dated 2 June 1966 relating to D8500-85, included the following:

'Paxman have completed, at their expense, a number of modifications to the engines but are not prepared to accept the Board's contention that a further nine are necessary, or in any way their responsibility. The main items are replacement of alloy cylinder heads and crankcases by cast iron to overcome cracking. Paxman are prepared to meet their contractual responsibility to rectify defects occurring during the first twelve months of service but have resisted any further liability.'

The Memorandum then goes on to say that, as part of an overall package of compromises involving liquidated damages for late delivery (reduced to actual losses) and oil contamination damage to traction motors (agreed to be BRB's liability), as well as engine modifications, Paxman had agreed:

'.....to supply all parts and materials required for certain specified disputed modifications, free of charge, and for certain other modifications at normal prices less 50%, and also to extend the guarantee period for crankcases to four years for the earlier locomotives in which the crankcase is suspect, and to replace any failures in the remainder in that extended period at 50% discount.'

At the Supply Committee Meeting on 9 June 1966 (Minute 443):

'The Committee approved the settlement of the dispute with Clayton Equipment Co. Ltd. and Davey Paxman & Co. Ltd.....on the terms set out in the Supplies Manager's memorandum dated 2 June 1966.'

On 24 October 1967, the Works and Equipment Committee (W&EC) (Min.3004, Item 7) authorised the replacement of aluminium cylinder heads on Paxman engines by 1,458 cast iron cylinder heads (i.e. covering all 243 engines); at the same time the Committee also authorised 132 cast iron crankcases, giving an overall estimated cost of £179,896. The W&EC authorisation for both the cylinder heads and the crankcases was at least partially retrospective.

Chapter 12
OTHER PERFORMANCE AND SERVICE ISSUES

12.1 Electrical Issues.

The P&SP Meeting minutes have, once again, been heavily researched with respect to electrical equipment issues as deployed on the Clayton fleet, although relative to the coverage of issues regarding the Paxman engine and associated equipment, references to electrical problems were comparatively scarce and very rarely to do with any actual quality issues associated with the GEC and Crompton Parkinson machines themselves.

The first P&SP Meeting minutes on 4 July 1967 make reference to Main Generator armature insulation deterioration due to the ingress of oily carbon deposits emanating from the exhaust system which passed through the generator compartment, making the comment: 'It is thought this will eventually lead to failure of all machines if allowed to persist'. This issue was addressed in Section 11.1.13.

Oil spillage as a consequence of cylinder head and crankcase fractures, plus general oil leakage from cylinder head joints led to damage to traction motors. Engine modifications already described substantially reduced the incidence of such costly electrical damage.

Minutes of the 10 January 1968 meeting recorded that:

'Workshops confirmed that difficulty in carrying out scheduled repairs to locomotives was being experienced owing to shortage of main generators. In fact, the 'float' of main generators were all away at various contractors for repair [to resolve armature insulation defects].'

Given the costs involved, repairs other than normal Works attention required special authority from the CM&EE (Scottish Region). Main generator attention by AEI, Le Marquand, etc, were strictly controlled, and, in fact, 'off-site' attention stopped altogether during the 1968/69 period. The P&SP Meeting Chairman frequently drew attention to the equipment available in the stored locomotives at Carlisle at this time.

Contractor repairs resumed in late 1969, although the 17 June 1970 Meeting minutes carried the following comment:

'A shortage of electrical machines, i.e. traction motors, main generators, was reported caused by long delays at repairers, allegedly due to shortage of copper. The repairer is most reluctant to quote delivery dates and the CM&EE, BRB representative indicated that delivery dates would be 26 weeks after clearance by BRB Inspector.'

By 13 January 1971 the situation had worsened considerably, with issues regarding main generator (MG) armature bearings increasing in significance:

'Glasgow Works representatives stated a serious problem existed in that a 62 weeks delivery was applicable to this item and they had had no spares available for the past 4 months. Three locomotives, namely 8540 and 8504 (both classified repairs) and 8574 (unclassified repair) had been fully repaired but could not be completed and returned to service due to these bearings not being available. They confirmed that all known sources were being approached and pressure will be maintained.

'BRB representatives hoped that the question of cost was not, in these times of difficulty, being allowed to jeopardise supplies.

'Scottish Region pointed out to BREL that the Main Generators remaining in some locomotives surplus to requirements would have bearings which were fit for further use. It was agreed that Scottish Region would liaise with Glasgow Works with a view to having the surplus locomotives made available again for removal of any such bearings.'

Whilst the use of bearings recovered from stored locomotives made a lot of economic and practical sense, there were down sides as the minutes from the 21 April 1971 meeting clearly illustrated:

'There were three further failures of MG bearings, and the incidence of this type of failure appears to be rising sharply. It is thought to be due to the use of old bearings. It was recognised, of course, that Works had no alternative but to use [such] bearings . . .owing to the extreme shortage of this component.'

By the 7 July 1971 meeting the screws were again being tightened on the Clayton fleet and as a consequence:

'The Chairman again stressed that no repairs to electrical machines are to be carried out without the authority of the CM&EE Scottish Region.'

The ingress of oil into the GEC and Crompton Parkinson main generators was clearly not a problem of the electrical equipment itself, and the bearing problems could be considered to be mechanical rather than electrical. As far as other electrical machinery was concerned, the minutes of the P&SP Meetings indicates that the availability of Crompton Parkinson traction motors was an issue from late 1969 onwards, as the 6 November 1969 minutes explained:

'No spare serviceable machines [traction motors] of this type are available at the present time in Glasgow Works. It was pointed out that, owing to the previous run down of the fleet, machines had been left unrepaired but steps are now in hand to effect.

'The Chairman asked DOD [Director of Design] that an investigation be made into the possibility of using a Clayton-built bogie with GEC motors in Beyer Peacock locomotives. If so, we could obtain as many bogies as required with GEC motors, thus relieving the Crompton Parkinson spares position.

'. . .It was also considered that the interchange of power units [generators] should be included and this action was agreed.'

By the 21 January 1970 meeting there had been further developments:

'The Chairman stated it had been suggested a locomotive be withdrawn from service to provide spares of these machines [although this did not happen in a planned sense].

'Regarding inter-changeability between main components of Clayton and Beyer Peacock built locomotives, DOD representative said that bogies complete would possibly be interchangeable mechanically (a trial would prove this) but electrically i.e. use of Crompton Parkinson motors with GEC generator, etc., was not recommended. Power unit [generator] inter-changeability was possible physically but again was not recommended from an electrical aspect.

'So far as traction motors were concerned, the motors were not interchangeable owing to fan attachment differences but it was agreed that Glasgow Works, when the opportunity arose, would check inter-changeability of complete blowers.'

D8589 was reported at the 24 March 1970 Meeting as having suffered collision damage and the group latched onto the fact that it would provide a useful source of Crompton Parkinson spares (plus one cast iron engine crankcase). However, it was July 1970 when this locomotive was declared surplus to requirements enabling the release of urgently required spares. By the 7 October 1970 meeting it was reported that:

'Both power units complete, i.e. including CP main generators and turbo-chargers together with CP traction motors, were now to hand at Glasgow Works.'

12.2 Transport of Spares to Depots.

The P&SP Meeting minutes carried many references to transport issues relative to the delivery of spare parts between the St Rollox Works Stores and the various depots, particularly relating to long delivery times and locomotives stopped for extended periods awaiting materials.

12.3 Spares Recovery.

The recovery of spares from redundant locomotives was referenced in at least eight separate minutes during 1970 and 1971. The rundown of the fleet in 1968/69 together with associated spares clearly left a chronic shortage of spares when the decision was made to retain and expand the fleet in late 1969. The Chairman of the P&SP Meeting recognised the value of the redundant fleet, at Carlisle and Polmadie in particular, and urged the Works representatives to take appropriate action. The minute taker very eloquently recorded his views; for example at the 17 June 1970 Meeting:

> 'The Chairman pointed out again that there were 13 locomotives available for spares at Polmadie'.

Given the constant reminders from the Chairman, the impression from the Minutes is that recovery of spares by the Works contingent was not given a high priority, to the extent that brand new spares were procured instead. Again from the 17 June 1970 Meeting minutes:

> 'An order placed by Glasgow Works for cast iron cylinder heads had been stopped, it being pointed out that the LMR locomotives all had CI cylinder heads, which could be removed for spares purposes.'

As regards the five Scottish Region locomotives withdrawn in March 1971, the minute taker really went to town:

> 'Regarding the 5 surplus (*Scottish Region*) locomotives (*D8500/10/30/5, D8606*), 4 Clayton/Paxman and 1 Beyer Peacock/Paxman, the depots would remove all parts, turbo-blowers, AWS equipment, and a number of small parts . . .The locomotives will then be offered to Supplies Manager, BREL and Scottish Region suggested that BREL consider obtaining the cast iron cylinder heads, cast iron crankcases, connecting rods, crankshafts and engine details, traction motors, main and auxiliary generators and bogies as well as electrical control equipment. In other words, the locomotives were not to go scrap merchants with anything useable but the maximum amount taken off first of all. By doing so, the availability position in future will be assisted.
> In the past, considerable difficulty has been experienced in obtaining Crompton Parkinson spares as fitted to Beyer Peacock locomotives and Scottish Region suggested that all relays, auxiliary machines and main electrical machines be removed from this locomotive, together with all control equipment.'

By August 1971, an embargo was placed on the purchase of major spares for these locomotives, together with the discontinuation of scheduled repairs.

12.4 P&SP Effectiveness.

At the final meeting of the P&SP on 8 September 1971:

> 'The Chairman . . .reminded those present that the first meeting was held in mid-1967 when defects and difficulties were numerous and performance in terms of Availability and Reliability were low - Availability being about 60% and Miles per Casualty 4,000. The work of the Committee and others concerned led to improvements such that a reasonably satisfactory performance was achieved and held over the period since then.'

The deployment of a multi-functional team to improve the performance was without doubt highly laudable, but, as already suggested, the successful application matrix-style management in large highly bureaucratic organizations can be very challenging. The P&SP group certainly seemed very capable of identifying the key issues to focus on, and were very effective in identifying potential solutions. The key issue was one of implementation, against the backdrop of conflicting Department objectives and where the full value of improvements was not fully understood in a cost/benefit sense. It is unbelievable that substantial performance benefits, expressed in thousands of pounds, was held up for so long by delays in the fitment of soft-iron cylinder head gaskets, costing pence.

In my view, for what it is worth, the Clayton performance could have been improved to a higher level, and faster, if the Chief Mechanical Engineer chairmanship had continued into the P&SP meeting group after the cessation of the BRB/Davey Paxman Liaison Meetings in 1967. The consequent clout and over-arching perspective would have ensured a greater impetus to progress. Instead, the P&SP operated at a fairly local level and floundered somewhat as a consequence.

12.5 Maintenance
12.5.1 Locomotive Maintenance Conditions.

Early maintenance conditions for the Claytons, along with other diesel classes, were less than ideal. D8500-53 were initially allocated to the major steam depot at 66A Polmadie; D8554 onwards were allocated to 64B Haymarket from October 1963 and although this depot had closed to steam by this point, the locomotives operated extensively from 64A St Margarets and 64C Dalry Road, both still very much open to steam. As the April 1964 *Railway Observer* noted:

'Several [Claytons] are now to be seen in and around St Margarets where one would not have thought the prevailing filth would have been conducive to diesel maintenance.'

A Technical Report to the BRB dated 17 March 1967 specifically recognised that the maintenance of the Clayton fleet was seriously affected by rebuilding work at Haymarket and conditions at Polmadie depot. Conditions improved later for the Clayton fleet, although for a number of years Thornaby, Gateshead and Polmadie remained as up-graded, but still steam-age, facilities. It was only at Carlisle Kingmoor, from January 1968, that the Claytons enjoyed facilities which were specifically designed for diesel traction from the outset. The June 1985 edition *Rail* magazine carried an article on Kingmoor depot which included an interview with the Chief Traction Maintenance Supervisor, John Duncan, who made the statement:

'Do you realise that if the Claytons had remained at Carlisle, they would still be running today . . .It's the same old story; if things are looked after they give good service but by the time we sorted them out the Class was on the doomed list. We were rectifying the faults but it was too late.'

The article continued:

'He makes this statement on the premise that had they been looked after properly, they would not have gained a poor reputation and added to the list of class withdrawals under the Traction Plan . . . John devised an ingenious method of clearing blocked turbo-chargers by attaching a compressed air supply to blow out the passages to stop oil passing the seals and to prevent seizure of the turbo with carbon build-up. This was undertaken automatically when a Class 17 was visiting

An unidentified Clayton in the 'prevailing filth' of St Margarets, Edinburgh, undated. It is hard to disagree with the *Railway Observer* correspondent! Is it any wonder that the GEC generators suffered insulation deterioration! (griffith_p (flickr))

the Fuel & Inspection depot in Kingmoor yard. The failure of this component plummeted from thirty a month to just two. And this was just the turbo-chargers!'

12.5.2 Working Conditions for Fitters.
The propensity for the Claytons to leak oil led to poor working conditions for fitters; as David Clough described in his book *British Rail Standard Diesels of the 1960s* (2009):

> 'With oil deposited everywhere in the engine compartment, maintenance was an unpleasant job, made worse because much of the work involved lying on top of the engine, which could only be done when it had cooled down. It was therefore unsurprising that there are allegations that depot staff were not as thorough in their work as they should have been and minor problems became major failures in consequence.'

M. McManus, St Rollox Apprentice (Classic Diesels & Electrics No.8) comments:

> 'The Clayton was a class designed for drivers, not fitters. The cab was vast in comparison to the engine room. Although much of the work could be carried out from the running plates through the engine room doors, certain jobs were not catered for. To set valve clearances, for example, you had to be both ambidextrous and a midget!'

Cost of Modifications.

Modification	Estimate spread of outstanding expenditure (£'000)			
	1967	1968	1969	1970
Authorised:				
1. General Mechanical	1	0	0	0
2. General Electrical	3	1	3	0
3. Engines	1	0	0	0
4. Main Generators	0	0	0	0
5. Traction Motors	0	0	0	0
6. Bogies	9	10	0	0
Total	14	11	3	0
To Be Authorised:				
1. General Mechanical	11	14	34	54
2. General Electrical	2	0	0	0
3. Engines	79	85	153	165
4. Main Generators	0	0	0	0
5. Traction Motors	0	0	0	0
6. Bogies	0	0	0	0
Total	92	99	187	219
Grand Total	106	110	190	219

12.6. Costs of Improvements.
12.6.1 Maintenance Costs.
Various archive documents make reference to the additional costs associated with maintaining two engines rather than one. It has been impossible to quantify this assertion given that costings were severely masked by the high incidence of unscheduled repairs.

12.6.2 Cost of Modifications.
An Addendum to the Technical Report to the British Railways Board produced by the Chief Engineer (Traction & Rolling Stock) dated 17 March 1967, included a section covering the cost of modifications, both authorised and anticipated as at 31 December 1966 (see table above). Although these figures inevitably changed, they illustrate the scale of expenditure envisaged to resolve the problems of the Class 17s.

12.7 Driver's Views.
The Claytons were very popular with drivers and secondmen and were ideally to suited for local trip workings due to their excellent all-round visibility. The driver's outlook was second to none, with the visibility of the nearside buffer from the driver's seat making the coupling-up to

trains a straightforward and safe process. The driver's cab desk has been described as 'superb' and 'well laid out' and the ability to change combinations of engines/generators and traction motors was considered by many to be 'technically brilliant'.

One down side, during the summer months, was that the cabs could get quite hot given the large number of windows. Clearly you couldn't please everyone! And what would it have been like if a train heating boiler had been installed?

'Clingon' (messaging on the WNXX website Forum) made the comment that:

'One [Motherwell] driver in particular told me that if you kept them at 3/4 throttle they would run all day without an issue but when pushed, they would invariably overheat and an engine would shut down, forcing the driver to run the other flat out and then it was game over.'

H. Friend in his article 'These We Have Loved: A BR Driver's Guide to Main-Line Traction' (*Traction* magazine, Issue No.2) commented:

'I liked them personally. There were many in evidence during the 1960s in the Tyneside and Teesside area in which I worked . . .they were delightful engines to work on, comfortable and pleasantly warm in winter.

'They were used extensively for freight work and this proved to be their undoing as they were flogged to an early grave simply because they were forced to haul far heavier loads than they were designed for.

'Their best feature was the fact they possessed two engines. If one failed, at least you could get home, even if it meant dumping the load somewhere.'

NBL Type 2 D6126 and an unidentified Class 17/1 heading south through Coatbridge Central, 1965. The Class 21 appears to be rescuing the Clayton (maybe after failing and being returned to Polmadie for repairs). Some sort of irony here. . .an NBL providing rescue assistance! Another photograph exists of D6126 passing northbound through Coatbridge Central, but whether the picture was taken before or after this image is unknown; given the locations of Eastfield and Polmadie, probably after with D6126 returning to its home depot. (Doug Kirk)

Chapter 13
ACCIDENT AND FIRE DAMAGE

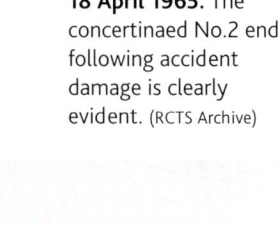

D8508, 66A Polmadie, 18 April 1965. The concertinaed No.2 end following accident damage is clearly evident. (RCTS Archive)

13.1 Accident/Derailment Damage.

D8507

D8507 was noted at 64B Haymarket on 19 May 1970 with damage to one end (*Railway Observer*, July 1970), this having occurred on or before 5 May 1970. D8507 was subsequently released from an Unclassified repair at St Rollox Works on 6 June 1970.

D8508

The date and location of the incident which affected D8508 is unknown, but will have been on or before 12 April 1965. The DLRC for D8508 indicates a release from Unclassified repairs at St Rollox Works on 26 June 1965.

D8545

The December 1963 edition of the *Railway Observer* reported the following:

'D8545 came to grief at Ayr in the early morning darkness of 22nd October (*1963*) when with a brake van it crashed through the buffers of the head-shunt at Falkland Jct., coming to rest canted towards the down main line throughout the ensuing day, causing all down trains to be heavily cautioned. The diesel was re-railed during the early hours of the following morning by cranes from Polmadie and Hurlford.'

It was noted at St Rollox Works on 3 November 1963 and was released from a Light Casual repair on 5 December 1963 after remedial work.

D8555

A report from the *Railway Observer*, this time the February 1964 edition, reported that:

'One of the new Clayton Type 1 diesels became a casualty when on 4th December (*1963*) D8555 (64B) on E58 trip ran away on the Loanhead branch with 36 coal wagons and caused a line blockage . . .after it became derailed at Millerhill.'

The Railway Magazine (April 1964) added:

'On December 18 (*1963*), D8555 was involved in an

Accident and Fire Damage • 245

D8567, Glasgow Works, October 1969. (Dave Thorpe)

D8570, Bescot (en route to Bird's, Long Marston scrap yard), Undated.
Tarpaulins cover the accident damage suffered by the No.1 end bonnet.
(Stephen Fisher)

accident at Millerhill Yard, when the engine, with a train of fully loaded coal wagons, became derailed and ploughed into an embankment.'

It will be noted that two different dates are quoted. The Diesel Locomotive Record Card for D8555 shows Unclassified Works treatment at St Rollox between 5 December 1964 and 31 January 1964. Even if D8555 spent the early part of this period 'on depot awaiting Works' the clear inference is that the RCTS date for the incident is correct.

D8567

According to the November 1969 edition of *The Railway Magazine*:

'. . .(D8567) was derailed on September 4 (*1969*) when working a train of scrap metal from Sighthill Yard to the Hallside steelworks at Newton. The train was within yards of its destination when the locomotive and three wagons overturned, bringing down the overhead gantries.'

D8567 was released from St Rollox Works on 3 November 1969 after Unclassified repairs.

D8570

The *Railway Observer* (January 1969) reported a further incident:

'. . .On 7th November (*1968*), a train carrying scrap metal from Motherwell headed by blue 8570 ran through the catch points at the end of the loop at Ross Jct. and almost ended up on the M74, stopping at the top of the embankment above the motorway. The six derailed wagons were soon removed but 8570 remained until 17th November . . .'

D8570 was stored unserviceable on 10 November 1968 and condemned on 23 November 1968 as a result of this accident.

D8581

The photograph in Section 10 of D8581 at 64B Haymarket in May 1969 illustrates the accident damage sustained by the locomotive although the exact date and whereabouts of the mishap are not known. The Clayton was officially stored during w/e 17 May 1969 pending a decision; after about 4½ months in store at Haymarket, D8581 was repaired receiving a full Classified treatment at St Rollox

D8616, 64B Haymarket, 30 August 1970. Accident damage to No.2 end sustained on or before 7 August 1970. (Exe-Rail [G. Lendon])

Works between 29 October and 15 November 1969.

D8585
No location or date details have been found for this locomotive; however it was stored unserviceable on 23 October 1968, reportedly due to collision damage, and condemned very soon afterwards on 27 October 1968. Although withdrawn, the damage may not necessarily have been particularly serious; several other undamaged locomotives during this period were being discarded as surplus to requirements.

D8589
Collision damage at Derwenthaugh resulted in D8589's withdrawal on 6 July 1970. No further details are known, although the minutes of the P&SP meeting of 24 March 1970 did make specific reference to the accident damage sustained by this locomotive.

D8590
The SLS magazine lists D8590 as withdrawn on 27 March 1971 due to collision damage. Further details concerning date, location and the circumstances of the incident are unknown.

D8599
The DLRC for D8599 includes an entry for Classified and Collision Damage repairs at St Rollox Works over the period 19 December 1966 to 21 January 1967. D8599 was sighted at the Works on 16 December 1966.

D8611
This locomotive was reported to be in Darlington Works on 6 June 1965 'after colliding with a lorry at a level crossing' (*Railway Observer*, July 1965).

D8613
A DLRC works entry lists D8613 as out of traffic between 27 August and 10 September 1969 for collision damage repairs at St Rollox Works.

D8616
Details surrounding D8616's accident are not known; it was released from St Rollox Works 24 October 1970 following Unclassified repairs.

13.2 Out of Control Trains.
A memo from the Movements Manager, Glasgow to Chief Operating Officer, London, 1 July 1965 included a statement of instances of out of control freight trains on the Scottish Region during 1963 and 1964. Trains involving Claytons are listed below:

Out of Control Freight Trains.

Date	Location	Locomotive	Load	Gradient (Falling)	Remarks
31/08/1963	Langloan Weighs	D8514	29=36	1 in 75	Locomotive overload.
11/09/1963	Fullwood Junction	D8506/D8528	55=68	1 in 86	Driver failed to comply with A.W.B. Instructions.
11/12/1963	Loanhead branch	D85xx	36=45	1 in 55	Driver failed to comply with A.W.B. Instructions.
20/12/1963	Wishaw South	D8510	37=43	1 in 102	Driver failed to comply with A.W.B. Instructions.
03/04/1964	Bridge of Earn	D85xx	14=53	1 in 75	Fitted train. Heavy braking necessary to control train on falling gradient caused excessive heating of tyres and brake blocks.
14/05/1964	Holytown Junction	D8564	24=48	1 in 180	Driver failed to comply with A.W.B. Instructions.
13/08/1964	Fullwood Junction	D8651 (sic)	28=54	1 in 86	Driver failed to comply with A.W.B. Instructions.
10/09/1964	Millerhill	D85xx	30=50	1 in 75	Driver misjudged distance in addition to which the air pressure was only 45/45 as against 60p.s.i.
02/10/1964	Holytown Junction	D8563	23=35	1 in 180	Driver failed to comply with A.W.B. Instructions.
29/10/1964	Craiglockhart	D85xx	21=60	1 in 143	Fitted train. Engine overloaded.

Notes:
1. Load: '29=36' for the first incident listed above means 29 wagons represented by an equivalent load of 36 16-ton mineral wagons. The latter number allows direct comparison between consists of different wagon types.
2. A.W.B. = Advanced Warning Board.

A separate report regarding the Craiglockhart incident provides additional information, as follows:

<u>Craiglockhart Junction, 26/10/64</u> [Note the different date but identical location/train details].
'A Type 1 Clayton diesel loco hauling 21 loaded wagons and brake van ran out of control when approaching Kingsknowe and diverted into a sand drag protecting a junction. Loco and leading 14 wagons derailed.
    'Train overloaded due to inaccurate assessment of train load by guards. Driver also failed to note during the course of journey at difficulty in handling the train.'

The locomotive involved might have been D8580; the DLRC for this locomotives shows Unclassified repairs at St Rollox Works for the period 27 October 1964 to 31 December 1964.

13.3 Fire Damage.

BR had serious problems with diesel locomotives catching fire. In an attempt to address the issue, and the associated safety and cost implications, BR set up regional reporting of fires so that the reasons and severity could be ascertained on a class-by-class basis. This commenced in 1961 and over the following years the reports on *Fires on Diesel Train Locomotives* produced by the Locomotive Performance and Efficiency Development Unit, Derby, contributed greatly to the understanding of the problem, and quickly identified potential solutions to the problems (locomotive modifications (e.g. re-routed pipework, spark guards), fire-fighting equipment, improved maintenance and cleaning routines, etc).

The full listing of Class 17s which were reported in the British Rail fire reports are listed below.

Source of Ignition:
A Brake block sparks.
B Hot engine parts, including exhaust blows.
C Electrical overloads or arcs.
D Mechanical seizures or failures.
E Train heating boiler burners.

Combustible material:
1 Oil impregnated dirt or waste.
2 Fuel or lubricating oil sprays or leaks.
3 Electrical insulation.
4 Not known.

Degree of damage:
S Severe (i.e. a fire which necessitated works attention).
NS Not severe.
NK Not known.

Class of train:
Initial digit of train reporting code, or, F (freight), LE (light engine), and T (trip freight).

Fire Damage.

Year	Loco No.	Date	Source of Ignition	Combustible Material	Degree of Damage	Class of Train	Comments
1965	D8516	11/05/1965	A	1	NS	8	
	D8562	14/05/1965	B	2	NS	8	
1966	D8518	12/01/1966	A	1	NS	6	
	D8562	19/01/1966	A	1	NS	4	
	D8548	12/03/1966	A	1	NS	8	
	D8578	17/03/1966	B	2	NS	4	
	D8570	30/03/1966	A	1	S	8	
	D8517	14/04/1966	A	1	S	8	
	D8580	19/04/1966	B	2	NS	8	
	D8563	23/04/1966	B	2	NS	4	
	D8549	23/04/1966	A	1	NS	F	
	D8501	26/04/1966	A	1	S	8	
	D8574	03/05/1966	B	2	NS	6	
	D8572	11/05/1966	A	1	NS	4	
	D8531	24/05/1966	B	2	NS	8	
	D8545	24/05/1966	B	2	NS	8	
	D8536	30/05/1966	A	1	NS	F	
	D8568	17/06/1966	B	2	NS	8	
	D8562	22/06/1966	B	2	NS	8	
	D8568	16/07/1966	B	2	NS	1	
	D8563	21/07/1966	B	2	NS	?	
	D8543	27/07/1966	A	1	NS	7	
	D8508	15/08/1966	A	1	NS	1	
	D8520	19/08/1966	A	1	NS	T	
	D8538	06/10/1966	A	1	NS	8	
	D8565	03/11/1966	B	1	NS	4	
	D8573	08/12/1966	B	2	NS	4	
	D8515	12/12/1966	A	1	NS	?	
	D8585	28/12/1966	C	2	NS	8	

Accident and Fire Damage • 249

Year	Loco No.	Date	Source of Ignition	Combustible Material	Degree of Damage	Class of Train	Comments
1967	D8565	06/01/1967	C	3	NS	LE	
	D8567	28/01/1967	B	2	NS	T	
	D8579	30/01/1967	C	3	NS	F	
	D8581	30/01/1967	B	1	NS	4	
	D8585	03/02/1967	D	2	NS	4	
	D8566	07/02/1967	C	3	NS	8	
	D8585	10/02/1967	D	2	NS	F	
	D8562	13/02/1967	D	2	NS	T	
	D8571	23/02/1967	D	2	NS	T	
	D8573	09/03/1967	D	2	NS	4	
	D8538	14/04/1967	A	1	NS	8	
	D8585	10/05/1967	D	2	NS	F	
	D8605	31/05/1967	D	2	NS	F	
	D8605	01/06/1967	D	2	NS	6	
	D8615	26/06/1967	D	2	NS	T	
	D8587	07/09/1967	B	2	NS	8	
	D8587	30/12/1967	D	2	NS	F	
	D8584	30/12/1967	B	2	?	F	
1968	D8527	02/10/1968	B	2	?	F	See Note (9).
1969	D8573	02/09/1969	A	1	NS	F	
	D8574	06/12/1969	B	2	?	F	
1970	D8600	04/05/1970	A	3	S	F	
	D8594	02/11/1970	D	2	NS	F	
1971 (Jan-Jun)	D8602	04/01/1971	B	1	S	F	

Notes:
1. Full reports seen for 1962-67, 1970 and 1971 (first six months). There were no Clayton fires reported during the period 1962-64.
2. No information seen for January-March 1968.
3. Summary reports seen for April-June and July-September 1968; one LMR Class 17 casualty was reported in each quarter (locomotive identities unknown).
4. Summary report seen for October-December 1968 mentioning one ScR casualty (presumably D8527 as reported in BR internal memo).
5. Summary report seen for January-March 1969; one ER Class 17 casualty reported (identity unknown).
6. Summary reports seen for April-June and July-September 1969; one ScR Class 17 casualty reported in each quarter (identities unknown).
7. Summary report seen for October-December 1969; two ER Class 17 reported (locomotive identities unknown) and one ScR casualty (presumably D8574 as reported in BR internal memo).
8. No information seen for July-December 1971.
9. The locomotive number for the 2 October 1968 incident is probably an error. D8527 was reported on fire at Lochmuir whilst working the 1352 Thornton to Aviemore freight; however, in October 1968 D8527 was allocated

to D10 Preston Division. The locomotive concerned may have been D8575 which was stored unserviceable on 13 October 1968 and subsequently withdrawn on 27 October 1968; in addition, the Performance & Service Problems Meeting minutes lists the reason for withdrawal of D8575 as 'Fire Damage'.
10. The St Rollox Shopping Summary (1969) shows D8602 out of traffic from 17 October 1969 and in Works from 20 October and 13 November 1969 for fire damage repairs, although nothing was reported as such on the DLRC.
11. Reported incidents show a high emphasis on Haymarket locomotives. Was this a reporting issue or were Haymarket locomotives more susceptible (differing maintenance standards, cleaning regimes, etc.)? No Eastern Region Claytons were reported until 1969.
12. There was a massive 'spike' in fire incidents in 1966 and first-half 1967, significantly reducing from 1968 onwards.

	Class 17	Class 20
1965	1 fire/annum per 57.0 locos.	1 fire/annum per 64.0 locos.
1966	1 fire/annum per 4.3 locos.	1 fire/annum per 76.5 locos.
1967	1 fire/annum per 6.9 locos.	1 fire/annum per 106.5 locos.

In 1965, the Claytons were the most susceptible locomotives with respect to 'B2' incidents (hot engine parts igniting fuel or lubricating oil).

D8612, 66A Polmadie, 28 May 1971. Exhibiting fire damage of the 'Fuel or lubricating oil sprays or leaks' variety, although this particular incident was not reported! (Colour-Rail)

Chapter 14
OPERATIONS: HIGH-LEVEL SUMMARY

14.1 Initial Intentions.

The Glasgow North, Glasgow South and Fife Area Schemes were envisaged to require forty-seven, sixty-five and thirty-four Type 1 locomotives respectively. In terms of construction, the 146 locomotives were split into two batches i.e. 58 English Electric Type 1's within the 1961 Building Programme and 88 Clayton Type 1's in the 1962 Programme.

D8070-99 and D8100-16 were delivered new to 65A Eastfield between June 1961 and February 1962 to fulfil the Glasgow North allocation. D8117-27 were delivered to 66A Polmadie between February and July 1962 as part of the Glasgow South allocation.

The English Electric deliveries left fifty-four locomotives outstanding for Glasgow South and thirty-four for the Fife Area Schemes. D8500-53 (allocated new to 66A Polmadie) and D8554-87 (to 64B Haymarket) when delivered between September 1962 and February 1965 fulfilled the outstanding Glasgow South and Fife requirements.

In reality, D8554-87 initially worked off 64A St Margarets, 64B Haymarket and 64C Dalry Road, plus Millerhill and Leith Central stabling points. Based on locomotive sighting information, it took until early 1967 for the Claytons to make a concerted effort to broaden their horizons north of the Forth with operating bases at 62A Thornton Junction, 62C Dunfermline and Alloa.

D8588-D8616 were initially justified for use within three Area Schemes (i.e. Holbeck (three), Hull (thirteen) and Sheffield (thirteen). On delivery, these twenty-nine Claytons were actually delivered to 51L Thornaby (four), 52A Gateshead (twelve) with thirteen to the Sheffield area as intended.

14.2 Polmadie.

Clayton duties included:

- General trip and transfer freight traffic in the Glasgow South, Parkhead, Motherwell, Carstairs, etc., areas.
- Raw materials, semi-finished, and finished steel traffic involving Clyde General Terminus, Ravenscraig, Dalzell, Clyde Ironworks, Clydesdale, Gartcosh, Glengarnock, etc.
- Inter-Regional freight duties: Mossend-Carlisle via the West Coast Main Line route (via Beattock) or the Glasgow & South Western route (via Dumfries).
- Glasgow Central station pilot duties.
- Prior to electrification of routes south of the Clyde, Claytons were used fairly extensively on the Glasgow commuter services to Gourock, Wemyss Bay and East Kilbride.
- During the 1963 summer months the Claytons operated 'holiday extras' from Glasgow to the Ayrshire coast and Isle of Man boat trains to Ardrossan.
- Engineering duties.

The disposition of the 66A Polmadie fleet on two dates in 1965 illustrate two features. Firstly, approximately two-thirds of the operating Polmadie fleet were based at their 'home' depot at weekends, with the other one-third at 66B Motherwell. Secondly, regarding the Motherwell locomotives, they appear to have been selected from the lowest and highest numbered locomotives of the Polmadie allocation; this was not just a feature of the two dates selected below but a general feature of the 1963-68 period as the sightings testify.

18/04/65	66A	D8507/8/10/4/6-20/2-4/6-38/40/2/5	28	
	66A (Repair Bay)	D8501/12/3/25/39/41/7	7	
	66B	D8500/3/5/9/11/43/4/8-52	12	47
08/08/65	66A	D8506/9/12/6-22/4-6/8/30/2-9/43/5/6/9/53	28	
	66A (Repair Bay)	D8504/23/7/9/31/40/4/8/51	9	
	66B	D8500-3/5/7/8/11/3-5/41/7/50/2	15	52

14.3 Haymarket.
Clayton duties included:

- Edinburgh area trip and transfer freight duties from 64A St Margarets, 64B Haymarket, 64C Dalry Road, Leith and Millerhill and later also out-stationed at Bathgate. Dalry Road was closed on 4 October 1965 with locomotives and men transferred to St Margarets; in turn, St Margarets closed on 22 April 1967 with resources transferred to Millerhill and Haymarket.
- From early 1967 and with the progressive demise of steam the Claytons operated further afield to include Fifeshire, based at 62A Thornton Junction, 62C Dunfermline and Alloa, operating trip and transfer freights as far north as Dundee and Perth. Thornton Junction depot closed on 8 September 1969 and locomotives were stabled in the Yard; 62C Dunfermline closed on 22 September 1969 with locomotives subsequently stabling at the Townhill Wagon Works.
- Inter-Regional freight duties: Millerhill-Berwick (ECML), and Millerhill-Carlisle (Waverley route via Hawick up to closure in early-1969). Double-headed Claytons were regular performers on the Bathgate-Kings Norton car trains.
- Edinburgh Waverley and Princes Street station pilot duties.
- Hawick pilot duties.
- Passenger duties, including to Callander on the Stirling-Crianlarich route.
- Holiday trains from Dundee/Edinburgh to Blackpool (as far as Carlisle).
- Engineering duties, including track lifting duties on the Waverley route.

14.4 Aberdeen Kittybrewster.
A few of the Beyer Peacock built Claytons in the D8604-16 range spent a short period operating from 61A Kittybrewster in 1966 with D8604/9/10/3/4/6 (at least) being noted in the Aberdeen area between late-May and mid-September 1966. The locomotives were always officially on Haymarket's books although the BLS *Railway Locomotives* magazine (March 1967) lists D8616 at 63A Perth in February 1967 carrying 61A shed plates. Duties included trip freight work in North-East Scotland.

14.5 Gateshead.
Clayton duties included:

- Blyth & Tyne coal traffic. N.B. Trials on the Consett iron-ore duties from Tyne Dock in 1963 proved the unsuitability of the Claytons on this very onerous traffic.
- Wearside coal traffic, including export coal via Sunderland South Dock.
- General trip freight around the Northumberland and Durham areas.
- ECS workings between Newcastle and Heaton.
- Engineering duties.

Double-headed Claytons or a Clayton with a Diesel Brake Tender were regularly rostered for the heavier mineral trains in the North-East to ensure adequate braking capability.

14.6 Thornaby.
Clayton duties included:

- Teesside area trip freight work, both north and south of the river, with a heavy concentration on coal, steel and chemical movements.
- Engineering duties.

14.7 Barrow Hill.
Clayton duties included:

- Nottinghamshire/Derbyshire trip-freight activities (coal, coal derivatives and traffic associated with the Staveley iron works, etc.).

14.8 Carlisle Kingmoor.

Minute 67/166(b) of the Scottish Railway Board Meeting of 1 November 1967 indicated that the transfer of Claytons to the LMR was 'made possible by the rationalisation of services and improvements effected in utilisation'; Minute 68/30(b) of the 6 March 1968 meeting mentions 'Continuing improvement in diagramming and utilisation'.

Clayton duties included:

Polmadie Duties.

- Trip and transfer freight traffic around Carlisle and the West Cumberland/Furness route to Workington, Barrow and Carnforth, including mineral traffic, rail trains from Workington, nuclear flask traffic associated with the Windscale/Sellafield re-processing facility.
- Carnforth-Skipton trip workings.
- Shap banking duties involving up to four Claytons (1 January to 5 May 1968).

It is alleged that the LMR completed staff training on the Claytons in July 1968 with withdrawal of the fleet commencing in October!

14.9 Other.

In September 1963, D8501/36 undertook trials over the East and West Yorkshire Union Railway route, based at 56B Ardsley; however they were no further developments.

D8505, Bridge of Earn, 31 March 1970.
(RCTS Archives)

254 • THE CLAYTON TYPE 1 BO-BO DIESEL-ELECTRIC LOCOMOTIVES – BRITISH RAILWAYS CLASS 17

D8538, Glasgow Central, 25 June 1967. Station pilot duties. Recently ex-works.
(Colour-Rail)

D8513, Sandyford, 31 May, 1967.
(RCTS Archives)

D8514 and D8503, Beattock, 10 August 1963. (Rail-Online)

D8536 and D8552, Kingshill, 27 August 1971. (RCTS Archive)

D8567, Cowlairs Works, 5 July 1966. Being shunted by D8567 is withdrawn Gresley A3 Pacific 60041. (Colour-Rail)

Haymarket Duties.

D8568, Dalmeny Junction, 27 March 1964. (RCTS Archive)

D8575 and D8555, North Queensferry, 21 August 1965. The Forth Bridge is clearly visible in the background. This location is reputed to have caused many problems for the flat Paxman engines as they were fully 'opened-up' immediately after the Forth Bridge speed limits, particularly with heavy loads as depicted here.
(Rail-Online)

D8587, Gorgie East, 30 July 1967. Gorgie East station was closed in 1962, when passenger rail services were withdrawn from the Edinburgh Suburban line.
(RCTS Archives)

D8572, Edinburgh Princes Street, 7 August 1965. Princes Street station closed on 6 September 1965, almost exactly one month after this picture was taken. Passenger services were subsequently routed into Waverley station.
(Author's Collection)

D8587 plus Stanier Black 5 44952, Edinburgh Princes Street, 3 July 1965. Superb architecture apparently totally destroyed by 1970, in the name of progress.
(Author's Collection)

D8530, Lady Victoria, Colliery, undated. (Colour-Rail)

D8606 and D8601, Steele Road (south of Riccarton Junction on the Waverley Route), April 1965. A very attractive shot of two Claytons not really doing what they were originally designed for. Centre-cab locomotives on Anglo-Scottish trunk services were not essential but would certainly have given the driver a very comfortable ride! There is some doubt about the date of this photograph and also the identity of the second locomotive. D8606 wasn't re-allocated from Barrow Hill to Scotland until May 1966, and D8601 followed much later from Gateshead in June 1971 after the closure of the Waverley route in January 1969. The second machine is, however, a Beyer Peacock specimen given the duplicate works plates. (Colour-Rail)

D8586, Edinburgh Princes Street Gardens, 1970 (Slide processed June 1970). (TOPticl Digital Memories)

Gateshead Duties.

D8592, Monkwearmouth Bridge, circa 1966. Mixed freight crossing the River Wear at Sunderland. (Rail-Online)

D8601, Newcastle, undated. Gunpowder wagons, perhaps?
(Transport Topics)

D8601, Location unknown, undated (Slide processed November 1970).
(TOPticl Digital Memories)

D8596, Durham Gilesgate, 23 August 1966. (RCTS Archive)

Carlisle Kingmoor Duties.

D8521 and D8537, Light engines passing 12A Carlisle Kingmoor depot, 18 April 1964. This photograph was taken prior to the Claytons being allocated to Carlisle with both locomotives allocated to Polmadie at this time. Differing fortunes for these two locomotives: D8521 was allocated to Carlisle from late 1967, later seeing an extended if sedentary life in Departmental service; D8537, on the other hand, was the first Clayton withdrawn and cut-up in 1968. (Author's Collection)

Operations: High-Level Summary • 263

D8531, Carlisle Citadel, undated. (Colour-Rail)

D8523 and D8515, Carnforth, 3 August 1968. One day before the end of steam on BR, notwithstanding the specials on 11 August. The coaling tower in the background was very close to redundancy but it still survives today, which is more than can be said for 116 of the Claytons. Surprising as it might seem, blue-liveried D8523 only lasted for two more months before withdrawal, with green D8515 lasting until October 1971 following re-allocation back to Scotland. (Author's Collection)

D8500 and D8522, plus Metrovick D5711, 10A Carnforth, 24 May 1968. (Author's Collection)

Railtours.

D8613, NCB Gladwell Colliery, 16 October, 1965. Assisting Fowler 4F 43953 on rail tour duties. (Colour-Rail)

D8510, NCB Beckermet Colliery, 2 March 1968. Brake-van excursion around West Cumberland. (Rail-Online)

Chapter 15
DETAIL DIFFERENCES

15.1 Works Plates.
15.1.1 D8500-87.
Two Works plates were fitted, one on each locomotive cab side. A 'full' Works Plate (with Serial No.) was also positioned on the underside of the main frame at one end and on one side only.

Clayton Cab-side Works Plate. (Alan Whincup)

A rarely seen example of a Clayton Mainframe Works Plate. Note the Serial No. format i.e. 4365/U 84 rather than 4365U/84 or 4365U84 as published in railway publications to date. (Alan Whincup)

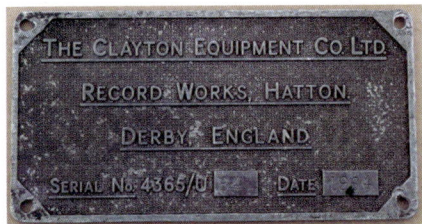

15.1.2 D8588-99, D8600-16.
Two cabside plates were fitted on the Beyer Peacock-built locomotives (on each cab side):

Beyer Peacock Works Plate from D8613. Approximate size 12in. x 5in. (Alan Whincup)

D8607, 16C Derby, 18 October 1964. Brand new and, perhaps, part way through a test run from Manchester. The light catches the two works plates on the cab side very well. (Colour-Rail)

Additional plate indicating building under licence. Approximate size 9in x 3¾in. (Alan Whincup)

15.2 Rolls-Royce engine (D8586/7).
The Rolls-Royce engined pair were immediately identifiable by their raised bonnet roofs, and associated roof hatches. Several photographs of both D8586 and D8587 in this book illustrate their distinctive appearance.

15.3 Multiple Working.
D8500-87 (Classes 17/1 and 17/2) were built with Red Diamond multiple working equipment, and D8588-D8616 (Class 17/3) with Blue Star. These two systems were not compatible due to electrical differences between the sub-classes (for example, weak-field changes for the GEC and Crompton Parkinson

Detail Differences • 267

equipment occurred at differing speeds); however the Red Diamond to Blue Star conversion was a relatively straightforward process and was ultimately undertaken at both Depots and Works.

To standardise the fleet, most if not all of the D8500-87 fleet was converted to Blue Star under Modification MB37/5, although some early withdrawals may have gone to the scrap yard prior to such modification.

The September 1966 edition of the *Railway Observer* recorded D8565/D8609 and D8582/D8604 operating together in May and June 1966 respectively, although these may have been operating in tandem. However the December 1966 *Railway Observer* specifically reported that 'D8540/1/53 now have the 'Blue Star' coupling code instead of the previous 'Red Diamond'.

Appendix A to the Works & Equipment Committee Meeting minutes of 07 November 1967 entitled 'Technical Modifications to Traction and Rolling Stock Authorised by the Chief Engineer (Traction & Rolling Stock) under Delegated Powers during 6 Months Ended 30th September 1967' included the following *retrospective* item:

Item No.	Description	Estimated Cost (£)
56	Modify train line functions on 88 Clayton 900hp locomotives	1,936

The P&SP minutes of the Meeting held on 4 July 1967 included the comment that 'Scottish Region reported that four locos had been modified from Red Diamond to Blue Star at depots, in accordance with modification MB37/5'. This could be construed to mean that the earliest conversion work was undertaken in Main Works. The meeting held on 25 September 1969 reported that all conversion work had been completed.

15.4 Boiler Pipes.

All of the Clayton Equipment built locomotives (D8500-87) were delivered with steam pipes. However, as regards the D8588-D8616 fleet, the Beyer Peacock 'Maintenance

Unidentified Clayton, 51L Thornaby, 1964. Probably one of the D8588-91 batch allocated to Thornaby at this time. Blue-Star coupling code. Note the absence of both a steam pipe and a hole drilled in the buffer beam for associated 'through' pipe-work. (Rail-Online)

Instructions' specifically state that steam heating pipes were fitted to D8604 to D8616 only; D8588-99, D8600-3, deployed in the North-East of England, were built devoid of such through steam pipe-work.

In later life many of the fleet built with steam pipes were modified at least to the extent that all or part of the buffer beam equipment was removed (see photograph on page 269). Some locomotives may also have had the under floor pipe work removed completely during Classified works attention.

15.5 Tablet Catchers.
Tablet catcher recesses were provided on all 117 locomotives, ready for the fitment of automatic tablet exchange equipment. In practice no locomotives were ever fitted with such equipment.

15.6 Lamp Irons.
The Beyer Peacock locomotives carried slightly taller buffer beam lamp irons compared with their earlier Clayton-built counterparts.

15.7 Spark Guards.
The July 1966 *Railway Observer* reported that 'It is understood that deflector plates are being fitted to the bogies of the Claytons to prevent sparks from the brake blocks igniting adjacent dirt.' To put this comment in perspective, the P&SP Meeting Minutes for 4 July 1967 indicated that spark guards had already been fitted to the Beyer-Peacock (Class 17/3) locomotives and that 'similarly designed guards are to be fitted to Clayton type locos (*D8500-87*) as per modification MB36/45' authorised on 26 January 1967. Workshops were asked to fit the guards whenever locomotives were lifted. By the 19 September 1967 meeting it was reported that none of the D8500-87 fleet had yet been fitted but that the 'required material was almost complete and fitting would begin in the near future'.

15.8 Automatic Warning System (AWS).
The Diesel Locomotive Record Cards provide information, albeit incomplete, regarding the retro-fitting of AWS to locomotives in the range D8540 to D8585 (see Section 10) during the period 1964/65. In general, work was undertaken at St Rollox Works for Polmadie locomotives and at Inverurie Works for the Haymarket fleet.

Presumably locomotives in the range D8500-39 were also retro-fitted but no evidence has been found to support this proposition. The DLRCs are silent with respect to AWS fitment to D8586/7 and the Beyer Peacock locomotives.

15.9 Ergonomic Seats.
The minutes of the 19 September 1967 P&SP meeting reported that 'four locomotives had been fitted with one ergonomically designed seat each, this being under experiment MDL/1696'. The locos concerned were D8505/8/41/9.

15.11 Demisters.
Cab windscreen demister apertures were frequently blocked-off by drivers to prevent the ingress of exhaust gas via the traction motor blower fan which fed the demisters. Official trials were subsequently undertaken on Haymarket locomotives with proper blanking plates placed in the air ducts. No safety issues were experienced and all locomotives were subsequently modified; Glasgow Works was instructed to leave the blanking plates in place when locomotives passed through Works.

15.12 Snow Ploughs.
The Claytons were fitted with miniature snow ploughs from time to time. Examples are illustrated in Section 10 (see photographs of D8542/3/81, D8606-8).

15.13 Depot Plates.
Steam-age oval shed plates were fitted to most new locomotives by Polmadie, Thornaby and Gateshead on arrival, but not by Haymarket. Photographs of Barrow Hill locomotives show no sign of shed plates being fitted. When fitted, the plates were attached to the cab sides immediately below the running number.

Ultimately all shed plates were removed and replaced either by transfers or by direct painting. Removal of the plates left two 'tell-tale' bolts on the cab sides to indicate the previous fitment of plates.

A significant number of Claytons in the D8554-79 range exhibited these bolts suggesting the fitment of 66A plates on re-allocation from Haymarket to Polmadie.

Interestingly, several locomotives in the D8580-7 and D8606-16 range also carried the bolts (e.g. D8581/3/6, D8607/8/12/3/6). D8580-7 were always allocated to Haymarket and the D8606-16 batch spent most of the lives at Haymarket after departing Barrow Hill in May 1966. The bolts attached to the latter batch could be explained by the fact that a number of the locomotives were noted operating around the Kittybrewster/Aberdeen area during the May-September 1966 period, with at least D8616 noted carrying 61A plates (see Section 14.4).

The bolts on D8581/3/6 may provide evidence to suggest that at least some Haymarket locomotives carried 64B plates.

15.14 25kV Signs.

The Claytons received a standard complement of ten signs which appeared to remain unchanged throughout their lives. A few signs went 'missing in action' but nothing like the 'chopping and changing' seen on the Metrovick and NBL Type 2s!

The standard positions were:

- Two signs on the bonnet sides adjacent to the radiator grille and traction motor grilles each side (4).
- Two signs on each cab front below the windows (4).
- One sign on each exhaust cowling between the cab front windows (2).

15.15 Data Panels.

Classification data panels were applied to the cab sides of the Claytons from 1969. The details listed on the panels were as follows:

VB Class 17
Weight tons 68
Brake force tons 35
RA 4
Max speed mph 60

Unidentified Class 17/1, Coatbridge Central (south end of sation), 1965. (Doug Kirk)

Chapter 16
LIVERIES

15.1 General.
As with the subject of appearance design and locomotive styling (Section 6) there is little archive literature on the subject of the livery styling prior to fleet introduction. Material that is available is provided below.

Mr John Barnes (Design Consultant) of Messrs. Allen-Bowden Ltd. had his first meeting with Clayton Equipment on 30 June 1961. Barnes discussed some initial sketches with Mr G. Williams (BTC Design Officer) on 18 August 1961 and Mr E.S. Cox (Assistant Chief Mechanical Engineer) on 25 August.

Barnes' sketches included the recommendation for a light colour for the solebar to offset the overall green of the locomotive body. Cox was critical of the suggestion and Williams had suggested red as an alternative. In a letter to Williams dated 30 August 1961, Barnes also suggested the application of light green to the front of the cab and the exhaust cover.

On 1 September Cox visited Clayton Equipment and discussions covered, inter alia, appearance design and locomotive livery. A letter from Williams to Barnes dated 7 September highlighted the fact that Cox would not accept any change of colour at solebar level and that it had been decided that the whole body would be finished in Locomotive Green with the exception of the roof which would be charcoal grey.

During November, Williams asked Barnes for a painting schedule for the Clayton locomotives and on 15 November 1961 Barnes responded:

'Thank you for your letter of November 14 regarding the painting schedule for the above locomotives. I understand from your previous letter on the subject that we can only use standard Locomotive Green with a charcoal grey cab roof and therefore it seems that all I can do is to instruct Clayton Equipment Co. Ltd. on these lines . . .

'If you think that there is any possibility at all of introducing a third colour onto this locomotive I would be glad to have an opportunity to do so, although I think it is already agreed that the use of a third colour is unacceptable to the Commission.'

Williams responded to Barnes on 22 November 1961:

'I agree with you that an all Locomotive Green livery and charcoal grey is going to look a bit dismal. I believe on one of your sketches it appeared that you had used a lighter colour for the upper part of the driving cab. I cannot remember whether I discussed this with you or whether you said that this, on your sketch, was merely a highlight to illustrate the 'tumble home' of the bodyside at this point, but you may care to think of using a mid grey at this point if you think it would be worthwhile and if it could be effectively carried round the front without disturbing the shape of the windscreen, etc.

'In any case, if you have any recommendations for the use of a third colour, one or two alternative sketches, however rough, might be worthwhile having to discuss with Mr Cox . . .'

Barnes' reply to Williams dated 27 November 1961 was enthusiastic:

'I visited Clayton Equipment Ltd. on Friday last and discussed outstanding details with Mr Boast.

'I discussed the prospect of a third colour and they are keen to do this if we can get approval from Mr Cox so, at your suggestion, I am preparing a further rough coloured elevation for you to submit to Mr Cox and I feel sure that this will add considerably to the appearance of the locomotive. As soon as this is ready I will post it on to you, when no doubt you will let me know whether you have been successful with your plea.'

This was followed by another letter from Barnes on 4 December 1961:

'I now have pleasure in forwarding herewith a rough drawing showing a proposed colour scheme.

'It may be that the colours are not strictly accurate as we have had to do this from memory, but I would ask that if you do get approval from Mr Cox perhaps you will let me have the numbers or names of the standard railway colours so that I can definitely tie this up with Clayton Equipment Co. Ltd.'

Williams eventually responded to Barnes on 15 February 1962:

'I am sorry to have been so long in writing to you again about the livery of the above locomotives. I have now had an opportunity of discussing this with the Chief Mechanical Engineer and he has agreed the basic scheme outlined in your colour drawing.

'The main body of the locomotive will be British Railways Locomotive Green and the cab sides and front should be finished in Sherwood Green No.177 - 25321 in the manner indicated on your drawing.

'I shall be grateful if you will now prepare an outline side and front elevation as quickly as possible, indicating the lines of demarcation of these two colours

and any other detail you feel is necessary and to support it with a brief written instruction. I will then forward this to the Chief Mechanical Engineer for distribution to the Regional sponsoring office and Messrs. Clayton Equipment Co. Ltd.'

Barnes dutifully forwarded two drawings to Williams on 21 February 1962; Williams forwarded these drawings to Cox on 26 February.

Whether an 'official' painting schedule was ever produced is unknown; certainly the document was not found amongst available archive material. However, the schedule (without the number positioning and crest details) would have looked something like this:

British Transport Commission - General Instructions for Painting Main Line Diesel Locomotives.

Type and Make of Locomotive: 900hp Type1 Clayton Equipment Co.

Item	Colour and Reference
1. Body	Standard Locomotive Green, B.R. Spec.30.
Except cab sides (above crease) and cab fronts including exhaust stacks	Sherwood Green Spec.177-25321
2. Cab Roof	Mid-Grey B.S.S. 2660 9-101
3. Bogies	Black
4. Undergear	Black
5. Buffer casings and beams	Red B.S.S. 2660 0-005

All locomotives carried numbers which were prefixed by a serif-style D.

16.2 GSY

All 117 Claytons were delivered in two-tone GSY livery, with running numbers prefixed by a serif 'D'.

Buffer beam numbers were applied at Barrow Hill to D8604-16.

D8571, 66A Polmadie, 25 July 1969. GSY livery. Serif-D prefix. Withdrawn and already showing signs of cannibalization. Note the miniature snow-ploughs. (Dave Thorpe)

D8605, Location unknown, but presumably Scotland by this date, June 1966. Number on buffer beam, typical of the Barrow Hill fleet. '1' rolled-up in the third position of the headcode box; typical for the Scottish Region although '1' can't have been too common for the Claytons. (TOPticl Digital Memories)

16.3 BFY
16.3.1 Class 17/1 Blue Livery (D8500-85).
BFY carried by: D8500-3/7/10/20/2/3/5-7/9/32/4/5/8/40/2/3/5/50/6/7/64-7/70/3/4/7/80/2-5 (Total: 37).

D8583, Haddington, 18 September 1967. BFY livery. Asymmetric numbers and large emblems, with Red Diamond coupling codes retained at this point. (TOPticl Digital Memories)

Thirty-seven Class 17/1s received blue livery with full yellow ends and asymmetric numbers (and sans-serif 'D'). All repaints were undertaken as part of Works classified repairs; details are provided below:

	BFY	Relevant St Rollox Works details:
	Repaint Date	DLRC Works Details, supported where necessary with SRSS & P&SP Information and/or Sightings
D8500	06/67 (we24/06/67)	DLRC not seen. Released we24/06/67 after I repair. 28/05/67
D8501	11/67	DLRC not seen. Ex-works 14/11/67 presumably after I repair. 24/10/67, 04/11/67
D8502	c11/67	DLRC not seen. 24/10/67, 04/11/67
D8503	Date unknown (post 10/67)	DLRC not seen. Carlisle Kingmoor: 08/10/67 (still GSY), 66A Polmadie: 31/12/67 (livery unknown but, as an LMR loco, indicative of a move to/from St Rollox Works)
D8507	06/67 (we10/06/67)	02/05/67-09/06/67 I
D8510	11/67 (we18/11/67)	12/10/67-18/11/67 I
D8520	07/67 (we08/07/67)	25/05/67-07/07/67 I
D8522	11/67 (we04/11/67)	13/09/67-04/11/67 I
D8523	07/67 (we05/08/67)	14/06/67-31/07/67 I
D8525	01/68 (we06/01/68)	01/12/67-06/01/68 I
D8526	10/67 (we21/10/67)	04/09/67-21/10/67 I
D8527	12/67 (we09/12/67)	01/11/67-09/12/67 I
D8529	10/67 (we07/10/67)	18/08/67-07/10/67 C
D8532	01/68 (we20/01/68)	12/12/67-20/01/68 I
D8534	08/67 (we19/08/67)	19/06/67-19/08/67 I
D8535	12/67 (we16/12/67)	09/11/67-16/12/67 I
D8538	06/67 (we17/06/67)	15/05/67-17/06/67 I
D8540	12/67 (we30/12/67)	27/11/67-30/12/67 I
D8542	11/67 (we11/11/67)	29/09/67-11/11/67 I
D8543	10/67 (we14/10/67)	28/08/67-14/10/67 C
D8545	10/67 (we28/10/67)	13/09/67-27/10/67 I
D8550	12/67 (we09/12/67)	01/11/67-09/12/67 I
D8556	09/67 (we09/09/67)	10/07/67-08/09/67 I
D8557	09/67 (we16/09/67)	05/08/67-16/09/67 I
D8564	12/67 (we23/12/67)	17/11/67-23/12/67 I
D8565	10/67 (we14/10/67)	06/09/67-14/10/67 C
D8566	09/67 (we09/09/67)	02 or 09/08/67-09/09/67 I
D8567	06/67 (we17/06/67)	06/05/67-16/06/67 I
D8570	01/68 (we13/01/68)	08/12/67-13/01/68 I
D8573	07/67 (we01/07/67)	30/05/67-01/07/67 I
D8574	08/67 (we12/08/67)	17/06/67-12/08/67 C

	BFY Repaint Date	Relevant St Rollox Works details: DLRC Works Details, supported where necessary with SRSS & P&SP Information and/or Sightings
D8577	09/67 (we23/09/67)	11/08/67-23/09/67 I
D8580	11/67 (we25/11/67)	12/10/67-25/11/67 I
D8582	07/67 (we15/07/67)	07/06/67-14/07/67 I
D8583	08/67 (we26/08/67)	26/06/67-26/08/67 I
D8584	12/67 (we23/12/67)	15/11/67-16/12/67 I
D8585	12/67 (we09/12/67)	11/10/67-02/12/67 I

Notes:
1. Other Class 17/1s with known 1967 classified repairs (with release from Works in GSY):
   D8505/8/13/6/7/9/21/30/6/41/9/51/60/2/3/8/9/78/9 (all pre 31/05/67).
2. Last and First:
   Last GSY release: D8563 (27/05/67)
   First BFY release: D8507 (09/06/67)
   Last BFY release: D8532 (20/01/68)
3. Emblem Size:
   Small emblems were applied to the early blue repaints (up to end-July 1967); large emblems replaced the small emblems from August with D8574 being the first (St Rollox Works release date 12/08/67). Thus:
   Small emblems: D8500/7/20/3/38/67/73/82 (Total: 8)
   Large emblems: D8501-3/10/22/5-7/9/32/4/5/40/2/3/5/50/6/7/64-6/70/4/7/80/3-5 (Total: 29).
4. D8512/21 subsequently carried blue livery during Departmental service (see Section 22).

16.3.2 Class 17/2 Blue Livery (D8586/7).
No Class 17/2s carried blue livery.

16.3.3 Class 17/3 Blue Livery (D8588-99, D8600-16).
Only one Class 17/3 carried blue livery prior to final withdrawal.

	BFY Repaint Date	Relevant St Rollox Works details: DLRC Works Details
D8606	01/68 (we20/01/68)	18/12/67-20/01/68 I

Notes:
1. D8606 carried large emblems.
2. D8598 subsequently carried blue livery during Departmental service (see Section 22).

16.3.4 General Comments.
1. All blue livery applications were undertaken between June 1967 and January 1968.
2. Although numerous Classified repairs were carried out after January 1968, repainting was restricted to 'touch-up and varnish' of the green livery, plus application of full yellow ends.

16.4 GFY
GFY carried by: D8504/6/8/14/5/28/36/9/41/6/8/9/51/2/5/8/9/61-3/8/79/81/6/9/92/3/7/8, D8601/2/4/7/8/10/2-6, plus possibly D8594-6/9 (Total: 40, possibly 44)

Clayton classified repairs dried-up during 1968 due to impending withdrawal; however, an upturn

D8563 65A Eastfield, 28 May 1971. Ex-works after a Classified repair at St Rollox and awaiting transfer back to Haymarket. Such a late repair will have helped survival until end-December 1971, the end of 'squadron' service on BR. 'Touch-up and varnish' paint job with newly applied full yellow ends. (Colour-Rail)

in traffic in 1969/70 resulted in the resumption of repairs. This did not mean the renewed application of blue livery, however; instead all repairs were followed up by 'touch-up and varnish' treatment, with the only livery modification being the application of full yellow ends. Original style numbers were retained through to withdrawal.

Claytons known from sightings and/or photographic evidence to have received GFY livery are listed below, together with any post-May 1967 DLRC shopping details (excluding rectification work) and/or St Rollox Shopping Summaries. On the basis that all repaints were undertaken on Works due to Trades Union agreements, these works dates, where sufficient information is available, give a strong indication as to when the full yellow ends were applied.

	Relevant St Rollox Works details: DLRC Works Details (post-May 1967, Classified or long-duration (>1 week) Unclassified repairs), P&SP Information and Relevant sightings	**Likely date of acquiring GFY livery** (marked ***)
D8504	DLRC not seen. 10/01/71	**Insufficient information.**
D8506	**160270-140370 U**	*** Listed GFY (RO1270).
D8508	**310371-260671 C**	***
D8514	No Intermediate repair shown on DLRC. 07/01/68	**Insufficient information.**
D8515	No repairs shown on DLRC after 05/06/65. Ex-Works xx0268 C	*** Listed GFY (RO0968).
D8528	061167-021267 U 18/07/68 **300970-141170 C**	***
D8536	170867-010967 U 310768-170868 U **130370-170470 C**	*** Listed GFY (RO1270).
D8539	021067-211067 U **160170-070270 C**	*** Listed GFY (RO1270).
D8541	**281070-281170 C**	***
D8546	**300170-070370 C**	*** Photographed GFY 07/03/70 recently ex-works.
D8548	221267-130168 U **131069-311069 C/I**	*** Listed GFY (RO1270).
D8549	**061170-051270 C**	***
D8551	290367-190567 I **171069-061169 U** 19/06/71	Possible date for application.
D8552	110368-230368 U **250171-030371 U**	Possible date for application. Listed GFY (RO1270).
D8555	**011070-311070 I (?)** 19/05/71	***
D8558	240767-170867 U **201170-311270 I**	***
D8559	**281169-191269 C/I**	*** Photographed GFY 24/02/70.
D8561	04/10/70 **220171-170271 U**	Possible date for application.
D8562	080367-050567 I **211169-131269 C/I**	*** Listed GFY (RO1270).
D8563	190467-270567 I 010970-160970 U **040571-220571 C**	*** Photographed GFY 28/05/71 (ex-Works at 65A).
D8568	**230969-181069 C** 160270-040370 U	***

	Relevant St Rollox Works details: DLRC Works Details (post-May 1967, Classified or long-duration (>1 week) Unclassified repairs), P&SP Information and Relevant sightings	**Likely date of acquiring GFY livery** (marked ***)
D8579	300367-260567 I	
	020667-160667 U	
	230470-230570 C	*** Noted GFY at 65A ex-works in 06/70 (RO0870).
D8581	250368-060468 U	
	291069-151169 C	*** Listed GFY (RO1270).
D8586	**080868-290369 C**	*** Listed GFY (RO1270).
	081269-221269 U	
D8589	DLRC not seen.	
	24/08/67	
	11/05/68, 25/05/68	*** Photographed GFY 16/06/68 ('freshly applied ').
D8592	160469-090569 U	
	040869-130869 U	
	xxxxxx-290870 C	***
D8593	**xxxxxx- 100868 C**	*** Photographed GFY 05/03/69.
D8597	**171269-140270 C/I**	*** Photographed GFY 04/05/70.
D8598	**xxxxxx-101070 C**	***
D8601	xxxxxx-060668 U	
	xxxxxx-141170 C	*** Listed as GFY (RO1270).
D8602	171069-131169 U	
	xxxxxx-250470 C	*** Photographed GFY 17/05/70 (recently ex-works).
D8604	xxxxxx-081267 U	
	xxxxxx-140968 C	***
	xxxxxx-091068 U	
D8607	**221267-030268 I**	*** Listed as GFY (RO1268).
	220469-020569 U	
	200270-210370 U	
	301170-301270 U	
D8608	**060268-160368 I**	***
D8610	**070268-160368 I**	***
D8612	**050968-280968 C**	*** Photographed GFY 04/01/69.
D8613	250867-150967 U	
	290967-131067 U	
	160468-180568 C	*** Photographed GFY 20/05/68 (ex-works).
D8614	**040168-100268 I**	***
	150969-041069 U	Listed GFY (RO1270).
D8615	**240668-130768 I**	***
	200669-270669 U	
	020270-210270 U	
D8616	**160268-300368 I**	*** Photographed GFY 02/05/68.

In addition the following Claytons *may* also have carried GFY livery:-

	Relevant St Rollox Works details: DLRC Works Details (post-May 1967, Classified or long-duration (>1 week) Unclassified repairs), P&SP Information and Relevant sightings	**Likely date of acquiring GFY livery** (marked ***)
D8594	211166-231166 U	Photographed GSY 16/06/68
	xxxxxx-191270 C	***
D8595	xxxxxx-040568 C	***
	01/08/68, xx/08/68	
D8596	xxxxxx-25 or 281167 ?	**Insufficient evidence.**
D8599	xxxxxx-300171 C	***
	xxxxxx-120271 U	

16.5 GSY at Withdrawal.

GSY was carried by at withdrawal: D8505/9/11-3/6-9/21/4/30/1/3/7/44/7/53/4/60/9/71/2/5/6/8/87/8/90/1, D8600/3/5/9/11 (Total: 35)

Given the 38 BFY and 44 GFY Claytons listed above (assuming that D8594-6/9 did indeed receive GFY), then this leaves 35 of the fleet which potentially carried GSY throughout their lives.

Photographic evidence of twenty-one of the thirty-five proves GSY at withdrawal: D8505/11-3/6/9/21/30/1/3/60/71/2/6/8/87/8, D8600/3/5/11.

Whilst the DLRC cards for D8544/54/90 have not been seen, for the remainder there are no post-May 1967 classified repair dates shown on the DLRCs (i.e. D8509/17/8/24/37/47/53/69/75/91, D8609), supporting the belief that the GSY was retained throughout.

It should be noted that D8537 was withdrawn prior to the full yellow-end edict being issued for green locomotives.

16.6 Summary: Liveries Carried and D-Prefix Position at Withdrawal.

Loco. No.	Liveries Carried			D or No-D at Withdrawal	Comments
D8500	GSY		BFY	?	D8500 on 05/09/70.
D8501	GSY		BFY	D	
D8502	GSY		BFY	D	
D8503	GSY		BFY	D	
D8504	GSY	GFY		No-D	
D8505	GSY			D	
D8506	GSY	GFY		No-D	
D8507	GSY		BFY	No-D	
D8508	GSY	GFY		No-D	See Note (1).
D8509	GSY			D	
D8510	GSY		BFY	D	
D8511	GSY			D	
D8512	GSY			D	BFY (8512) in Departmental service.

Loco. No.	Liveries Carried			D or No-D at Withdrawal	Comments
D8513	GSY			D	
D8514	GSY	GFY		D	
D8515	GSY	GFY		D	
D8516	GSY			D	
D8517	GSY			D	
D8518	GSY			D	
D8519	GSY			D	
D8520	GSY		BFY	D	
D8521	GSY			D	BNY (S18521) in Departmental service.
D8522	GSY		BFY	D	
D8523	GSY		BFY	D	
D8524	GSY			D	
D8525	GSY		BFY	D	
D8526	GSY		BFY	D	
D8527	GSY		BFY	D	
D8528	GSY	GFY		No-D	
D8529	GSY		BFY	D	
D8530	GSY			D	
D8531	GSY			D	
D8532	GSY		BFY	D	
D8533	GSY			D	
D8534	GSY		BFY	D	
D8535	GSY		BFY	D	
D8536	GSY	GFY		No-D	
D8537	GSY			D	
D8538	GSY		BFY	No-D	
D8539	GSY	GFY		No-D	
D8540	GSY		BFY	No-D	
D8541	GSY	GFY		No-D	
D8542	GSY		BFY	D	
D8543	GSY		BFY	?	D8543 on 11/08/70
D8544	GSY			D	
D8545	GSY		BFY	No-D	
D8546	GSY	GFY		No-D	
D8547	GSY			?	Early withdrawal (02/69).
D8548	GSY	GFY		D	
D8549	GSY	GFY		No-D	
D8550	GSY		BFY	D	
D8551	GSY	GFY		No-D	

Loco. No.	Liveries Carried			D or No-D at Withdrawal	Comments
D8552	GSY	GFY		No-D	
D8553	GSY			?	Early withdrawal (10/68).
D8554	GSY			D	
D8555	GSY	GFY		No-D	
D8556	GSY		BFY	?	Early withdrawal (02/69).
D8557	GSY		BFY	D	
D8558	GSY	GFY		No-D	
D8559	GSY	GFY		D	
D8560	GSY			D	
D8561	GSY	GFY		No-D	
D8562	GSY	GFY		No-D	
D8563	GSY	GFY		No-D	
D8564	GSY		BFY	D	
D8565	GSY		BFY	No-D	
D8566	GSY		BFY	?	Early withdrawal (12/68).
D8567	GSY		BFY	No-D	
D8568	GSY	GFY		D	
D8569	GSY			D	
D8570	GSY		BFY	D	
D8571	GSY			D	
D8572	GSY			D	
D8573	GSY		BFY	D	
D8574	GSY		BFY	D	
D8575	GSY			?	Early withdrawal (10/68).
D8576	GSY			D	
D8577	GSY		BFY	D	
D8578	GSY			D	
D8579	GSY	GFY		No-D	
D8580	GSY		BFY	D	
D8581	GSY	GFY		No-D	
D8582	GSY		BFY	?	Early withdrawal (01/69).
D8583	GSY		BFY	D	
D8584	GSY		BFY	?	Early withdrawal (10/68).
D8585	GSY		BFY	?	Early withdrawal (10/68).
D8586	GSY	GFY		D	
D8587	GSY			D	
D8588	GSY			D	
D8589	GSY	GFY		?	

Loco. No.	Liveries Carried		D or No-D at Withdrawal	Comments
D8590	GSY		?	D8590 on 30/09/70.
D8591	GSY		?	Early withdrawal (10/68).
D8592	GSY	GFY	No-D	
D8593	GSY	GFY	D	
D8594	GSY	GFY?	?	
D8595	GSY	GFY?	D	
D8596	GSY	GFY?	D	
D8597	GSY	GFY	No-D	
D8598	GSY	GFY	No-D	BFY (8598) in Departmental Service.
D8599	GSY	GFY?	No-D?	
D8600	GSY		D	
D8601	GSY	GFY	No-D	
D8602	GSY	GFY	No-D	
D8603	GSY		D	
D8604	GSY	GFY	D	
D8605	GSY		D	
D8606	GSY	BFY	D	
D8607	GSY	GFY	D	
D8608	GSY	GFY	D	
D8609	GSY		?	Early withdrawal (10/68).
D8610	GSY	GFY	D	
D8611	GSY		D	
D8612	GSY	GFY	D	
D8613	GSY	GFY	D	
D8614	GSY	GFY	D	
D8615	GSY	GFY	D	
D8616	GSY	GFY	D	

Notes:
1. D8508 cab ends were painted in dark Locomotive Green by withdrawal, leaving only the side window area in Sherwood Green.
2. All BFY repaints included D-prefixed numbers at the time of repaint.

Chapter 17
STORAGE AND WITHDRAWAL

17.1 Storage 1968-69: Internal BR Reporting.
Over the critical period May 1968 to October 1969, the Scottish Region circulated weekly memoranda regarding storage of their motive power stock; these documents were for internal consumption only and were never released into the public domain. As a consequence these stock changes were never published in contemporary railway magazines or society journals.

Week Ending	Cumulative Total	Stored (serviceable (s) / unserviceable (u))	"Drawn from Store" (DfS)
04/05/1968	0		
11/05/1968	1	D8537 S(u) 05/05/68 (Surplus/Engine Defect))	
18/05/1968	1		
.....			
20/07/1968	1		
27/07/1968	0		D8537 DfS 21/07/68 (condemned 21/07/68)
03/08/1968	0		
.....			
28/09/1968	0		
5/10/1968	1	D8553 S(u) 05/10/68 (Surplus)	
12/10/1968	1		
19/10/1968	2	D8575 S(u) 13/10/68 (Surplus)	
26/10/1968	4	D8584 S(u) 23/10/68 (Surplus), D8585 S(u) 23/10/68 (Collision damage)	
02/11/1968	0		D8553/75/84/5 DfS 27/10/68 (condemned: D8553/75 on 27/10/68, D8584/5 on 29/10/68)
09/11/1968	0		
16/11/1968	1	D8570 stored (u) 10/11/68 (Collision damage)	
23/11/1968	0		D8570 DfS 23/11/68 (condemned 24/11/68)
30/11/1968	0		
.....			
11/01/1969	4	D8560 S(u) 07/01/69 (Condition), D8544/54/71 S(s) 05/01/69 (Surplus)	

Week Ending	Cumulative Total	Stored (serviceable (s) / unserviceable (u))	"Drawn from Store" (DfS)
18/01/1969	5	D8547 S(u) 12/01/69 (Surplus)	
25/01/1969	2		D8544/54/71 DfS 19/01/69 (all reinstated - Traffic requirements)
01/02/1969	2		
08/02/1969	5	D8556/64/77 S(u) 02/02/69 (Condition)	
15/02/1969	0		D8547/56/60/4/77 DfS 09/02/69 (all condemned 16/02/69)
22/02/1969	0		
01/03/1969	1	D8572 S(u) 23/02/69 (Condition)	
08/03/1969	1	D8571 S(u) 02/03/69 (Condition)	D8572 DfS 02/03/69 (reinstated - Traffic requirements)
15/03/1969	2	D8554 S(u) 02/03/69 (Condition) (late entry)	
22/03/1969	2		
29/03/1969	3	D8572 S(u) 23/03/69 (Condition)	
05/04/1969	3		
12/04/1969	3		
19/04/1969	5	D8562/78 S(u) 13/04/69 (Condition)	
26/04/1969	5		
03/05/1969	5		
10/05/1969	5		
17/05/1969	6	D8581 S(u) 11/05/69 (Collision damage)	
24/05/1969	7	D8551 S(u) 18/05/69 (Condition)	
31/05/1969	7		
07/06/1969	3		D8554/71/2/8 DfS 18/05/69 (late entry) (all condemned 18/05/69)
14/06/1969	3		
.....			
05/07/1969	3		
12/07/1969	2		D8562 DfS 06/07/69 (reinstated - Traffic requirements)
19/07/1969	2		
.....			
04/10/1969	2		
11/10/1969	1		D8551 DfS 05/10/69 (reinstated - Repair)
18/10/1969	1		
25/10/1969	0		D8581 DfS 19/10/69 (reinstated - Repair)
01/11/1969	0		

Notes:
1. Many of the storage dates (and indeed subsequent 'drawn from store' dates) quoted were the Sunday of the relevant week-ending (Sunday-Saturday), presumably reflecting accepted reporting norms.
2. This listing does NOT include those locomotives which moved direct from an operational allocation (i.e. Polmadie or Haymarket) to withdrawal from traffic.
3. D8551/62/81 were reinstated as part of the process to provide sufficient resources to meet traffic requirements envisaged from late-1969.

17.2 Storage 1971: Official BR Reporting.

D8506/31/43/83/6 were the only Claytons externally reported as stored, the first four in July 1971 and D8586 in August 1971.

17.3 Withdrawal Rationale: National Traction Plans.

The 1967 and 1968 National Traction Plan requirements for the Claytons were as follows:-

1967 National Traction Plan - Type 1 Requirements.

	Stock @ 31/07/67	Forecast Year-ending Stock Requirement							
		1967	1968	1969	1970	1971	1972	1973	1974
Total	444	455	406	326	289	249	208	147	81
ScR	168	154	139	111	100	90	77	57	45
ER	164	166	108	105	101	93	87	68	36
Class 17	117	117	117	88	61	21	0	0	0
Class 20	228	228	228	228	228	228	208	147	81

1968 National Traction Plan - Type 1 Requirements.

	Stock @ 05/10/68	Forecast Year-ending Stock Requirement						
		1968	1969	1970	1971	1972	1973	1974
Total	390	326	283	263	253	240	220	196
ScR	134	126	112	100	90	77	57	45
ER	100	91	56	48	48	48	48	36
Class 17 (ScR)	63	55	47	35	21	0	0	0
Class 17 (LMR)	35	0	0	0	0	0	0	0
Class 17 (ER)	18	18	8	0	0	0	0	0
Class 17 (Total)	116	73	55	35	21	0	0	0
Class 20	228	228	228	228	228	228	220	196

Notes:
1. The 1967 NTP assumed a 90% availability level and a utilisation of 12hr per day (hauling trains or shunting).
2. The 1968 Plan included some Class 17 adjustments as follows:
   - Availability: Reduced from 90 per cent to 80 per cent.
   - Utilisation: Increased from 12 to 12.5hrs hauling trains or shunting per day.

The actual Class 17 stock at the end of December 1968 was seventy-one (fifty-seven ScR, fourteen NER), reducing to sixty-four (fifty ScR, fourteen NER) at the end of February 1969. Five locomotives were reinstated on 18/05/69 (D8500/4/5/8/13) ostensibly as replacements for five withdrawals.

Minutes of the 'Performance &Service Problems of Clayton Type 1 Locomotives' Group meeting recorded the following information:-

Min.103, Meeting 25 September 1969:-
'The Chairman referred to the recent changes in the National Traction Plan which meant that the previously agreed NTP forecasts were unlikely to

be achieved. This came about by subsequent significant traffic levels and a need to improve the level of movement resources in order to meet the requirements of the Freight Plan in the future. So far as Class 17 locomotives are concerned, this means that 60 will be retained until the end of 1969, reducing to 50 by the end of 1971 and thereafter a gradual withdrawal . . .

'To meet the foregoing, Scottish Region had arranged for 6 locomotives to be given Intermediate repairs before the end of 1969 and 14 Intermediate repairs have been included in the 1970 budget. Owing to certain Regional plans, the Scottish Region fleet was to be temporarily increased in the near future by taking locomotives out of store on the LMR and putting these back into service.

'Regarding the Eastern Region locomotives, their representative stated that there would be 14 retained in service for an indefinite period and 6 would be given Intermediate repair in 1970, probably followed by 4 more in 1971.

'It should be noted that, although intermediate repairs are proposed for the locomotives, this classification does not apply to the power unit which will receive a 'general' in each case.'

Min.112, Meeting 6 November 1969:-
'Eight locomotives, Nos. 8503, 8506, 8507, 8510, 8516, 8528, 8529 and 8531, stored unserviceable by LM Region have been transferred to Scottish Region; these are now . . .being repaired and placed in service at Scottish Region depots.'

The reinstatement of D8530 in October 1969 was not mentioned. By the end of 1969 the actual Clayton fleet stock increased to 73 (59 ScR, 14 NER).

Min.122, Meeting 21 January 1970:-
'The Chairman . . .stated that proposals had been made by Scottish Region, not yet agreed by BRB, to increase the Intermediate repairs in 1970 budget up from the agreed 14 to 22, with a further 22 to be done in 1971 . . .These repairs are intended to ensure that there would be sufficient Class 17 locomotives to meet requirements in the period up to 1974.

'Eastern Region representatives confirmed that 6 locomotives are budgeted for scheduled repair in 1970 and a further 8 are planned for 1971.'

Min.131, Meeting 24 March 1970:-
'Eastern Region reported locomotive 8589 had been in a collision and was at present stored unserviceable waiting decision.'

The withdrawal of D8589 in July 1970 reduced the fleet to seventy-two (59 ScR, 13 NER).

Min.159, Meeting 13 January 1971:-
'As things stand, 11 Scottish Region locomotives are scheduled for classified repair this year . . .'

In March 1971 D8500/10/30/5, D8606 were withdrawn reducing the Scottish complement to 54.

Min.165, Meeting 21 April 1971:-
'The total Scottish Region complement will in due course rise from 54 to 62 and these locomotives will require to be maintained for some considerable time - at least until 1974 . . .

'Five locomotives of this class . . ...will remain in Eastern (*Region*).'

As late as April 1971, at least as far as the Chairman of the P&SP group was concerned, the Clayton fleet was seen to continue into 1974.

Minute 165 indicated the transfer of eight of the remaining Eastern Region Claytons to Scotland, leaving a nominal five in the North-East of England. However, the meeting had apparently missed the withdrawal of D8590 in March 1971, which rendered the mathematics incorrect! In the event all of the remaining twelve North-Eastern Claytons were transferred to Scotland during the April-June 1971 period giving a total Scottish fleet size of sixty-six.

An item in the 'Home News' section of the September 1971 edition of *Modern Railways* indicated 'BR to scrap 225 locomotives' which, given publication lead-times, suggests a decision taken in June/July. 'Reduction in freight traffic and the expense of maintaining the locomotives concerned have led BR to withdraw five electric and 220 now non-standard diesel locomotives' were given as the reasons and included the remaining sixty-six Claytons.

Min.172, Meeting: 7 July 1971:-
'Since the last meeting, there has been a complete withdrawal of this class of locomotive from Eastern Region and 12 Beyer Peacock locomotives have been transferred to Scottish Region.

'The Chairman indicated that there is a proposal to withdraw 8 class 17 locomotives from Scottish Region and the 8 locomotives would be those which are due to go through works for scheduled repair in 1971. The 8 locomotives in question are Nos. 8505, 8543, 8557, 8516, 8542, 8529, 8531, 8550.

'BRB, London, representative intimated that there are plans afoot which would place an embargo on the purchase of major spares for these locomotives, with the possibility that no more scheduled repairs would be carried out.'

As it turned out, no Claytons were withdrawn between the July and September P&SP meetings, although D8506/31/43/83/6 were placed in store.

Min.180, Meeting 8 September 1971:-
'The Chairman intimated that following an exercise relative to Locomotive Resources, a number of locomotives were to be dispensed with and so far as the Scottish Region was concerned the whole of the Class 17 fleet was to be eliminated. This would take place in two stages. The first stage by 31 October, 1971, by which time 56 would be withdrawn and the second stage by 31 January, 1972 by which time the remaining 10 would be taken out of service.

'It was stated that the 10 locomotives which would be retained up to 31 January, 1972 are 8504/07/08/38/40/51/52/61/63/65.'

In the event the batch of ten locomotives which benefitted from a short extension in traffic were initially D8504/7/8/36/48/52/8/61/3/5 although D8504/61/5 were subsequently withdrawn and substituted by D8529/74/98 before final withdrawal on 31 December 1971. At this point, apart from the RCD locomotives, Class 17 became operationally extinct.

17.4 Withdrawal Dates: Chronological Listing.

Summary of Withdrawals and Re-instatements:

Summary of Withdrawals and Re-instatements:

Date	Locomotives Withdrawn/Re-instated	Stock Change	Remaining Stock — Class 17 Total	Remaining Stock — Classes 17/1+17/2	Remaining Stock — Class 17/3
Position end-06/68			117	88	29
1we 27/07/68 (21/07/68)	D8537 (ex-store)	-1	116	87	29
1we 12/10/68 (12/10/68)	D8500/1/4-9/11-3/6-24/6/8-31/3/4	-27	89	60	29
1we 19/10/68 (13/10/68)	D8609/11	-2	87	60	27
1we 02/11/68 (27/10/68)	D8553/75 (both ex-store)	-2	85	58	27
4we 16/11/68 (28/10/68)	D8605	-1	84	58	26
1we 02/11/68 (29/10/68)	D8584/5 (both ex-store)	-2	82	56	26
1we 02/11/68 (02/11/68)	D8503/14 (both ex-store)	0	82	56	26
1we 02/11/68 (02/11/68)	D8512/29 (initially re-instated to store, reinstated to traffic 04/11/68)				
1we 30/11/68 (24/11/68)	D8570 (ex-store)	-1	81	55	26
1we 07/12/68 (04/12/68)	D8566/9	-2	79	53	26
2we 28/12/68 (22/12/68)	D8591/5/6	-3	76	53	23
2we 28/12/68 (28/12/68)	D8510/2/27/9/32	-5	71	48	23

Date	Locomotives Withdrawn/Re-instated	Stock Change	Remaining Stock		
			Class 17 Total	Classes 17/1+17/2	Class 17/3
1we 11/01/69 (05/01/69)	D8582	-1	70	47	23
1we 22/02/69 (16/02/69)	D8547/56/60/4/77 (all ex-store)	-5	65	42	23
1we 01/03/69 (28/02/69)	D8576	-1	64	41	23
1we 07/06/69 (18/05/69) (late entry)	D8544, plus D8554/71/2/8 (all ex-store)	0	64	41	23
1we 24/05/69 (18/05/69)	D8500/4/5/8/13 (re-instated)				
1we 11/10/69 (05/10/69)	D8530 (re-instated)	+1	65	42	23
1we 08/11/69 (02/11/69)	D8503/6/7/10/6/28/9/31 (re-instated)	+8	73	50	23
1we 11/07/70 (06/07/70)	D8589	-1	72	50	22
5we 06/03/71 (03/03/71)	D8530	-1	71	49	22
4we 27/03/71 (27/03/71)	D8590	-5	66	46	20
4we 03/04/71 (27/03/71)	D8500/10/35, D8606				
4we 09/10/71 (21/09/71)	D8506/31/43/83/6 (all ex-store) D8513/49/55/7/92/4, D8616	-12	54	37	17
4we 09/10/71 (04/10/71)	D8579/88/93, D8604	-4	50	36	14
4we 09/10/71 (05/10/71)	D8502/15/28/40/1/50/1/99, D8607/8/10/2/4/5	-14	36	29	7
4we 09/10/71 (06/10/71)	D8503/5/16/25/9/38/9/42/5/6/59/62/7/8/73/4 D8580/1/7/97/8, D8600-3/13	-26	10	10	0
4we 27/11/71 (07/11/71)	D8504/61/5	0	10	9	1
4we 27/11/71 (07/11/71)	D8529/74/98 (re-instated)				
4we 01/01/72 (31/12/71)	D8507/8/29/36/48/52/8/63/74/98	-10	0	0	0

Notes:

1. An internal memorandum from the Scottish Region General Manager to the BRB Chief Operating Officer entitled 'National Traction Plan' and dated 25 April 1968 stated (with reference to the Clayton fleet):

'Confirming telephone conversation between our representatives, Flint/Hamilton, the following is a preview of the proposed transfers to apply between 4th and 6th May, 1968, as far as the Scottish Region is concerned.

8533, 8534	To Carlisle Kingmoor	4.5.68
8535, 8536, 8537	Surplus to Requirements	4.5.68

'Will you please say if you require the Claytons....shown surplus to our requirements.'

D8533 and D8534 were the last of thirty-five Claytons transferred to Carlisle (LMR). There were clearly a further three Claytons surplus to the Scottish Region's requirement and the next three locomotives numerically (D8535-7) were offered up for further use or withdrawal. In the event all three were retained on the Scottish Region with subsequent surplus locomotives withdrawn on the basis of condition rather than numerical simplicity.
2. Locomotives stored, then subsequently reinstated, in 1968/69 are excluded from the above analysis.
3. The table excludes the reinstatement and immediate withdrawal of D8506/30 in May 1969.
4. D8592 was possibly also withdrawn from stored status as implied by P&SP Meeting (08/09/71).

5. Excluding withdrawals from Departmental stock, D8500/3-5/7/8/10/2/3/6/28/31/74/98 were withdrawn twice, D8506/30 three times (although both locomotives were reinstated and immediately withdrawn on 18/05/69), and D8529 four times.
6. The December 1971 *Railway Observer* reported that D8587 and D8613 were the last locomotives of Classes 17/2 and 17/3 respectively to be withdrawn from traffic in October 1971, although D8598 was subsequently reinstated. When D8598 was withdrawn in December, this locomotive acquired the honour of being the last Class 17/3 withdrawn.
7. According to the March 1972 edition of *Railway Observer*: 'The last Clayton to be withdrawn was 8574 on 31st December and it was used on the last regular duty a few days before this - working from Rutherglen p.w. depot.'
8. The locomotives with the shortest lives were:-

Clayton-built: D8585 with 4 years, 5½ months (collision damage).
Beyer Peacock built: D8611 with just less than 3 years, 10 months (due first Intermediate repair).

9. The longest continuous lifespan (excluding the 'Parkhead stoppage') was D8502 from 6 October 1962 to 5 October 1971 - virtually exactly nine years! However, in terms of total elapsed time, the lifespans of D8507/8 were longer at 9 years and 2 months because of their December 1971 withdrawal; however both spent considerable amounts of time out of traffic at Carlisle in 1968/69.

17.5 Withdrawal Administration.
17.5.1 Financial Protocol.
BR auditing procedures and financial protocols required formal authorisation of locomotive withdrawals; much of this was undertaken retrospectively but facilitated 'keeping the books in order'. Some of the documents covering the 1968-71 Clayton withdrawals are reproduced below in abridged form.

17.5.2 1968 Withdrawals.

Memorandum to Investment Committee, 7 August 1969
National Traction Plan: Locomotive Condemnations - 1968.

'The purpose of the submission is to seek formal approval to the condemnation of the locomotives withdrawn from stock up to the end of 1968, not covered by earlier authorisation, and which conform to the proposals contained in the National Traction Plan.

'The locomotives have been, or will be, handed over to the Supplies Manager for disposal after recovery of the serviceable parts.

'It is recommended that formal authority be given for the withdrawal of 158 diesel train locomotives, 251 diesel shunting locomotives and 12 electric locomotives (*total 421*) at a total scrap value of £0.5m.'

Signed: Chief Operating Officer, Chief Finance Officer, Chief Engineer (T&RS), Chief Officer (New Works).

The 158 diesel locomotive withdrawals included 46 Claytons (40 Class 17/1s and 6 Class 17/3s); financial details of the Clayton withdrawals as appended to the memorandum were:

	Original Cost	Written-down Book Value	Residual scrap value
D8500/1/3-14/6-24/6-34 D8537/53/66/9/70/5/84/5	£2,302,864	£1,693,023	£50,500 (*39x£1,250*) plus £1,750 for D8570)
D8591/5/6, D8605/9/11	£ 366,533	£ 293,226	£ 7,808 (*Av. c£1,300*)

Such was the lateness of this documentation relative to the actual withdrawal dates of the locomotives concerned that adjustments could be made allowing for the reinstatement of 14 Claytons during 1969 (i.e. D8500/3/4-8/10/3/6/28-31), plus Western Region diesel-hydraulic D6330, together with the transfer of two Claytons at scrap value to the Research Department, Derby (D8512/21). As a consequence a new memorandum was issued in February 1970 reducing the number of withdrawals requiring authorisation to 404:

Memorandum to Investment Committee, 3 February 1970.
National Traction Plan: Locomotive Condemnations - 1968.
'The purpose of this memorandum is to seek formal approval to the condemnation of the locomotives withdrawn from stock up to the end of 1968, not covered by earlier authorisation . . .

'It is recommended that formal authority be given for the withdrawal of 404 locomotives at a total scrap value of £0.5m., and approval to the transfer of 2 diesel train locomotives to Departmental stock . . .'
Signed: Executive Director (Systems & Operations), Executive Director (Finance).

Authorisation by the Investment Committee followed on 17 February 1970 (Min.3412, Item 5).

17.5.3 1969 Withdrawals.

Memorandum to Investment Committee, August 1970.
National Traction Plan: Locomotive Condemnations - 1969.
'The purpose of this memorandum is to seek formal authorisation to the condemnation of the locomotives withdrawn from stock during 1969, not covered by earlier authorisation and which conform to the proposals outlined in the National Traction Plan dated December, 1968 . . .

'The number of locomotives withdrawn during 1969 is 158 [46 diesel locomotives, 111 diesel shunting locomotives and 1 electric locomotive] . . .

'It is recommended that formal authority be given to the withdrawal of 158 locomotives during 1969 at a total scrap value of £0.2m.'
Signed: Executive Director (Systems & Operations), Executive Director (Finance), Chief Officer (New Works).

The forty-six diesel locomotives withdrawn in 1969 included twelve Claytons, all Class 17/1s, as follows:

	Original Cost	Written-Down Book Value	Residual scrap value
D8544/7/54/6/60/4/71/2/6/7/8/82	£690,864	£538,872	£21,425 (*11x£1,750, plus £2,175 for D8572*)

17.5.4 1970 Withdrawals.

Memorandum to Chief Executive (Railways), 26 April 1971.
Locomotive Condemnations: 1970.
'This memorandum seeks formal authority for condemnation of locomotives withdrawn from stock during 1970 . . .

'It is recommended that formal authority be given to the withdrawal of 64 locomotives during 1970 (4 *diesel locomotives, 55 diesel shunting locomotives and 5 electric locomotives*) at a total scrap value of £0.07m.'
Signed: Executive Director (Systems & Operations), Executive Director (Finance), Chief Officer (New Works).

The 1970 withdrawals included one Clayton locomotive (Class 17/3):

	Original Cost	Written-Down Book Value	Residual scrap value
D8589	£ 56,088	£ 42,627	£ 185

17.5.5 1971 Withdrawals.

Memorandum to Chief Executive (Railways), August 1972.
Locomotive Condemnations: 1971.
'This memorandum seeks formal authority for the condemnation of locomotives withdrawn from stock during 1971 which arose principally from the decision taken early in the year to reduce freight service provision. The reduction in the locomotive fleet was aimed principally at the elimination of the diesel-hydraulic locomotives and other less reliable classes, thereby achieving considerable cuts in depot maintenance and workshop costs.

'It is recommended that formal authority be given to the withdrawal of 362 locomotives during

1971 [5 electric, 247 diesel and 110 diesel shunting locomotives]....' Signed: Executive Director (Systems & Operations), Executive Director (Finance). This submission included 72 Claytons (48 Class 17/1s, 2 Class 17/2s and 22 Class 17/3s):

	Original Cost	Written-Down Book Value	Residual scrap value
D8500/2-8/10/3/5/6/25/8-31/5/6/8-43/5/6 D8548-52/5/7-9/61-3/5/7/8/73/4/9-81/3 D8586/7 D8588/90/2-4/7-9, D8600-4/6-8/10/2-6	£3,336,323	£2,357,907	£88,128 *(Av. £1,224)*

Memorandum to Investment Committee, 18 September 1972.
<u>Locomotive Condemnations: 1971.</u>
1. The attached submission seeks formal authority for the condemnation of 362 locomotives [5 electric, 247 diesel and 110 diesel shunting locomotives] withdrawn from stock in 1971 . . .
   Signed: Chief Executive (Railways).

D8531, D8516 and D8580, St Rollox Works, 26 September 1974. (Anthony Sayer)

Chapter 18
STORAGE LOCATIONS: DEPOTS AND YARDS

18.1 12A Carlisle Kingmoor (Steam)
Selected sightings:

27/07/68	D8508	
08/09/68	D8508	
09/10/68	D8505/8/12/21/3/6/34	
24/10/68	D8500/3/4/6-9/11/4/6-24/8/30/1/3/4	
07/12/68	D8500/1/3/4/6/7/9/11/3/4/6-24/6/8/30/1/4	D8508/33 not listed. D8501/13/26 added.
31/12/68	D8500/1/3-9/11/3/4/6-24/6/30/1/3/4	D8528 not listed. D8505 added.
03/01/69	D8500/1/3-9/11/3/4/6-24/6/8/30/1/3/4	
11/01/69	D8500/1/3-9/11/3/4/6/8-24/6-8/30/2-4	Two separate listings!
11/01/69	D8500/1/3-9/11/3/4/6-24/6/8/30/1/3/4	

Unidentified Clayton and D5719, 12A Carlisle Kingmoor (Steam), 4 January 1969. The withdrawn locomotives seen at the Steam Depot on this date (in order recorded) were: D8529/32/10/12/27/07/31/13/21/05/20/26/34/04/23, D5707, D8511/22, D5719, D8508, D5716, D8533, D5702, D8501, D5712, D8509, D5708, D8517, D5717, D8500/24, D5714/01, D8514/06/16, D5711/06, D8530/28/03/18.
(John Grey Turner)

Two magazine extracts are pertinent:

'Since the closure of Kingmoor and Upperby MPD's on 1st January 1968, diesel locomotives are stabled at various points within the locality. The steam shed at Kingmoor, despite the loss of connections with the main and goods lines at Etterby last February, still contains stored and condemned locomotives, access being gained through the ground frames on the up through siding adjacent to the shed. 44767, 44932 and 45262 remain amongst numerous Clayton Class 17's.' (*Railway Observer*, February 1969)

'Demolition of the former Kingmoor Steam shed appears to have now commenced (*circa April 1969*) and all of the withdrawn Clayton and Met-Vick diesels, which were stored there have now been moved to a point in the northern end of Carlisle New Yard.' (*Railway Locomotives*, June 1969)

18.2 12A Carlisle Kingmoor (Diesel)

Selected sightings:

03/01/69	D8510/2/27/9/32
11/01/69	D8510/2/27/9/32
26/05/69	D8512/21
14/09/69	D8530
20/09/69	D8530

D8510 and D8512, plus two others, 12A Carlisle Kingmoor (Diesel), 4 January 1969.
(John Grey Turner)

18.3 Carlisle Kingmoor Yard

Selected sightings:

12/05/69	D8501/3/6/7/9-12/4/6-24/6-9/31-4	D8530 not listed; recording omission?
26/05/69	D8501/3/4/6/7/9-11/3-20/2-4/6/8-34	D8504/13 re-instated on 18/05/69.
		D8512/21 not listed; to 12A (Diesel) (sighted 26/05/69).
		D8515 incorrectly included in list.
		D8527 not listed; recording omission?
06/07/69	D8501/3/5/8-11/4/6-20/2-4/6-34	D8505/8 incorrectly included and D8506/7 omitted; recording errors?
20/08/69	D8501/3/6/7/9-11/4/6-20/2-4/6-34/77	
24/08/69	D8501/3/6/7/9-11/4/6-9/22-4/6-34.	D8520/77 not listed; recording omissions?
20/09/69	D8501/3/6/7/9-11/4/6-20/2-4/6-9/31-4/77	D8530 not listed; to 12A (Diesel) (first sighted 14/09/69).
30/03/70	D8501/9/14/8-20/2/6/7/32-4	D8503/6/7/10/6/28/9/31 not listed; re-instated 02/11/69.
		D8511/7/23/4 not listed; to 66A (first sighting 14/12/69).
		D8577 not listed; to Bird, Long Marston (first sighting 11/10/69)
xx/04/70	D8501/9/14/8-20/2/6/7/32-4	
25/05/70	D8532-4	D8501/9/14/8-20/2/6/7not listed; D8501/9/14/8-20/7 to 66A (first sighting 24/05/70), D8522/6 to St Rollox Works (first sighting 25/05/70)
		D8532-4 first sighted at 66A on 13/06/70.

Note:

1. 'Abington: On 27th April (*1970*), 1739 (D05) hauled 8501/9/14 from Kingmoor to Polmadie as 8Z42, and 8518/9/20 followed later as 8Z43. On the 28th, 8Z42 consisted of 8522/6/7 hauled by 6839 (66A), and 8Z43 was 8532/3/4.' (*Railway Observer*, July 1970).

   There is doubt surrounding the D8532-4 move on 28/04/70, given their sighting at Carlisle Yard on 25/05/70 and their absence at 66A Polmadie on 24/05/70.

Map of Carlisle Kingmoor Yard. From Ordnance Survey 1:2500; 1967. The two large fans of sidings at the top centre of the map are the Up Sorting Sidings, and the pair at the bottom centre are the Down Sorting Sidings. The most easterly lines constitute the West Coast Main Line (WCML) and the more widely spaced fan of sidings at the bottom right of the map between the WCML and the Down Sorting Sidings are the Up Departure Sidings.

During 1969/70 withdrawn locomotives were stored in the fan of six dead-end sidings immediately to the north of the Down Sorting sidings, with the longest of the six utilised for storing the Metrovicks during 1969 (marked in yellow).

Claytons were stored on at least four of the five adjacent shorter dead-end roads.

(© Crown Copyright and Landmark Information Group Ltd 2020 [www.Old-Maps.co.uk Ref.787653137])

D8534 and D8527, Carlisle Kingmoor Yard, 21 August 1969. Three days later the following twenty-four locomotives were sighted here (in order recorded): D8503/29/01/26/18/14/06/07/09/23/11/30/16/17/27/22/33/24/19/10/32/34/31/28. (S. Blencowe)

D8511, Carlisle Kingmoor Yard, 3 May 1969. (Rail-Online)

Storage Locations: Depots and Yards • 295

D8519, Carlisle Kingmoor Yard, 3 May 1969. (Rail-Online)

18.4 66A Polmadie

Polmadie became a major dumping ground for stored and withdrawn Claytons spanning the period from 1968 to 1975.

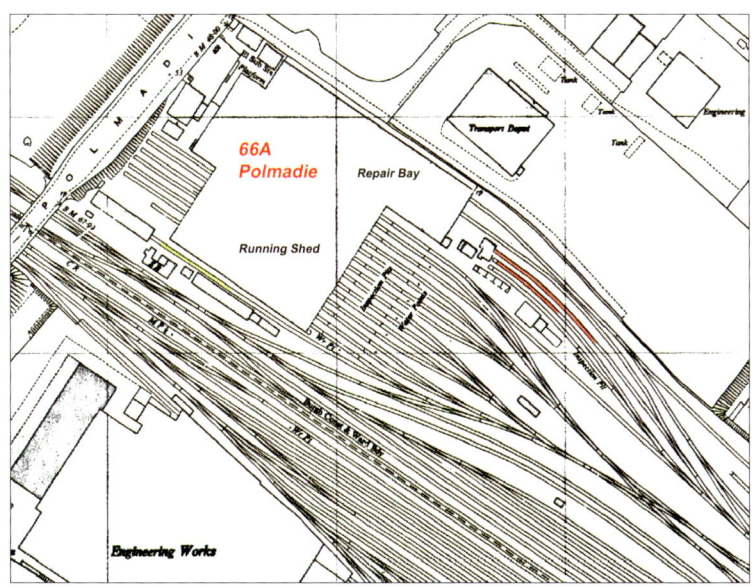

Map of 66A Polmadie depot. From Ordnance Survey 1:2500; 1971.

The main dumping areas for the stored and withdrawn Claytons at Polmadie were as follows:

> on the dead-end roads to the south-east of the Repair Bay (highlighted red), on the through roads inside the Running Shed, and, during 1971, on the track along the south-west wall of the Running Shed (sometimes known as the 'Dead Road') (highlighted yellow).

(© Crown Copyright and Landmark Information Group Ltd 2020 [www.Old-Maps.co.uk Ref.63751930])

D8511, D8523 and D8524, 66A Polmadie, 30 December 1969.
(Author's Collection)

D8503, 66A Polmadie, August 1970. Not officially stored but D8503 is clearly not a runner at this point. The November 1970 *Railway Observer* listed it as 'in store' in a visit report dated 4 August 1970, but the Clayton actually survived until the October 1971 cull.
(Grahame Wareham)

Storage Locations: Depots and Yards • 297

D8511, 66A Polmadie, August 1970. (Grahame Wareham)

D8530 and D6137, 66A Polmadie, 5 August 1971. (Author's Collection)

D8567 and D8557, 66A Polmadie, 10 September 1971. Three locomotives on one of the dead-end roads near the Polmadie Repair Bay presumably awaiting attention. The second locomotive, D8557, was withdrawn on 21 September 1971, just eleven days after this photograph was taken; whether it worked between the 10th and the 21st appears somewhat unlikely. D8567, on the other hand, survived into October. D8567 with small logo, and D8557 with large logo for comparison. (Rail-Online)

D8613, 66A Polmadie, 1972. Stored in the familiar Clayton storage location on the dead-end roads immediately to the south-east of the Polmadie Repair Bay. Note the diesel locomotive bogie-cleaning area in the foreground. (Laurie Mulrine)

Storage Locations: Depots and Yards • 299

D8548, 66A Polmadie, 9 November 1972. Within three days of arriving back at Polmadie after exile at the old steam shed at Ardrossan; based on Laurie Mulrine's sighting information, D8548, together with D8504, were the last Claytons to depart from the Ardrossan 'dump', which at its peak totalled twenty-eight locomotives. The Claytons suffered a considerable amount of vandalism and parts theft whilst at Ardrossan.
(Laurie Mulrine)

D8573, 66A Polmadie, 1973. Repair Bay visible to the right of picture.
(Rail-Online)

D8616 flanked by D8612 and D8542, 66A Polmadie, 28 September 1974.
(Anthony Sayer)

D8607, with D8546 and D8563, 66A Polmadie, 28 September 1974.
(Anthony Sayer)

Storage Locations: Depots and Yards • 301

Selected sightings:
N.B. ZH = St Rollox Works:

17/08/68	D8537	
06/10/68	D8537	Unknown if D8553 was present.
12/01/69	D8537/44/7/53/6/60/6/9/70/1	
13/03/69	D8547/56/60/4/6/9/71/6	D8537/53 not listed; moved to J. MacWilliam, Shettleston.
		D8544 not listed; reinstated to traffic.
		D8554/85 not listed; not yet arrived at 66A?
		D8570 not listed; recording omission?
29/03/69	D8547/54/6/60/4/6/9/70-2/6/85	
30/03/69	D8547/54/60/4/6/9/70/2/6/85	D8556/71 not listed; recording omission?
13/07/69	D8544/54/71/8	D8547/56/60/4/70/2/6 not listed; to Bird, Long Marston.
		D8566/9/85 not listed; to J. MacWilliam, Shettleston.
02/08/69	D8544/54/71/8	
xx/10/69	D8544/54/71/8	
14/12/69	D8511/7/23/4/44/54/78	D8511/7/23/4 added; ex Carlisle Yard (last sighted 20/09/69).
		D8571 not listed; to ZH (first sighted 11/01/70).
11/01/70	D8517/24/44/54/78	D8523 not listed; to ZH (first sighted 11/01/70).
		D8511 not listed; recording omission?
29/03/70	D8511/7/24/44/54/78	
11/04/70	D8511/7/24/44/54/78	
24/05/70	D8501/9/11/4/7-20/4/7/44/54/78	D8501/9/14/8-20/7 added; ex Carlisle Yard (last sighted xx/04/70).
13/06/70	D8501/11/4/7-20/4/32-4/44/54/78	D8532-4 added; ex Carlisle Yard (last sighted 25/05/70).
		D8509/27 not listed; recording omission?
14/06/70	D8501/9/11/4/7-20/4/7/32-4/44/54/78	
03/07/70	D8501/9/11/4/7-20/4/7/32-4/44/54/78	
30/07/70	D8511/4/7-9/24/32-4/44/54/78	D8501/9/20/7 not listed; recording omission?
31/07/70	D8511/4/7-9/24/32-4/54/78	D8501/9/20/7/44 not listed; recording omission?
02/08/70	D8501/9/11/4/7-9/24/7/32-4/44/78	D8520/54 not listed; recording omission?
04/08/70	D8501/9/14/8-20/7/32-4/44/54/78	D8511/7/24 not listed; to J. MacWilliam, Shettleston.
16/08/70	D8509/18-20/33/4/44/54/78	D8501/14/27/32 not listed; to ZH (first noted 22/08/70).
04/10/70	D8509/18-20/33/4/44/54/78	
07/11/70	D8509/18-20/33/4/44/54/78	
10/01/71	D8518-20/33/4/44	D8554/78 not listed; to J. MacWilliam, Shettleston.
		D8509 not listed; recording omission?
14/02/71	D8509/18-20/33/4	D8544 not listed; to J. MacWilliam, Shettleston.
11/04/71	D8509/18-20/30/3-5	D8530/5 added.
25/04/71	D8500/9/18/9/20/30/3-5	D8500 added.
28/05/71	D8500/10/8/30/4/5	D8509/19/20 not listed; to ZH (first sighted 17/04/71).
		D8533 not listed; to ZH (first sighted 17/04/71).

07/08/71	D8500/10/8/30/4/5	
21/08/71	D8500/10/8/30/4/5	
29/08/71	D8500/6/10/8/30/1/4/5/43	D8506/31/43 added.
26/09/71	D8506/13/31/4/43/9/55/7	D8500/10/8/30/5 not listed; to ZH (D8500/10/8/35 first sighted 27/09/71, D8530 cut up on arrival).
		D8549/55/7 added.
27/11/71	D8502/4/15/6/25/31/4/8-42/5/6/50/1 D8557/9/62/7/73, D8607/8/10/2-6	D8506 not listed; to ZH (first sighted 28/11/71). D8513 not listed; to ZH (cut up on arrival). D8543/9/55 not listed; to ZH (first sighted 21/11/71). D8502/4/15/6/25/38-42/5/6/50/1/9/62/7/73, D8607/8/10/2-6 added (withdrawals).
03/01/72	D8507/8/25/9/36/45/52/62/7/8/74/98	D8502/4/15/6/31/4/8-42/6/50/1/7/9/73, D8607/8/10/2-6 (plus D8548/58/63 (withdrawn 12/71)) not listed; to Ardrossan (first sighted 03/01/72). D8507/8/29/36/52/74/98 added (12/71 withdrawals). D8568 added, ex-Motherwell (last sighted 27/11/71).
09/01/72	D8507/8/25/9/36/45/52/62/7/8/74/98	
xx/02/72	D8507/8/25/9/36/45/52/62/7/8/74/98	
27/02/72	D8507/8/25/36/68, plus D8529/62 working at 66A, D8552 at 66B and D8574 at 66C	D8545/67 not listed; to GEC Stafford (first sighted 24/02/72). D8598 not listed; to Derby RCD (first sighted 12/03/72).
30/03/72	D8507/8/25/9/36/52/62/8/74	
17/08/72	D8507/8/25/9/36/52/62/8/74	
06/09/72	D8507/8/25/9/36/52/62/8/74	
16/09/72	D8507/8/25/9/36/52/62/74	D8568 not listed; to Clitheroe (first sighted 11/09/72).
14/10/72	D8508/25/9/36/62	D8507/52/74 not listed, to ZH (first sighted 14/10/72).
30/10/72	D8539/42/62, D8607/12	D8508/25/9/36 not listed, to ZH (first sighted 22/10/72). D8539/42, D8607/12 added; ex Ardrossan (last sighted 06/09/72).
06/11/72	D8502/31/3/9-42/6/50/1/7/8/62/3/73, D8607/8/10/2/3/6	D8502/31/3/40/1/6/50/1/7/8/63/73, D8608/10/3/6 added; ex Ardrossan (last sighted 06/09/72).
09/11//72	D8502/4/31/4/9-42/6/8/50/1/7/8/62/3, D8573, D8607/8/10/2/3/6	D8504/48 added; ex Ardrossan (last sighted 06/09/72).
19/11/72	D8502/4/31/4/9-42/6/8/50/1/7/8/62/3, D8573, D8607/8/10/2/3/6	
14/01/73	15 Claytons	
10/02/73	D8504/39/42/6/8/50/1/7/63/73, D8607/8/12/3/6	D8502/31/4/40/1/58/62, D8610 not listed, to ZH (first sighted 10/02/73, presumably by 14/01/73).
18/11/73	D8504/39/42/6/8/50/1/7/63/73, D8607/8/12/3/6	

10/08/74	D8504/39/42/6/8/50/1/7/63/73, D8607/8/12/3/6	
25/08/74	D8504/42/6/8/50/1/7/63/73, D8607/8/12/3/6	D8539 not listed; to 66B (first sighted 13/08/74), then 65A (first sighted 22/09/74).
26/01/75	D8504/42/6/8/50/1/7/63/73, D8607/8/12/3/6	
08/03/75	D8504/42/6/8/50/1/7/63/73, D8607/8/12/3	D8616 not listed; to Dundee (first sighted 17/03/75).
27/04/75	D8504/42/6/8/50/1/7/63/73, D8607/8/12/3	
17/05/75	No Claytons listed	All 13 Claytons transferred to ZH (first sighted 17/05/75).

Notes:
1. Why was it so long before early-withdrawn D8544/54/78 were moved away to scrapyards for disposal?
2. D8522/6 may have stayed at Polmadie for a short time in May 1970 before moving on to St Rollox Works. D8522/6/7 were moved north from Carlisle on 28/04/70.
3. D8534: LMR-withdrawn D8534 managed to linger long enough at Polmadie to be included in the 28-strong contingent of Claytons sent to Ardrossan for storage in December 1971.
4. D8562: The 20/10/72, 06/11/72 and 09/11/72 visit reports shows the arrival/return of the remaining 22 Claytons from Ardrossan to Polmadie; all the previous Clayton residents, with the exception of D8562, had been removed to St Rollox Works by that time to make space.
5. On 1 April 1975 BR announced the construction of a new carriage servicing depot at Polmadie; to assist building work all movements into and out of Polmadie were restricted to the east end of the depot with effect from 4 May 1975. Prior to this restriction coming in to place the Claytons were removed from Polmadie to St Rollox Works.

Four Claytons (at least) at Rutherglen, 25 May 1975. Something of mystery, this photograph. Stated as being Rutherglen and with the appearance of being taken from a moving train. My initial thought is that the Claytons were in transit from Polmadie to St Rollox Works, but the quoted date doesn't quite fit. The latest sighting found for the final thirteen Claytons at Polmadie is 27 April 1975, with a subsequent first sighting at St Rollox Works on 15 May! (Colour-Rail)

18.5 Ex-67D Ardrossan

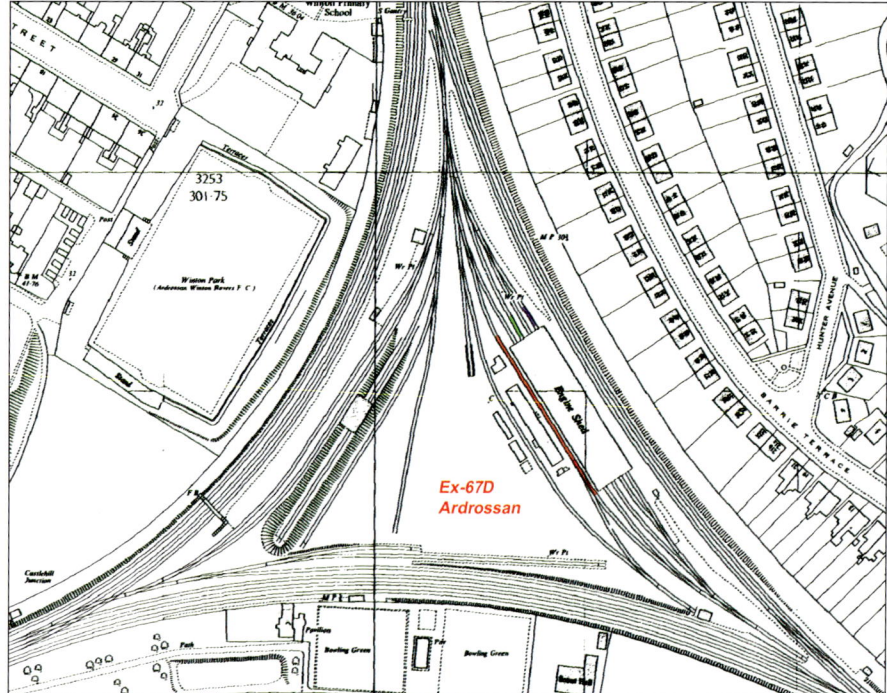

18019. Map of Ex-67D Ardrossan. From Ordnance Survey 1:2500; 1965.

At its peak, twenty-eight Claytons were dumped at the old steam depot at Ardrossan. D8548 (highlighted in purple) and D8531 (green) were parked outside to the north-west of the depot building. D8515/6/38/59 and D8614/5 were also parked outside immediately adjacent to the south west wall of the depot building (with D8516 furthest south) (highlighted in red). The remaining twenty were stored inside on the four through roads, five locomotives per road.
(© Crown Copyright and Landmark Information Group Ltd 2020 [www.Old-Maps.co.uk Ref.949848286])

D8548 and D8531, Ex-67D Ardrossan, 17 August 1972. The depot building was capable of holding twenty Claytons undercover (if such protection were needed). By August 1972 there were twenty-two locomotives dumped here, so two braved the sea-side elements outside to the north-west end of the shed building. When there were twenty-eight Claytons at Ardrossan, the additional six were parked adjacent to the south-west side of the depot (on the line to the extreme right in the photograph).
(Anthony Sayer)

Storage Locations: Depots and Yards • 305

D8548, Ardrossan, 17 August, 1972.
(Anthony Sayer)

D8531, Ardrossan, 17 August 1972.
(Anthony Sayer)

The March 1972 edition of the *Railway Observer* records: '…Those (*Class 17's*) stored at Polmadie were removed during mid-December (*1971*) to the old steam depot at Ardrossan, which is otherwise disused…'
Selected sightings:

03/01/72	D8502/4/15/6/31/4/8-42/6/8/50/1/7-9/63/73, D8607/8/10/2-6
26/02/72	D8502/4/15/6/31/4/8-42/6/8/50/1/7-9/63/73, D8607/8/10/2-6
14/05/72	D8502/4/15/6/31/4/8-42/6/8/50/1/7-9/63/73, D8607/8/10/2-6
17/07/72	D8502/4/31/4/9-42/6/8/50/1/7/8/63/73, D8607/8/10/2/3/6
17/08/72	D8502/4/31/4/9-42/6/8/50/1/7/8/63/73, D8607/8/10/2/3/6
06/09/72	D8502/4/31/4/9-42/6/8/50/1/7/8/63/73, D8607/8/10/2/3/6

Notes:
1. The last sighting of D8502/4/15/6/31/4/8-42/6/50/1/7/9/73, D8607/8/10/2-6 at Polmadie prior to their movement to Ardrossan was 27/11/71. All except D8534 were withdrawn from Scottish Region books during the period September-November 1971; D8534, however, was withdrawn by the LMR in October 1968, having spent over three years in store around Carlisle and at Polmadie!
2. D8548/58/63 were withdrawn on 31/12/71 and presumably must have arrived at Ardrossan very early in January 1972.
3. *British Railways Illustrated* in June 2007 (Vol.16, No.9, p413) carried a letter from Ian D. Osborne in which he stated: '…..regarding the stored Clayton diesels at Ardrossan…..on 26th February 1972. The numbers of the Claytons I recorded (28) at the old shed were as follows, in the order I recorded them: 8515, 8615, 8614, 8538, 8559, 8516, 8608, 8542, 8557, 8539, 8541, 8612, 8558, 8607, 8610, 8573, 8531, 8548, 8563, 8504, 8534, 8502, 8546, 8540, 8613, 8550, 8551, 8616…..' (N.B. My colour coding, with locomotives highlighted in red outside).

My sightings on 17 August 1972 in order were:

D8531, D8573, D8610, D8607, D8558, D8541, D8612, D8539, D8557, D8542, D8608, D8616, D8551, D8550, D8613, D8540, D8546, D8502, D8534, D8504, D8563, D8548

4. D8515/6/38/59, D8614/5 were not listed on 17/07/72 having moved to Glasgow Works (first sighted there on 08/06/72 (D8515/38, D8614/5 - photographed in Carriage Sidings) and 18/06/72 (D8516/59 - sighted in the Top Yard)).
5. After the 06/09/72 Ardrossan visit, the 22 remaining Claytons were **not** noted at Polmadie on 14/10/72 but D8539/42, D8607/12 were back at Polmadie by 30/10/72, D8502/31/3/40/1/6/50/1/7/8/63/73, D8608/10/3/6 by 06/11/72 and D8504/48 by 09/11/72.

18.6 64B Haymarket
Selected sightings:
(N.B. ZH = St Rollox Works)

24/11/68	D8575/84	
02/01/69	D8575/84	
12/01/69	D8575/82/4	
30/03/69	D8582/4	D8575 not listed; to J. MacWilliam, Shettleston (sighted 09/03/69).
xx/xx/69	-	D8582/4 not listed; to J. MacWilliam, Shettleston (sighted 25/05/69).
25/04/71	D8510, D8606	
16/05/71	D8606	D8510 not listed; to 66A (sighted 28/05/71).

20/06/71	D8606	
28/08/71	D8606	
27/09/71	D8583/6/92/4	D8606 not listed; recording omission?
		D8583/6/92/4 added.
09/10/71	D8528/81/3/92-4/9, D8604/6 (plus D8578 (sic D8586?))	D8528/81/93/9, D8604 added.
27/11/71	D8560 (*sic*)/1/80/7, D8601-3	D8580/7, D8601-3 added.
		D8528/81/6/93 not listed; to Millerhill Yard (sighted 27/11/71).
		D8583 not listed; to Millerhill Yard (sighted 03/01/72). N.B. Not sighted at MH on 27/11/71.
		D8599 not listed; to ZH (sighted 16/10/71).
		D8592/4, D8604/6 not listed; to ZH (first sighted 21/11/71).
19/12/71	D8587	D8561 not listed; to Millerhill Yard (sighted 03/01/72).
		D8602/3 not listed; to ZH (sighted 09/01/72).
		D8580, D8601 not listed; recording omission?
03/01/72	D8505/80/7, D8601	D8505 added; ex-Millerhill Yard (last sighted 27/11/71).
08/01/72	D8505/80/7, D8601	D8587, D8601 photographed still at 64B on 22/01/72.

D8606, 64B Haymarket, August 1971. (Grahame Wareham)

18.7 Millerhill Yard (Up Sorting Sidings).

D8583, etc., Millerhill Yard, 18 August 1972.
(Anthony Sayer)

D8587, etc., Millerhill Yard, 18 August 1972.
(Anthony Sayer)

Selecting sightings:

27/11/71	D8505/28/79/81/6/8/93/7, D8600	Stabling point or Yard?
03/01/72	D8528/61/5/79/81/3/6/8/93/7, D8600	D8561/83 added; ex-64B.
		D8565 added; ex-?
		D8505 not listed; to 64B.
08/01/72	D8528/61/5/79/81/3/6/8/93/7, D8600	
27/02/72	D8505/28/61/5/79-81/3/6-8/93/7, D8600/1	D8505/80/7, D8601 added; ex- 64B (last sighted 08/01/72 (all)/ 22/01/72 (only D8587, D8601 photographed)).
31/03/72	D8505/28/61/5/79-81/3/6-8/93/7, D8600/1	
03/06/72	D8505/28/61/5/79-81/3/6-8/93/7, D8600/1	
18/08/72	D8505/28/61/5/79-81/3/6-8/93/7, D8600/1	
02/09/72	D8505/28/61/5/79-81/3/6-8/93/7, D8600/1	

Notes:

1. On 14/04/72 G.Brookes recorded the 15 locomotives at Millerhill Yard in the following order (north to south):

   D8586, D8588, D8593, D8579, D8528, D8583, D8597, D8581, D8561, D8600, D8565, D8587, D8505, D8580, D8601.

   The order of locomotives recorded on 18/08/72 from north to south were as follows:

   D8587, D8505, D8580, D8601, D8597, D8581, D8561, D8600, D8565, D8586, D8588, D8593, D8579, D8528, D8583.

   Clearly in between the two visits the Claytons had been moved within the yard as three separate shunts.

2. After the last Millerhill Yard visit on 02/09/72, 13 of the 15 Millerhill Claytons were next noted at St Rollox Works on 14/10/72 i.e. D8505/28/61/5/79-81/3/6/93/7, D8600/1. D8587/8 were not listed although both were listed on a subsequent visit on 15/11/72; however D8578 was listed in error on the 14/10/72 visit and may have represented some sort of typographical error for D8587/8.

18.8 65A Eastfield.

D8539, 65A Eastfield, 28 September 1974. Other selected sightings of this locomotive here are 22/09/74, 08/03/75 and 15/11/75. Used for re-railing purposes; note the lifting-lug immediately above the nearest bogie.
(Anthony Sayer)

18.9 Thornton Junction

Sightings:

02/01/69	D8585, D8609/11
11/01/69	D8585, D8609/11
13/03/69	D8609

18.10 Dundee.

D8616, Dundee, June 1975. Other selected sightings at Dundee are: 17/03/75 and 04/08/75. Used to test re-railing equipment. (Colour-Rail)

Chapter 19

DISPOSAL: THE PRIVATE YARDS (1968–71)

19.1 Introduction.
Four companies were involved with the disposal of Claytons during the 1968-71 period i.e. J. MacWilliam at Shettleston (twenty), the Bird Group at Long Marston (seven), J. Cashmore at Great Bridge (one) and A. Draper at Hull (one), all at their own premises. In addition, W. Willoughby cut up the single 1970 withdrawal on site at 52A Gateshead (D8589).

19.2 J. MacWilliam, Shettleston (1969–71).
Summary of Claytons cut at J. MacWilliam, Shettleston (1969-71):
D8511/7/24/37/44/53/4/66/9/71/5/8/82/4/5/91/5/6, D8609/11 (Total: 20)

N.B. D8537/53/4/66/78 'Disposal not Proven'.

Sightings:

09/03/69		D8575
25/05/69		D8569 (being cut-up), D8582/4/5 (intact)
15/08/69	Photo	D8595/6 (cabs only)
23/08/70		D8517/24 (intact), D8511 (cab only)
17/04/71		D8544 (D8554/71/8 not listed)
xx/06/71		D8571

Comparative summaries of cutting-up information (Sources: AHBRD&E5 and 'D&ELfS):

	AHBRD&E5	D&ELfS
D8511	By 03/09/70	08/70
D8517	By 25/09/70	08/70
D8524	09/70	08/70
D8537	01/69	01/69
D8544	By 01/05/71	04/71
D8553	02/69	01/69
D8554	04/71	04/71
D8566	05/69	04/69

	AHBRD&E5	D&ELfS
D8569	By about 01/06/69	04/69
D8571	07/71	06/71
D8575	By 02/04/69	03/69
D8578	08/71	Disposal unknown.
D8582	By 19/06/69	05/69
D8584	By 19/06/69	05/69
D8585	By 19/06/69	06/69
D8591	By 30/08/69	08/69 ('requires confirmation')
D8595	06/69 (cab cut later)	08/69
D8596	06/69 (cab cut later)	05/69
D8609	06/69	06/69 ('requires confirmation')
D8611	Not recorded	06/69 ('requires confirmation')

Additional information from Hooper's book:

D8578	Cut-up at J. MacWilliam Shettleston, March 1971
D8591	Intact at J. MacWilliam, Shettleston, 9 August 1969

D8569, J. MacWilliam, Shettleston, 1969. (Transport Topics)

Cabs from D8595 and D8596, J. MacWilliam, Shettleston, 15 August 1969.
Note the lorry in the background with 'MacWilliam' emblazoned on the side (not the incorrect 'McWilliam' which most books incorrectly use for this scrap merchant).
(Stewart Blencowe)

19.3 Bird Group, Long Marston (and Cardiff?) (Bizle Group of companies) (1969/70).

Locomotives bought by the Bird Group inevitably resulted in some long distance movements, with locomotives stabled at very points on route. The following photographs illustrate Claytons at Bescot Yard and Worcester.

D8570, Bescot station, May 1969. (Stephen Fisher)

D8556/64/72, Bescot Yard, June 1969. (Stephen Fisher)

D8564 and D8572, Bescot Yard, June 1969. (Stephen Fisher)

D8560, D8576 and D8547, Worcester, 24 June 1969. (Transport Treasury (A. Swain))

D8560, Worcester, 24 June 1969. (Transport Treasury (A. Swain)

Summary of Claytons cut at Birds, Long Marston (1969/70): D8547/56/60/4/70/6/7 (Total: 7)

N.B. D8547 'Disposal not Proven' (Cardiff).

Reports and sightings of Claytons en route to Long Marston are listed below:

20/05/69	In train passing Bromford Bridge: D8570	
xx/05/69	In train passing Bescot: D8570	
24/06/69	Bescot Yard: D8547/60/76 (arrival date)	
24/06/69	Worcester: D8547/60/76	
27/06/69	Bescot Yard: D8556/64/72 (arrival date)	D8572 arrived in error; see Section 19.4.

Selected sightings at Bird's Long Marston:

22/05/69	D8570	
31/05/69	D8570	
28/06/69	D8547/60/70/6	D8547/60/76 added.
29/06/69	D8547/56/60/4/70/6	D8556/64 added.
05/07/69	D8556/60/4	D8547 not listed; recording error? D8570 not listed; cut-up?
26/07/69	D8547/56/60/4/76	
19/08/69	D8547/56/60/4/76	
11/10/69	D8547/56/64/76/7	D8560 not listed; cut-up? D8577 added.
11/01/70	D8547/56/64/76/7	
07/03/70	D8547/56/64/76/7	
30/03/70	D8556/64/77	D8547/76 not listed (see below).
23/08/70	D8577	

Comparative summaries of cutting-up information (Sources: AHBRD&E5 and 'D&ELfS):

	AHBRD&E5	D&ELfS
D8547	Shortly after 16/03/70 (at Cardiff)	03/70 (at Long Marston, 'requires confirmation')
D8556	By 30/04/70	04/70
D8560	12/69	12/69
D8564	By 30/04/70	04/70
D8570	By 27/06/69	05/69
D8576	By 28/03/70	03/70 ('requires confirmation')
D8577	By 30/04/70	Circa 07/70

Comments:
1. Sighting information suggests that D8560 was cut-up around September 1969.
2. Based on sightings, D8577 may have lasted beyond August 1970 (although this may have been in a part-cut state).
3. J. Hooper (2016) states that both D8564 and D8577 were cut up on 28 June 1970. The reference to D8564 probably relates to the final delayed disposal of the cab.

A 'Diesel Dilemma' (using the Peter Hall (RCTS) phraseology) exists with respect to the Claytons sold to Bird's. Whilst it is clear that all seven locomotives sold to the company arrived at the Long Marston site, what is not clear is whether all were actually cut there.

The Birmingham Locomotive Club Circular No.3/70 recorded:

'Movements of ex-BR diesel stock . . .on 16/3 (*1970*) 8547/56/65 (*sic*)/76 from Bird's Long Marston to Bird's Cardiff.'

Similarly the June 1970 edition of the *Railway Observer* reported:

'On 16th March (*1970*) D8547/56/64/77 were booked to be moved from Birds, Long Marston to Bird's, Cardiff.'

'Booked' is clearly the operative word as it is clear that at least D8556/64/77 did <u>not</u> move to Cardiff on the specified date given the sighting of these three locomotives on 30 March 1970 still at Long Marston. However, it is

D8564, Bird Group, Long Marston, August 1970. Withdrawn only 14 months after having received an Intermediate repair and repaint into blue livery. Problems with the Paxman engine precipitated its early demise. (Grahame Wareham)

quite possible that D8547 (and also D8576?) did move to Cardiff. Unfortunately no sightings of these two locomotives have been found after 7 March 1970, although Roger Harris (AHBRD&E5) does suggest that D8576 was cut up at Long Marston by 28 March 1970.

Further clarification is required on D8547/76, and indeed details of the exact disposal dates of D8556/60/4/77.

19.4 J. Cashmore, Great Bridge (1970).

As already mentioned in Section 19.3 above, D8572 was sent south from Scotland for scrap at Bird's, Long Marston arriving at Bescot Yard with D8556/64 on 27 June 1969, in error for D8577. Once the error was realised, D8572 was detached from the convoy pending further movement instructions. The September 1969 edition of the *Railway Observer* reported:

> '8572 was sent in error. 8577 should have arrived, but before it could be returned it

Disposal: The Private Yards (1968–71) • 317

D8572 with D7598, Bescot Yard, 1 September 1969. The Clayton looks in better external condition than the Class 25, but appearances deceive!
(John Chalcraft (Rail-Photoprints))

was damaged by children and considered unfit to travel back to Scotland.'

Following a tendering process, D8572 was sold to J. Cashmore, Great Bridge where it arrived on 12 June 1970, after virtually twelve months storage at Bescot Yard.

Selected sightings at Bescot Yard: 10/07/69, 19/11/69, 22/03/70, 31/05/70.

Summary of Claytons cut-up at Cashmore, Great Bridge (1969/70): D8572 (Total: 1)

Sightings of D8572 at Cashmore's yard: 21/06/70 and 29/06/70.

Comparative summary of cutting-up information (Sources: AHBRD&E5 and 'D&ELfS):

	AHBRD&E5	D&ELfS
D8572	08/70	08/70

19.5 A.Draper, Neptune Street, Hull.

Summary of Claytons cut at A.Draper, Hull (1969-71): D8605 (Total: 1)

Arrival dates for D8605 at Draper's, Hull vary from 5, 14 and 18 May 1969 (*Railway Observer, Railway World* and Scotney & Eagan (*Draper's Scrapyard, Hull*, Book Law Publications, 2015 (DSH))

respectively). On the basis of a sighting of D8605 at Thornaby on 10 May, the date of 5 May can be rejected. However, the 14 and 18 May dates may represent the difference between arrival in the Hull area as opposed to arrival in Draper's yard proper.

Selected sightings at Draper's: 18/05/69, 23/06/70, 12/08/70 (intact on all dates)

Comparative summaries of cutting-up information (Sources: DSH, AHBRD&E5 and D&ELfS):

	DSH	AHBRD&E5	D&ELfS
D8605	22/03/71	06/70	06/70

D8605, A. Draper, Hull, 29 June 1969. (Rail-Online)

The date discrepancies for the disposal of D8605 are by far the most extreme of all of the Clayton fleet. As regular visitors to the Neptune Street yard (and Sculcoates before), I will unreservedly go with Messrs. Scotney and Egan! They record D8605 as arriving at Neptune Street yard on 18 May 1969 and cut up on 22 March 1971, the delay in disposal apparently due to the fact that:

'...it was found that certain parts had been removed prior to its departure from Thornaby depot. Consequently a dispute arose whereby BR denied all knowledge of the absent bits and Draper's dug in until something was sorted out.'

19.6 W. Willoughby, Choppington.

Summary of Claytons cut by W. Willoughby on-site at 52A Gateshead: D8589 (1)

Selected sightings:

26/07/70	D8589
04/08/70	D8589 (accident damage, off bogies)
18/10/70	D8589

Comparative summary of cutting-up information:

	AHBRD&E5	D&ELfS
D8589	04/71 (all trace gone by 31/05/71)	04/71

Chapter 20
DISPOSAL: ST ROLLOX WORKS

20.1 Introduction.
Summary of Claytons cut at St Rollox Works:

D8500-3/5/6/9/10/2-5/8-23/6-8/30/2-5/8/40/1/3/5/9/55/8/9/61/2/5/7/79/81/3
D8586/7,
D8588/90/2-4/7-9, D8600-4/6/10/4/5   (Total: 61)

N.B. D8513/30 could be considered as 'Disposal not Proven' given the absence of any known sightings at St Rollox Works.

Scrapping of the Claytons at St Rollox Works started with D8530 in October 1971 and suddenly ceased in April/May 1974 with two Claytons left in a partially cut-up condition; these two locomotives (believed to be D8505/59) remained in this condition until at least June 1975.

The whole process of accurately understanding what happened at St Rollox Works during the six years spanning 1970 to 1975 is fraught with difficulty for the following reasons:

- Lists of available visit reports vary significantly depending on the time of year with noticeably fewer during the winter months, with January and December being particularly poorly represented.
- Available sightings vary significantly in terms of the areas covered during the Works visits. Invariably the Works Test House, Erecting Shop and Erecting Shop Yards were visited, but depending on the willingness of the guide, pressure brought to bear by enthusiasts to visit the more far-flung parts of the Works and time constraints, such areas as the Carriage Sidings and the Top Yard were occasionally missed out from the visit itinerary. The Scrapping Area was the area most frequently ignored. Listings were, therefore, often understated.
- Many enthusiasts, unfortunately disregarded withdrawn locomotives (particularly if they had disappeared from their ABC's) and, as such, they went unrecorded. For the more frequent visitors to St Rollox Works, the Claytons almost became permanent fixtures and, for this reason as well, were often ignored.
- The Scrapping Area, when it was visited, was frequently treated in a very cavalier fashion by enthusiasts. Scrapping of locomotives at St Rollox Works was notorious with respect to the time it took for the cut-up remains of locomotives to be removed from site. As a consequence 'flame-cut' cab side panels could linger for months but were still recorded by enthusiasts, so that, many years down the line, these numbers became to reflect

extant locomotives rather than cut-up remains. In contrast to comment above, locomotive listings in the Scrapping Area effectively became overstated!

- Most listings lacked any anecdotal information, such as ' Carriage Sidings' or 'Scrapping Area' (to indicate where locomotives were situated within the works), 'stored' or 'withdrawn' (to describe a locomotive's status), or, 'intact' or 'part cut-up' (to indicate a locomotive's condition). In the relatively few instances where such information was provided, a clear indication of how particular Claytons were proceeding towards their ultimate demise was exceedingly helpful.
- One anecdotal piece of information used from time-to-time was the 'Dump' (or 'Clayton Dump'). This was actually very confusing because the main congregations of Claytons, the so-called 'Dump', actually migrated around the site over time!
- Accuracy of recording also played its part, more so in St Rollox Works than elsewhere it seems. During the 'busy' September 1971 to March 1974 period, if two (or more) separate lists of sightings were compared for the exact same date, there would be differences; without exception, and some very significant!

The migration of locomotives, combined with incomplete Works visits (in terms of geographic coverage) quite possibly led to the comment in the February 1974 edition of the *Railway Observer*:

'Several . . .instances have been recorded of certain Claytons disappearing from the dump at St Rollox, only to reappear several months later'.

The RO put this down to their 'having presumably visited some other depot to enable suitable spares to be removed prior to complete scrapping of the locomotive.'

Using D8507 as an example, this locomotive seemed to disappear completely during March 1975. The clue, however, had already been explained in the December 1974 *Railway Observer* when for a visit dated 5 October 1974 it commented:

'The only Class 17 on the 'Clayton Dump' (*actually the Carriage Sidings*) was 8507, although 8508/16/25/29/31/36/52/74/80 were in the cutting-up siding.'

So whilst the group of nine Claytons had been moved to an area much closer to the main Works complex, D8507 became stranded over a quarter of a mile to the east of the Main Works buildings. My contention is that once Claytons arrived at St Rollox Works after the end of 1971 they never left (not until the final exodus of September 1975, that is), but simply got missed!

The accompanying maps show the area east of the main Shops at St Rollox Works; a 1963 Ordnance Survey plan has been used to give the required level of detail, although it should be noted that in the mid-1960s the Works was substantially re-developed leading to some major alterations in the central area immediately to the east of the main Workshops.

The main areas used for the storage and scrapping of the Claytons are colour-marked, as follows:-

Erecting Shop Yard (blue)

Carriage Sidings Adjacent to the Top Yard embankment (2 lines) (orange)

Carriage Sidings Middle Area (pink)
Used temporarily in late-1972 during the heavy influx of Claytons from Millerhill Yard and Polmadie and again in mid-1975 when the remaining locomotives were marshalled ready for transfer to external scrap yards.

Old Paint Shop (Varnish Shop) Yard (brown)

North of Traverser (3 lines) (green)
The most northerly of these three lines by-passed the Scrapping Area and extended a considerable distance further west. All three lines were used from circa September 1974.

Siding immediately West of Scrapping Area (purple)
Part of the northerly line of the three sidings mentioned above. This area was used when the final influx of thirteen Claytons were received from Polmadie during April/May 1975.

Scrapping Area (red)

Disposal: St Rollox Works • 321

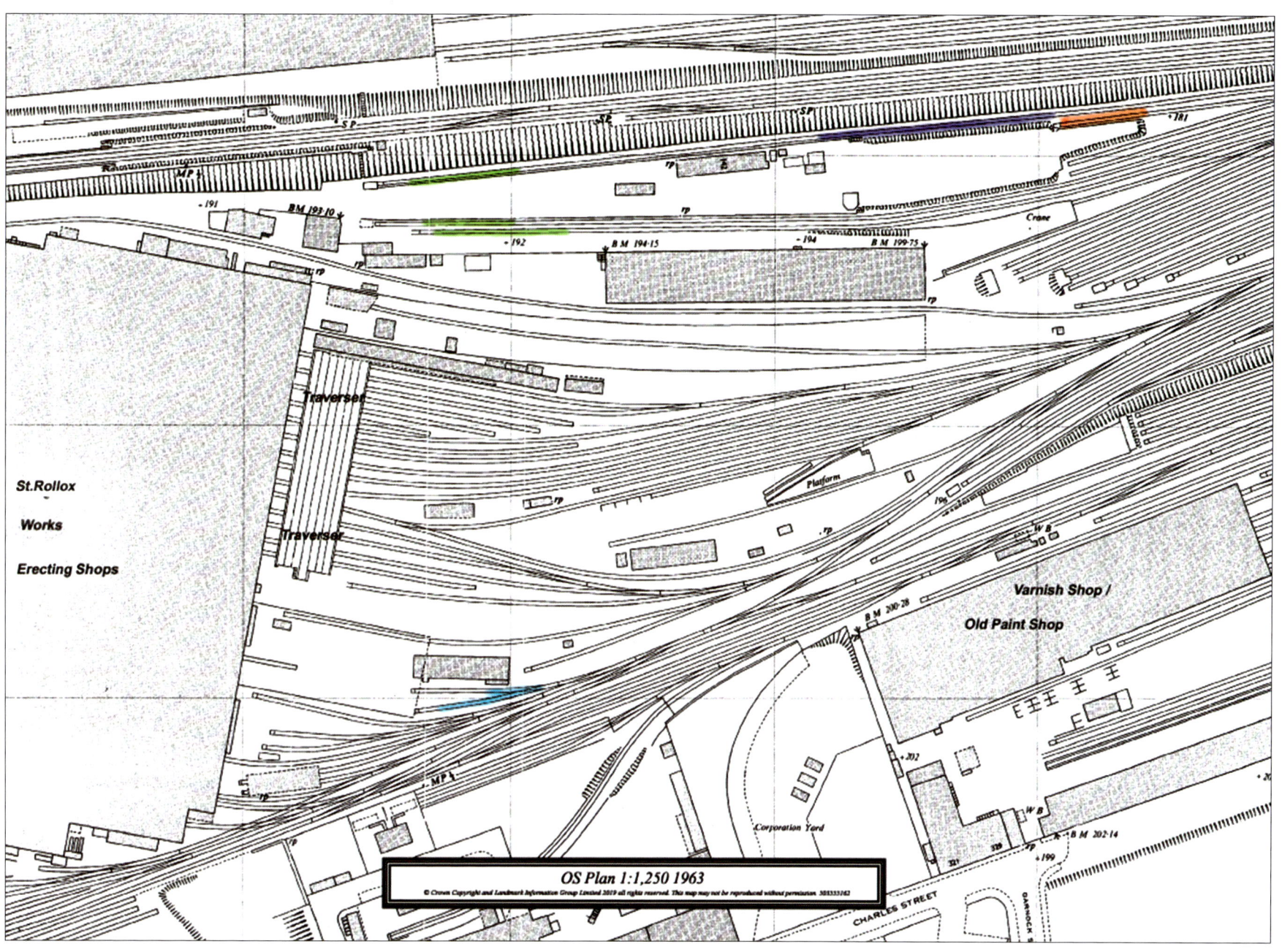

Above and next page: Maps of St Rollox Works east of the Main Erecting Shop, Machine Shop and Carriage Shops. Ordnance Survey 1:1250; 1963. (© Crown Copyright and Landmark Information Group Ltd 2020 [www.Old-Maps.co.uk Ref.503555162])

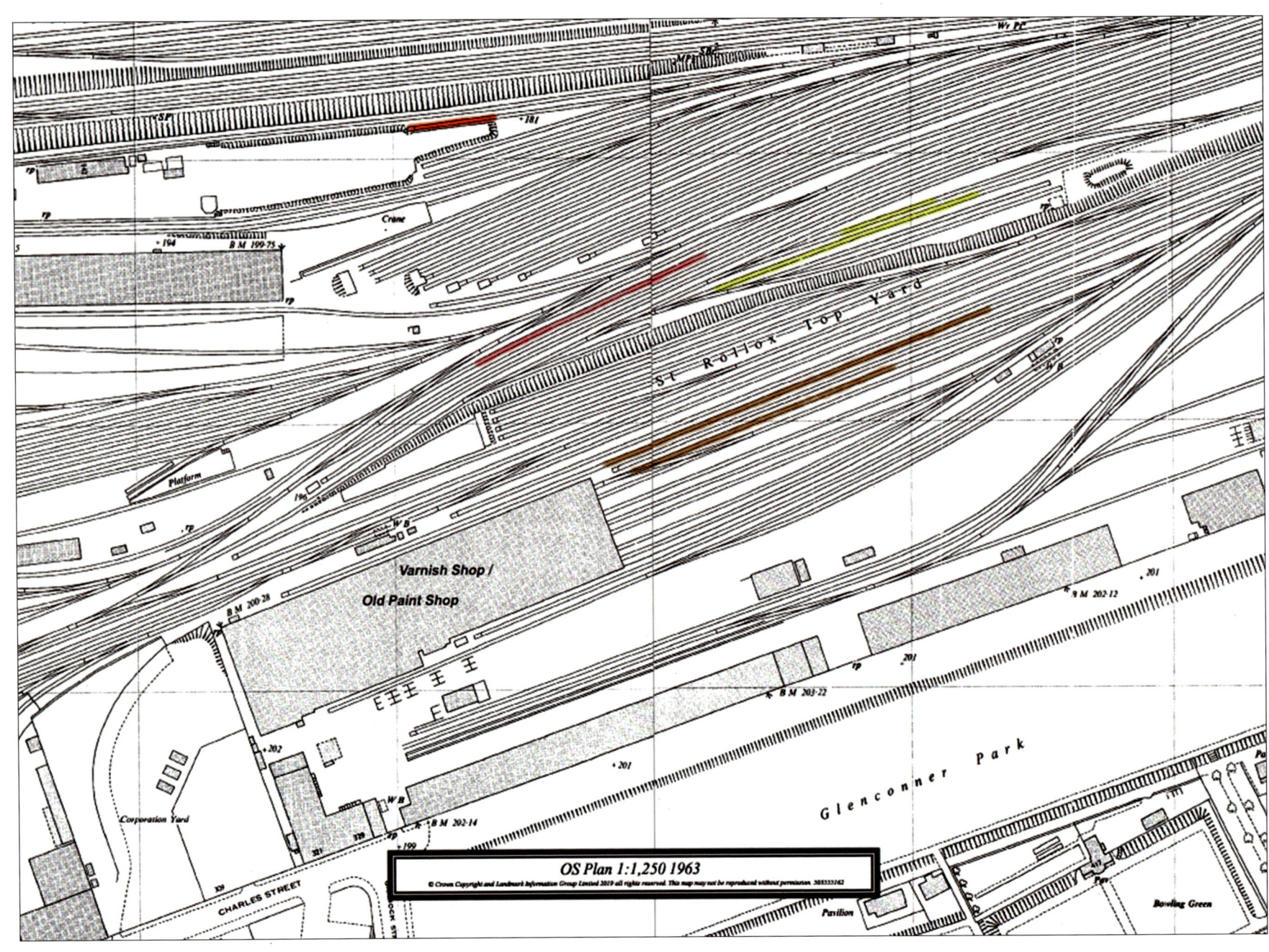

Disposal: St Rollox Works • 323

D8552, D8574, D8507, etc, St Rollox Works (Carriage Sidings), Undated (slide processed October 1972). Photograph taken from the Top Yard embankment looking north-east. The last four locomotives are believed to be D8583, D8528, D8579 and D8593 recently arrived from Millerhill Yard. (TOPticl Digital Memories)

D2424, D2416, D8507, D8574, D8552, D8536, D8508, St Rollox Works (Carriage Sidings), 22 April 1973. Looking west. (B.J. Nicolle [Rail Image Collections])

D8583 (plus D8593, D8528, D8586, D2436, D8561, D5114, etc) and D5067, St Rollox Works (Old Paint Shop Yard), 22 April 1973. Looking west. (B.J. Nicolle [Rail Image Collections])

D8552, D8574 and D8507, St Rollox Works (Carriage Sidings), 12 May 1973. Looking north-east. (Edward Bather)

St Rollox Works (Old Paint Shop Yard), 12 May 1973. Looking east. See the plan on page 338 for the locomotive identities. (Edward Bather)

D8559 and plenty more, St Rollox Works (Old Paint Shop Yard), 12 May 1973. Looking north-east. (Rail-Online)

D8545, St Rollox Works (Old Paint Shop Yard), 12 May 1973. Looking south-west from the top of the cab of a Clayton it would appear! (Rail-Online)

D8610, St Rollox Works (Old Paint Shop Yard), 12 May 1973. (Rail-Online)

D8529 and D8508, St Rollox Works (north of the old Works Traverser), 26 September 1974. Looking east. (Anthony Sayer)

Claytons dumped at St Rollox Works (north of the old Works Traverser), circa June 1975. Looking west, with: Back row (left to right): D8531, D8516, D8580; Middle row (left to right): D8529, D8525; Nearest row (left to right): D8508, D8536, D8552, D8574. (TOPticl Digital Memories)

20.2. Literature search.

Following my experiences with the quality of the NBL and Metrovick Type 2 storage information (see previous books in the Pen & Sword Locomotive Histories series), a similar analysis for the Clayton fleet was thought to be worthwhile. As before, the process starts with a literature search to round-up currently available information, this to be followed by primary-sourced sighting information (Section 20.3) and a comparison of the two to highlight any accuracy issues (20.4).

Starting with the literature search, four publications were reviewed:

Railway Observer (July 1976) (RO)
Allocation History of BR Diesels & Electrics Part 5, R.Harris (2005) (AHBRD&E5)
Modern Locomotives Illustrated No.186 (December 2010/January 2011) (MLI186)
Diesel & Electric Locomotives for Scrap, A.Butlin (2015) (D&ELfS)

Month	RO (Commencing month of cut-up)	D&ELfS (Date cut)	MLI186 (Disposal)	AHBRD&E5 (Date scrapped)
10/71	D8530	D8530	D8530	D8530
11/71	D8500/10/35/99	D8500/10/35/99	D8500/10/35/99	D8500/10/35/99
12/71	D8501/6/13/4	D8501/6/13/4/90	D8501/6/13/4	D8501/6/13/4/90
01/72	D8509/33	D8509/33	D8509/33	D8509/33
02/72	D8518	D8518	D8518	D8518
03/72	D8592/4	D8592/4	D8592/4	D8592/4
04/72	D8602/3/4/6	D8602/3/4/6	D8602/3/4/6	D8602/3/4/6
05/72	D8519/20/2/6/55	D8519/20/2/6/55	D8519/20/2/6/55/90	D8519/20/2/6/55
06/72	D8502/23/7/32/43/9	D8502/23/7/32/43/9	D8502/23/7/32/43/9	D8502/23/7/32/43/9
07/72				
08/72				
09/72			D8515	
10/72				
11/72		D8515		D8515
12/72	D8538, D8614	D8538, D8614	D8538, D8614	D8538, D8614
01/73	D8512/67	D8512/40/67	D8512/40/67	D8512/40/67
02/73		D8534/41/58	D8534/41/58	D8534/41/58
03/73			D8502	
04/73		D8502, D8600	D8600	D8502, D8600
05/73			D8565	
06/73		D8562		D8562
07/73		D8586/8/97	D8528/61/2/86/8/97	D8528/88
08/73		D8528/65/83, D8610	D8583, D8610	D8565/83/6/97, D8610
09/73		D8561/93	D8593	D8561/93
10/73				
11/73		D8545/79	D8545/79, D8601/15	D8545/79
12/73		D8615		D8615
01/74				
02/74		D8587, D8601	D8587	D8587, D8601

Disposal: St Rollox Works • 329

Month	RO (Commencing month of cut-up)	D&ELfS (Date cut)	MLI186 (Disposal)	AHBRD&E5 (Date scrapped)
03/74			D8559	
04/74		D8559		D8559
05/74		D8505/81	D8505/81	D8505/81
03/79		D8598	D8598	D8598
04/79		D8521	D8521	D8521

Notes:
1. D8515/40/90 were not included in the July 1976 *Railway Observer* listing.
2. Locomotives highlighted in red and blue are discussed later in Section 20.4.

20.3 Sightings of Claytons.

Selected St Rollox Works visits covering sightings of withdrawn Claytons:

11/10/69	Nil.
11/01/70	D8523/71
28/03/70	D8523/71
11/04/70	Erecting Shop Yard: D8523/71

D8613, St Rollox Works (Erecting Shop Yard), 11 April 1970. The Claytons behind D8613 are condemned D8523 and D8571. D8613 was in the Works for Unclassified repairs.
(Bill Hamilton)

25/05/70	D8522/3/6/71	D8522/6 added; ex-Carlisle Yard (last sighting 28/04/70 at Abington en route north).
04/08/70	D8522/3/6/71	
22/08/70	D8501/12/22/3/6/7/32/71. D8527/32 dumped beyond old carriage shop	D8501/27/32 added; ex-66A (last sighting 04/08/70). D8512 (sic D8514?, D8514 disappeared from 66A by 16/08/70)).
12/09/70	D8512/22/3/6/71	D8501/27/32 not listed; recording omission? D8512 (sic D8514?); D8512 seen at Longsight on 12/09/70.
03/10/70	Old Paint Shop Yard: D8501/14/22/3/6/7/32/71	D8514 added; ex-66A (last sighting 04/08/70).
10/10/70	D8501/14/22/3/6/7/32/71	
10/01/71	D8501/14/22/3/6/7/32/71	
20/02/71	D8501/14/22/3/6/7/32/71 (All 'Half a mile from the works')	
14/03/71	D8501/14/22/3/6/7/32/71/90	D8590 added; ex-52A (last sighting 13/02/71).
17/04/71	D8501/9/14/9/20/2/3/6/7/32/71/90	But N.B. D8509/19/20 sighted at 66A on 25/04/71.
19/05/71	Yard: D8533 Old Paint Shop Yard: D8501/9/14/9 D8520/2/3/6/7/32/71/90	D8509/19/20/33 added; ex-66A (last sighting 25/04/71).
31/05/71	D8501/9/14/9/20/2/3/6/7/32/3/71/90	

D8526, St Rollox Works (Old Paint Shop Yard), 21 August 1971.
(TOPticl Digital Memories)

19/06/71	D8501/9/14/9/20/2/3/6/7/32/71/90 ('...the usual dump of Claytons was visible about a quarter mile to the north (*sic east*)')	D8533 not listed; recording omission?
07/08/71	D8501/9/14/9/20/2/3/6/7/32/3/90	D8571 not listed; to J. MacWilliam, Shettleston (first sighting xx/06/71).
10/08/71	D8501/9/14/9/20/2/3/6/7/32/3/90 (in order recorded: D8522/26/23/19/09/20/32/27/ 14/33/90/01)	
16/08/71	D8501/9/14/9/20/2/3/6/7/32/3/90	
30/08/71	D8501/9/14/9/20/2/3/6/7/32/3/90	
27/09/71	D8500/1/9/10/4/8-20/2/3/6/7/32/3/5/90 'with another unidentified member of the Class reduced to frame level (8599)'	D8500/10/8/30/5 added; ex 66A (last sighting 29/08/71; also photographic evidence of D8535 at 66A on 07/09/71). D8599 not withdrawn by this date; unidentified loco assumed to be D8530.
16/10/71	Yard: D8510/8/99 Carriage Sidings: D8501/9/14/9/20/2/3/6/7/32/3/90 '...8518/99 were outside. Further locomotives added to the dump were D8519/20/7/33/90...'	D8599 added; ex-HA (last sighting 09/10/71). D8500/35 not listed; moved to Scrapping Area (Scrapping Area not visited?), or, cut-up?
21/11/71	D8501/9/10*/4/9/20/3/6/7/32/3/43/9/55 D8590/2/4/9*, D8604/6 * 'being cut-up'	D8543/9/55 added; ex-66A (last sighting 26/09/71). D8513 added but straight to Scrapping Area?; ex-66A (last sighting 26/09/71) and cut-up. D8592/4, D8604/6 added ex- 64B (last sighting 19/10/71). D8518/22 not listed; recording omission?
28/11/71	D8501/3/6/9/14/8-20/2/3/6/7/32/3/43/9/55 D8590/2/4/9, D8604/6	D8503 added; ex-?? D8506 added; ex-66A (last sighting 26/09/71). D8510 not listed; cut up. D8599 remains only?
09/01/72	D8503/18-20/2/3/6/7/32/43/9/55/92/4, D8602-4/6	D8501/6/9/14/33/90 not listed; moved to Scrapping Area (Scrapping Area not visited?), or, cut-up? D8602/3 added; ex-HA (last sighting 27/11/71).
12/02/72	D8500/3/18-20/2/3/6/55/92/4, D8602-4/6	D8527/32/43/9 not listed; moved to Scrapping Area (Scrapping Area not visited?), or, cut-up. D8500 listed; error?
27/02/72	D8503/19/20/2/3/6/55/92/4, D8602-4/6 In order recorded: D8503/55/23/19/20/26/22, D8606/02/04/03, D8592/94	D8518 not listed; moved to Scrapping Area (Scrapping Area not visited?), or, cut-up.
01/04/72	D8503/19/20/2/3/6/55	Scrapping Area not visited (see below).

D8602, St Rollox Works (Scrapping Area), April 1972.
(Colour-Rail)

D8603, St Rollox Works (Scrapping Area), April 1972.
(Colour-Rail)

02/04/72		D8503/19/20/2/3/6/55 (Works Yard), D8602/3 (Scrap Road, being cut-up), plus two unidentifiable frames, and cut-up remains of D8592	D8592 cut-up remains. D8594, D8604/6 not listed; cut-up, or, inidentifiable frames.
09/04/72		Old Paint Shop Yard: D8503/19/20/2/3/6/55 Scrapping Area: D8602/3	D8602/3 presumably still part-cut. In order recorded: D8503/55/23/19/20/26/22

An undated photograph on p128 of John Hooper's Clayton book (2016) illustrates seven locomotives awaiting scrap outside the Old Paint Shop. Nine locomotives are listed in the caption (D8503/55/23/19/20/26/22, D8603/12) and the first seven equate exactly to those listed on 9 April 1972, both in terms of the locomotives sighted and the order recorded. The other two locomotives listed in the caption, D8603/12, were probably D8602/3 which were located in the Scrapping Area in early-April; D8612 was located at Ardrossan during April 1972.

14/04/72	Photo1	Top Yard: D8503 (West end of Line) plus D8519/20/3 and 3 others	No photos of Scrapping Area.
	Photo2	Top Yard: D8522 (East end of line)	
25/04/72		D8503/19/20/3/55	D8602/3 not listed; cut-up. D8522/6 not listed; moved to Scrapping Area (Scrapping area not visited?), or, cut-up?
26/04/72		D8503/19/20/3/55	

Top Yard progressively filled up with Class 29s from April 1972.

14/05/72		D8503/23/55	D8519/20 not listed; moved to Scrapping Area (Scrapping area not visited?), or, cut-up?
28/05/72		D8503/19/23/55, D8606	D8519 part-cut/cut-up remains? D8520 not listed: cut-up. D8606 presumably cut-up remains!
10/06/72	List 1	No Class 17s listed	D8503/23/55 not listed; moved to Scrapping Area (Scrapping Area not visited?), or, cut-up?
10/06/72	List 2	Carriage Sidings: D8515/38, D8614 plus at least one other	D8515/38, D8614 and possibly D8615 added; from Ardrossan (last sighted 14/05/72). Were D8516/59 also present?
18/06/72		D8515/6/38/59, D8614/5	D8516/59 added (may have been present on 10/06/72).
06/08/72		D8515/6/38/59, D8614/5	
08/08/72		D8512/5/6/38/59, D8614/5	D8512 added: from Longsight (last sighted 05/07/72).
13/08/72		D8512/5/6/38/59, D8614/5	
17/08/72		D8515/6/38/59, D8614/5	D8512 not listed; recording omission?
		Carriage Sidings (W to E): D8515, D8614, D8615, D8538	
		Top Yard (W to E): Class 29s, D8516, D8559	

D8615, St Rollox Works (Carriage Sidings), July 1972. (Laurie Mulrine)

D8538, St Rollox Works (Carriage Sidings), 17 August 1972.
(Anthony Sayer)

D8515, D8614, D8615, D8538 (left to right), St Rollox Works (Carriage Sidings), 17 August 1972.
(Anthony Sayer)

Date		Locomotives	Notes
28/08/72		D8512/5/6/38/59, D8614/5 In order recorded: D8512...D8516,D8559... D8515, D8614, D8615, D8538	
02/09/72		D8512/5/6/38/59, D8614/5 Top Yard: D8516/59; Carriage Sidings: D8515/38, D8614/5	
14/10/72		Dump: D8505/7/15/6/28/38/52/9/61/5/74/9-81/3 D8586-8/93/7, D8600/1/14/5 Outer Yard: D8512	All 15 Claytons ex-Millerhill Yard added (last sighting 02/09/72). D8507/52/74 added; ex-66A (last sighting 16/09/72).

Top Yard filled up with Claytons after Class 29s had been removed for cutting up.

Date			Locomotives	Notes
20/10/72			D8505/7/8/16/25/8/9/36/52/9/61/5/74/9 D8580/1/3/6-8/93/7, D8600/1/14/5	D8508/25/9/36 added; ex-66A (last sighting 14/10/72). D8512/5/38 not listed; recording omission?
28/10/72		Photo	Top Yard (in order E to W): D8561/97/86, D8600, D8565, D8601, D8581/87/79/80/05/59/16	Includes 11 ex-Millerhill. Where were D8528/83/8/93? Presumably in Carriage Sidings - see photo on page 337.
05/11/72			D8505/7/8/12/5/6/25/8/9/36/8/45/52/9 D8561/5/7/74/9-81/3/6-8/93/7	D8600/1/14/5 not listed; magazine transcription error? D8545/67 added; ex-Stoke Cockshute (last sighting 01/10/72 (D8545 in transit 26/10/72).
19/11/72		List 1	D8505/7/8/12/6/25/8/9/36/45/52/9, D8561/5/7/74/9-81/3/6-8/93/7, D8600/1/14/5	D8515/38 not listed in either list; cut-up.
19/11/72		List 2	D8505/7/8/12/6/25/8/9/36/45/52/9 D8561/5/7/74/9-81/3/6/8/93/7, D8600/1/15	D8614 not listed; presumably in Scrapping area (not visited in second list).
14/01/73			33 Claytons	Presumably included D8502/31/4/40/1/58/62, D8610 ex-66A; last sighted 19/11/72. Only 15 Claytons at 66A on 14/01/73.
10/02/73		List 1	D8502/5/7/8/16/25/8/9/31/4/6/40/1/5/52/8/9 D8561/2/5/74/9-81/3/6-8/93/7, D8600/1/10/5	D8502/31/4/40/1/58/62, D8610 added; ex-66A (last confirmed sighting 19/11/72).
10/02/73		List 2	D8502/5/7/8/16/25/8/9/31/4/6/45/52/9 D8561/2/5/74/9-81/3/6-8/93/7, D8600/1/15 plus frames/bogies only of D8541/58	D8512/67, D8614 not listed; cut-up. D8540 cut-up remains? D8541/58 part-cut (completed by 01/03/73 per AHBRD&E5).
08/04/73			D8505/7/8/16/25/8/9/31/6/45/52/9/61/5 D8574/9-81/3/6-8/93/7, D8601/10/5	D8540/1/58 not listed; cut-up. D8502/34, D8600 not listed; moved to Scrapping Area (Scrapping Area not visited), or, cut-up). D8562 not listed; recording omission?

D8583, St Rollox Works (Carriage Sidings), either 4 February or 18 March 1973. Looking west with D8507, D8574, D8552, D8536 and D8508 beyond D8583. (Hugh Searle)

D8580, D8505, D8559, D8516, St Rollox Works (Old Paint Shop Yard), either 4 February or 18 March 1973. The Old Paint Shop (or Varnish Shop) forms the backdrop to the line of Claytons. (Hugh Searle)

D8565 St Rollox Works (Erecting Shop Yard / Goliath Crane area), April 1973. (Jonathan Martin)

21/04/73		D8505/7/8/16/25/8/9/31/6/45/52/9/61/2/5, D8574/9-81/3/6-8/93/7, D8601/10/5
25/04/73	Photo1	Old Paint Shop Yard (E to W): D5067 . . . D8583/93/28/86, D2436, D5114 . . .
	Photo2	Carriage Sidings (E to W): D2424/16, D8507/74/52/36/08

St Rollox Works: 12/05/73.

Old Paint Shop Yard (E to W)

Loco						5371	5364	5406	4097	5405	G Van	4095	Wagon	3284	5006	
Loco															8615	
Livery															GFY	Old Paint
Date C/U															12/73	Shop (Varnish Shop)
Loco						Wagon	5067	8588	8597	8545	Wagons	8610	8531	8525	8529	
Livery								GSY	GFY	BFY		GFY	GSY	BFY	XFY	
Date C/U							10/73	07/73	08/73	02/74		08/73	King	King	King	
Loco	8583	8593	8528	8586	2436	8561	5114	8601	8581	8587	8579	8580	8505	8559	8516	Wagon
Livery	BFY	GFY	XFY	GFY		GFY		GFY	GFY	GSY	GFY	BFY	GSY	GFY	GSY	
Date C/U	08/73	09/73	08/73	08/73	11/73	09/73	10/73	02/74	02/74	02/74	02/74	King	04/74	04/74	King	

Carriage Sidings (E to W)

Loco					2416	2424	8507	8574	8552	8536	8508
Livery							BFY	BFY	GFY	GFY	GFY
Date C/U					11/73	01/74	King	King	King	King	King

Loco		3932	8000	5121	5394

Scrapping Area.

Loco	8562 (nose ends removed)	Plates off 8600
Livery	G	
Date C/U	05/73	

Erecting Shop Yard

Loco	8565
Livery	BFY
Date C/U	08/73

Total No. of Claytons = 28.

N.B. Loco nos. in red = Nos. determined from observations immediately prior or subsequent to 12/05/73.

D8562, St Rollox Works (Scrapping Area), 12 May 1973. The short siding into the Scrapping Area is clearly visible. (Rail-Online)

Date	Locomotives	Notes
05/05/73	D8505/7/8/16/25/8/9/31/6/45/52/9/61/2/5, D8574/9-81/3/6-8/93/7, D8601/10/5	
12/05/73	D8505/7/8/16/25/8/9/31/6/45/52/9/61/2/5, D8574/9-81/3/6-8/93/7, D8601/10/5	
13/05/73	D8505/7/8/16/25/8/9/31/6/45/52/9/61/2/5, D8574/9-81/3/6-8/93/7, D8601/10/5	
28/05/73	D8505/7/8/16/25/8/9/31/6/45/52/9/61, D8574/9-81/3/6-8/93/7, D8601/10/5 'By 28th May 8565 had disappeared (*since 120573*), presumably cut up'	D8562 not listed; cut-up. D8565 not listed: moved to Scrapping Area (Scrapping Area not visited).

St Rollox Works: 28/05/73.
Indicative locomotive disposition based on Sean Greenslade's observations (in order recorded):

8516,8615,8559,8505,8529,8580,8525,8579,8531,8587,8610,8581,8601,8545,5114,8561,8597,
2436,8588,8586,5067,8528,8593,8583, ... 2416,2424,8507,8574,8552,8536,8508.

N.B. Colour-coding illustrates no change in location positions since 12/05/73.

09/06/73	Old Paint Shop Yard: D8505/16/25/8/9/31/45/59, D8561/79-81/3/6/7/93, D8601/10/5
	Carriage Sidings: D8507/8/36/52/74
	Scrapping Area: D8588/97

St Rollox Works: 09/06/73.
Indicative locomotive disposition based on Bill Hamilton's observations (in order recorded):

8615,8516,8559,8505,8529,8580,8525,8579,8531,8587,8610,8581,8601,8545,8561,8586,8528,
8593,8583, ... 8507,8574,8552,8536,8508, ... 8588,8597

N.B. Colour-coding illustrates no change in location positions since 28/05/73 other than movement of D8588/97 to Scrapping Area.

05/08/73	D8505/7/8/16/25/9/31/6/45/52/9/61, D8574/9-81/3/7/93, D8601/15	D8588/97 not listed; cut-up. D8528/86, D8610 not listed; moved to Scrapping Area (Scrapping Area not visited), or, cut-up).
11/08/73	D8505/7/8/16/25/9/31/6/45/52/9/61/74/9, D8580/1/3/7/93, D8601/10/5	D8528/86 not listed; cut-up. D8583, D8610 being cut-up in Scrapping Area.
	'. . .Of the Class 17 locomotives observed on the 'dump' on 12th May, 8528/86/8/97 had disappeared while 8583 and 8610 were being stripped . . .' (*edited*)	

St Rollox Works: 11/08/73.
Indicative locomotive disposition again based on Bill Hamilton's observations:

D8516,D8615,D8559,D8505,D8580,D8529,D8579,D8587,D8525,D8581,D8531,D8601,D8545,D8593,
D8561, ... D8507,D8574,D8552,D8536,D8508

N.B. Colour-coding illustrates no major change since 09/06/73 except D8610, and, D8528/61/83/6 preferentially dug out of the Old Paint Shop Sidings leaving D8545 and D8593 behind for later.

16/08/73	D8505/7/8/16/25/8/9/31/6/45/52/9/61/5/74/9, D8580/1/3/6/7/93/7, D8601/10/5	D8528/65/86/97 presumably cut-up remains? D8583, D8610 being cut-up in Scrapping Area.
15/09/73	D8505/7/8/16/25/9/31/6/45/52/9/61/74/9, D8580/1/7/93, D8601/15 Cutting started on D8561 (AHBRD&E5)	D8583, D8610 not listed; cut-up. D8561 being cut-up.

13/10/73	D8505/7/8/16/25/9/31/6/45/52/9/74/9-81/7, D8601/15.	D8561/93 not listed; cut-up.
	'At the 'dump' 8561/93 had disappeared since the visit in August (see p350, October R.O.) . . .'	

St. Rollox Works: 23/10/73.

Locomotive disposition based on Exe-Rail photographs:

<u>Carriage Sidings (W to E)</u>

D8508, D8536, D8552, D8574, D8507...?...D2424, D2416, D8587, D8581, D8601, D5149 . . .
  GFY   BFY   GFY

D8508 and D8536, St Rollox Works (Carriage Sidings), 23 October 1973. (D. Hunt [Exe-Rail])

D8574 between D8552 and D8507, St Rollox Works (Carriage Sidings), 23 October 1973. (D. Hunt [Exe-Rail])

D8587 (with D8581?), St Rollox Works (Carriage Sidings), 23 October 1973. Transferred across from Old Paint Shop Sidings. (D. Hunt [Exe-Rail])

D8601 (with D8581? and D5149), St Rollox Works (Carriage Sidings), 23 October 1973. (D. Hunt [Exe-Rail])

18/11/73		D8505/7/8/16/25/9/31/6/45/52/9/74/9-81/7, D8601/15	
13/01/74		D8505/7/8/16/25/9/31/6/45/52/9/74/9/80/7, D8615	D8581, D8601 not listed; moved to Scrapping Area (Scrapping Area not visited), or, cut-up). D8505 listed twice.
10/03/74		D8505/7/8/16/25/9/31/6/52/9/74/80	D8545/79/87, D8615 not listed; moved to Scrapping Area (Scrapping Area not visited), or, cut-up).
23/03/74		D8505/7/8/16/25/9/31/6/52/9/74/80 'D8505/59 and the frames and bogies of two others...on the scrap road.' (*presumably two of D8545/79/87, D8615*)	D8505/59 in Scrapping Area.
13/04/74		D8505/7/8/16/25/9/31/6/52/9/74/80	D8505/59 in Scrapping Area.
14/04/74		D8505/7/8/16/25/9/31/6/52/9/74/80	D8505/59 in Scrapping Area.
01/05/74	Photo1	Carriage Sidings: D8507/8/36/52/74	D8505 (*plus D8559*) in Scrapping Area.
	Photo2	Scrapping Area: D8505 (part-cut), plus another unidentified (D8559?), minimal remaining above frame)	

D8508, D8536, D8552, D8574, D8507, St Rollox Works (Carriage Sidings), 1 May 1974. (G.H. Taylor [Transport Treasury])

Disposal: St Rollox Works • 343

D8505, St Rollox Works (Scrapping Area), 1 May 1974. Possibly D8559 behind; see Section 20.4.
(G.H. Taylor [Transport Treasury])

28/05/74	D8505/7/16/25/9/31/6/52/74/80	D8508 not listed; recording omission?
		D8505 (*plus* D8559) in Scrapping Area.
15/06/74	'Dump': D8507/8/16/25/9/31/6/52/74/80	
	Scrapping Area: '. . .frames and bogies of two others' (*presumably D8505/59*)	
20/06/74	D8507/8/16/25/9/31/6/52/74/80	
24/08/74	D8507/8/16/25/9/31/6/52/74/80	
26/09/74	D8508/16/25/9/31/6/52/74/80	D8507 not listed; Carriage Sidings not visited?
	All located north of Traverser.	
05/10/74	Carriage Sidings: D8507	
	North of Traverser: D8508/16/25/9/31/6/52/74/80	
	'The only Class 17 on the 'Clayton Dump' was 8507, although 8508/16/25/29/31/36/52/74/80 were in the cutting-up siding.'	

26/01/75	D8507/8/15/25/9/31/6/52/74/80	
08/03/75	D8508/16/25/9/31/6/52/74/80	D8507 not listed; Carriage Sidings not visited?
20/03/75	D8508/16/25/9/31/6/52/74	D8507 not listed; Carriage Sidings not visited? D8580 not listed; recording omission?
29/03/75	D8508/16/25/9/31/6/52/74/80	D8507 not listed; Carriage Sidings not visited?
17/05/75	D8504/7/8/16/25/9/31/6/42/6/8/50-2/7/63, D8573/4/80, D8607/8/12/3	D8504/42/6/8/50/1/7/63/73, D8607/8/12/3 added; ex-66A (last sighting 27/04/75).

D8529 and D8508, St Rollox Works (North of Traverser), circa June 1975. (TOPticl Digital Memories)

Disposal: St Rollox Works • 345

D8574, St Rollox Works (North of Traverser), 17 May 1975. (Malcolm Best)

D8563 and D8546, St Rollox Works (Line west of Old Scrapping Area), 17 May 1975. Recently arrived Claytons from 66A Polmadie. The buffer stops in the foreground represent the end of the siding which formed the scrapping area for over sixty Claytons and nineteen of the twenty NBL Class 29s. The shallow cutting immediately behind the Claytons here was the old track-bed for trains into Glasgow Buchanan Street station which closed in 1966. (Malcolm Best)

D8504 and D8573, St Rollox Works (Line west of Old Scrapping Area), 17 May 1975. Recently arrived Claytons from 66A Polmadie. (Dave Lennon)

D8608, D8550 and D8542, St Rollox Works (Carriage Sidings), 17 May 1975. D8551 and D8613 were also in this line-up, just out of shot beyond D8542. (Malcolm Best)

Disposal: St Rollox Works • 347

D8507 and 26005, St Rollox Works (Carriage Sidings), 17 May 1975. The frequently missed Clayton during late-1974 and 1975 visits to the Works. I missed it myself on 26 September 1974! As the *Railway Observer* commented in October 1971: 'The blue painted Claytons on the dump are now getting very bleached'; certainly very true of D8507 four years later. (Malcolm Best)

St Rollox Works: 17/05/75 (Open Day).
Locomotive disposition based on the photographs of Malcolm Best and Dave Lennon, and Richard Boyd's notebook listing.

Carriage Sidings (adjacent to Top Sidings):	D8507
Carriage Sidings (central area) (W-E)	D8613, D8551, D8542, D8550, D8608
Area north of Old Traverser (W-E):	D8531, D8516, D8580
	D8529, D8525
	D8508, D8536, D8552, D8574
Line west of Old Scrapping Area (W-E)	D8557, D8607, D8612, D8504, D8573, D8548, D8546, D8563

27/05/75		D8504/7/8/16/25/9/31/6/42/6/8/50-2/7/63, D8573/4/80, D8607/8/12/3	
07/06/75		Crane Shop: D8542/51, D8613	
		Works Yard: D8550, D8608	
		'In addition to the five Claytons listed above . . . the following changes on the 'dump' were noted:	
		8541 - reappeared (sic), 8546/8/63 - disappeared.	
		'Dump' (deduced): D8504/7/8/16/25/9/31/6/52/7, D8573/4/80, D8607/12	
		'The unidentified remains of two Claytons were also seen (D8505/59?).'	
12/06/75	Photos	D8557, D8607+others	
Mid-07/75		Yard: D8504/8/36/42/6/8/50/1/2/63/73/4, D8607/8/12/3	D8507 not listed; Carriage Sidings not visited? D8557 not listed; recording omission?
		Dump: D8516/25/9/31/80	
10/08/75		D8504/7/8/16/25/9/31/6/42/6/8/50/1/2/7/63, D8573/4/80, D8607/8/12/3	
		In order recorded: D8504,D8551,D8542,D8550, D8508,D8536,D8552,D8574,D8548,D8546, D8563,D8613,D8608,D8557,D8607,D8612, D8573,D8507 (one line)	Red locos for Great Bridge (scrap). Blue locos for Norwich (scrap). Green locos to Shettleston (scrap). Clearly some preparatory sorting of locomotives had occurred by mid-August (and possibly by mid-July).
		then 8 DMU/EMU cars, then D8529,D8525 then 7 DMU/EMU cars, then D8531,D8516,D8580	

D8504, D8551, D8542, D8550, etc., St Rollox Works, 10 August 1975.
(Kevin Redwood)

D8521(S18521), St Rollox Works (Erecting Shop Yard), 25 November 1978. (Anthony Sayer)

12/08/75	D8504/7/8/16/25/9/31/6/42/6/8/50/1/2/7/63, D8573/4/80, D8607/8/12/3	
18/10/75	Nil.	D8507/8/16/25/9/31/6/52/74/80 not listed; to R.A. King, Norwich. D8557/73, D8607/8/12/3 not listed; to J. MacWilliam, Shettleston. D8504/42/6/ 8/50/1/63 not listed; to J. Cashmore, Great Bridge.
15/10/78	S18521, D8598	
25/11/78	Erecting Shop Yard: S18521 Scrapping Area: D8598 (dismantling just commenced)	
21/04/79	Scrapping Area: S18521 (baseframe/bogies only), D8598 (cut-up remains)	

D8598, St Rollox Works (Scrapping Area), 25 November 1978. (Anthony Sayer)

20.4 Analysis: Literature Search v. Actual Sightings.

Cutting up of the Claytons during the October 1971 to January 1973 period was reported in the July 1976 edition of the *Railway Observer* (RO). Sightings generally support the RO dates very closely and it appears that these dates provided the basis for information incorporated into subsequent publications. There are some exceptions, however:

D8515 and **D8590** for some reason, were never reported in the RO listing, and this presumably goes some way towards explaining the range of alternative dates found for these two locomotives in the literature search:

	Date range	Most likely based on sightings
D8515	09/72-11/72	11/72
D8590	12/71-04/72	12/71

D8527/32/43/9. The RO (and, therefore, subsequent publications) show these locomotives as being cut-up during June 1972. Sightings, however, suggest that they were all cut-up much earlier during February/March 1972.

After January 1973, and without further information supplied via the *Railway Observer*, disposal information for the Claytons became noticeably 'blurred'. Part of this will have been due to incomplete information caused by the St Rollox Works visits not always including the Scrapping Area, and, presumably also to people's differing definitions as to when a locomotive is considered cut. Thus variances between publications of a month either way is understandable and probably the best we can expect.

There are, however, between January 1973 and April 1974, three locomotives where quoted dates of disposal are greater than one month apart, as follows:

	Date range	Most likely based on sightings
D8561	07/73-09/73	09/73
D8565	05/73-08/73	05/73
D8601	11/73-02/74	01-02/74

There seems to be general agreement in the literature that **D8581** was cut-up in May 1974; however, sightings suggest that it was actually scrapped in January/February 1974. Similarly, **D8545** and **D8579** appear to have been cut-up in January/February 1974, rather than November 1973.

The Final Pair: **D8505** and **D8559**?

In terms of the 'final pair', it remains to be determined conclusively which they were. The May 1974 edition of the RO quotes '. . .D8505/59 and the frames and bogies of two others . . . on the scrap road' on 23 March 1974. Photographic evidence shows D8505 and another unidentifiable and substantially dismantled Clayton in the Scrapping Area on 1 May 1974 (see page 343). Subsequent RCTS visits to the Scrapping Area on 15 June 1974 and 7 June 1975 also make reference to the <u>unidentifiable</u> remains of two Claytons. D8505 seems certain, but was the other one indeed D8559? Or was the other unidentifiable Clayton one of the two 'frames, plus bogies' mentioned by the RO report of the 23 March 1974 visit?

Chapter 21

SCRAPPING: THE PRIVATE YARDS RETURN (1975–76)

21.1 Private Yard Disposal Recommences.

Cutting of the Claytons ceased at St Rollox Works in 1974, leaving twenty-five locomotives to be dealt with. In 1975 invitations were submitted to various scrap merchants to tender for these locomotives. Ultimately three yards were involved in the disposal of the outstanding Claytons, one Scottish (J. MacWilliam once again), and two in England. J. Cashmore at Great Bridge had featured before with the disposal of D8572, but R.A. King, Norwich was a 'new entrant' in the field of Clayton disposal.

After a further role-reversal, the final two 'cut-ups' (D8521/98 ex-Derby RCD) were disposed of at St Rollox Works in 1979 (see page 349).

21.2 J. Cashmore, Great Bridge (1975).

Summary of Claytons cut at J. Cashmore, Great Bridge: D8504/42/6/8/50/1/63 (Total: 7)

N.B. All 'Disposal not Proven'.

The November 1975 *Railway Observer* reported:

'The remaining Class 17 locomotives stored at Glasgow Works have all been sold for scrap and were despatched as follows:

'8542/8/50/1 to J. Cashmore, Great Bridge as 9X20 special train via Carlisle and Hellifield 16th September, with 8504/46/63 to the same destination the following day.'

In contradiction, NCTS magazine No.23 stated:

'On 10th September (*1975*) 25130 was seen hauling D8542/48/50/51 through

D8504, D8563 and D8546, Hunslet South Sidings, 18 September 1975. (Keith Long)

D8563 and D8546, Hunslet South Sidings, 18 September 1975.
(Keith Long)

Chesterfield and the following day 40019 pulled D8504/46/63 through Chesterfield.'

Sightings of D8504/46/63 at Hunslet, Leeds on 18 September 1975 would suggest that the *Railway Observer* report was correct.

Comparative summary of cutting-up information (Sources: AHBRD&E5 and D&ELfS):

	AHBRD&E5	D&ELfS
D8504	About 11/75	11/75
D8542	Not specified	09/75
D8546	About 10/75	09/75
D8548	Probably by 1/10/75	09/75
D8550	Not specified	09/75
D8551	Not specified	09/75
D8563	Not specified	09/75

21.3 J. MacWilliam, Shettleston (1975).

Summary of Claytons cut at J. MacWilliam, Shettleston (1975): D8557/73, D8607/8/12/3/6 (Total: 7)

N.B. D8557, D8607/8/13/6 'Disposal not Proven'.

The November 1975 *Railway Observer* again reports:

'The remaining Class 17 locomotives stored at Glasgow Works have all been sold for scrap and were despatched as follows: '8557/73, 8607/8/12/3 to J. McWilliams, Shettleston, 26th August as two special trains.

'In addition, 8616 from Dundee was sold to J. McWilliams, Shettleston and travelled from Dundee in 8X08 23.30 Aberdeen-Thornton Yard 19th August connecting into 8X35 05.40 Thornton Yard-Grangemouth 21st August thence 8X12 20.00 Grangemouth-Cadder the same day, being tripped to Shettleston shortly afterwards.'

Comparative summary of cutting-up information:

	AHBRD&E5	D&ELfS
D8557	09/75	08/75
D8573	Early 09/75	08/75
D8607	By 25/09/75	08/75
D8608	09/75	08/75
D8612	09/75	08/75
D8613	By 05/09/75	08/75
D8616	About 09/75	08/75

It will be noted that J. MacWilliam took all of the remaining Beyer-Peacock built Claytons.

D8612 and D8573, J. MacWilliam, Shettleston, October 1975. (Allan Trotter (Eastbank))

D8573 and D8612, J. MacWilliam, Shettleston, October 1975. (Allan Trotter (Eastbank))

D8612, J. MacWilliam, Shettleston, October 1975. (Allan Trotter (Eastbank))

21.4 R.A. King, Norwich (1975/6).

Summary of Claytons cut at R.A. King, Norwich (1975/6):
  D8507/8/16/25/9/31/6/9/52/74/80 (Total: 11)

N.B. D8531 'Disposal not Proven'.

The November 1975 *Railway Observer* once more provides:

'The remaining Class 17 locomotives stored at Glasgow Works have all been sold for scrap and were despatched as follows:
  '8507/8/16/25/9/31/6/52/74 to R.A. King, Norwich, as one train, special authority having been obtained for nine dead locomotives to travel together. They were conveyed in 9X20 03.35 special freight leaving Glasgow 4th September which was booked to arrive in March Up Yard at 17.55 8th September, to depart at 19.45 and arrive at Norwich at 23.30.
  '8580 to R.A. King, Norwich by 8X79 04.15 Cadder Yard to Millerhill, 11th September to connect with 8X21 03.48 Millerhill-Tyne Yard the following day.'

Furthermore, the February 1976 edition reported:

'Withdrawn Class 17 8539 was in the yard (*Tyne*) on 2nd (*December*), being hauled from Eastfield to Norwich as 9X20 special.'

Photographs and sighting information of Claytons in transit to Norwich are provided below:

Tyne Yard:
Sightings:

06/09/75	D8507/8/16/25/9/31/6/52/74
	In order recorded: D8574/52/36/08/07/16/31/25/29

D8574, Tyne Yard, undated (probably 6 September 1975).
(Rail-Online)

March:
Sightings:

08/09/75	D8507/8/16/25/9/31/6/52/74 (9X20 hauled by 47360)
07/12/75	D8539

Wensum Goods Yard, Norwich:

Map of Norwich Wensum Goods Yard. Ordnance Survey 1:10,560, 1970/71. Norwich Thorpe station top left of map. Route to Great Yarmouth and Lowestoft top right and route to Ipswich and London centre bottom of map. Storage area used for Claytons in Wensum Goods Yard highlighted in red. (© Crown Copyright and Landmark Information Group Ltd 2020 [www.Old-Maps.co.uk Ref.214475115])

21008. D8525, Wensum Goods Yard, Norwich, undated. Line up of six Claytons awaiting movement to King's scrap yard, believed to be (in order) D8529, D8525 (both BFY), D8531, D8516 (both GSY), D8508, D8536 (both GFY). The absence of D8580 probably dates this picture as mid-September 1975. (Transport Treasury/Ian C. Allan)

D8531 and D8516, Wensum Goods Yard, Norwich, undated. Both locomotives still in the GSY livery in which they were first delivered over twelve years previously; good quality paintwork! These were the only two GSY liveried Claytons to travel to Norwich. (Transport Treasury/Ian C. Allan)

D8531 and D8525, Wensum Goods Yard, Norwich, undated. Wensum Goods was used as a staging point before the Claytons were moved across to King's scrapyard in consists up to three locomotives at a time.
(Transport Treasury/Ian C. Allan)

Seven Class 17's awaiting their final journey, Wensum Goods Yard, Norwich, 27 September 1975. Locomotives as before with the addition of D8580 at the farthest end. Norwich Crown Point Depot now occupies this area.
(David Harlott)

Further sightings at Wensum Goods Yard:

24/10/75	D8508/16/25/9/31/6
18/12/75	D8525/9/39
27/12/75	D8525/9, 'together with a green-liveried member of the class with its number painted over (D8539).'

The Railway Magazine (December 1975) reported that: '8507 and 8552 were moved into Norwich shed on September 14 (*1975*), apparently for their fuel tanks to be drained'.

Locomotives were taken into King's scrapyard in batches; approximate dates of transfer from Wensum Goods or Norwich depot to the scrap yard were as follows:

D8507/52/74	By 27/09/75
D8580	By 24/10/75
D8508/36	By 31/10/75
D8516/31	By 27/12/75
D8525/9/39	21/01/76 (RO 0476)

Sightings at R.A. King's yard:

xx/10/75	D8507/52, plus remains of another believed to be D8574.
24/10/75	D8580 (partly stripped). D8507/52/74 'already demolished' (RW 0476)
31/10/75	D8508/36 (N.B. D8580 not listed; already cut-up?)
01/11/75	D8508/36
27/12/75	D8516, 'plus the frames of another member of the class visible' (possibly D8531)
10/04/76	D8525 (part-cut), D8539 (intact)
09/06/76	One Clayton 'with numbers painted out.' (*D8539*)

D8552 (left, in GFY livery) and D8507 (BFY), R.A. King, Norwich, October 1975. Note the partially cut-up remains of another Clayton (believed to be D8574) at the extreme left of the picture i.e. baseframe only on bogies. (Transport Treasury)

Comparative summary of cutting-up information:

	AHBRD&E5/(RW0176*)	D&ELfS
D8507	By 24/10/75*	10/75
D8508	Not specified	11/75
D8516	01/76	01/76
D8525	02/76	01/76
D8529	By 01/03/76	01/76
D8531	11/75	11/75
D8536	12/75	12/75
D8539	07/76	07/76
D8552	By 24/10/75*	09/75
D8574	By 24/10/75*	09/75
D8580	By 01/11/75	09/75

On the basis of the above, D8552/74/80 were cut-up in October and D8531 in December 1975. The photograph below indicates that D8525 lasted longer than suggested above, surviving in part-cut form until at least April 1976.

D8525, R.A. King, Norwich, 10 April 1976. Engines and generators plus the cab are all that remain on the baseframe. (Rail-Photoprints (R. Collen-Jones))

D8539, R.A. King, Norwich, 10 April 1976. The locomotive which frequently caused problems due to its number having been inexplicably painted out. D8525 in the distance. (Rail-Photoprints (R. Collen-Jones))

Chapter 22
POST-WITHDRAWAL

22.1 Generators.
22.1.1 GEC Stafford.

During early 1972, a national fuel crisis developed as a direct consequence of the miners' strike which commenced on 9 January; on 11 February a three-day working week was imposed on major commercial users of coal. As a way of offsetting the impact of electricity supply shortages a number of firms looked at alternative supply arrangements.

In the case of GEC at Stafford, the idea of deploying railway locomotives to supply power was hatched and in February three Claytons arrived to perform emergency electricity generation duties, with D8512 arriving from Longsight on the 23rd followed by D8545/67 from Polmadie on the 24th. All three were placed in a siding adjacent to the GEC Works on the up side of the main line just past the Queensville Curve

An offer of additional pay was accepted by the miners in a ballot on 25 February and they returned to work three days later, only four or five days after the arrival of the three Claytons. Whether any of the three Claytons actually generated any electricity for GEC is unknown.

On 10 March 1972, D8512/45/67 were towed to Stoke Cockshute depot; D8512 was noted

D8567, D8545 and D8512 (left to right), GEC Stafford, undated. Three 'generator' Claytons temporarily residing near the Queensville Curve. Note the three different sizes of BR double-arrow emblem. (Author's Collection)

back at Longsight in April, with D8545/67 subsequently transferred back to Scotland.

22.1.2 Glasgow Depots.
In a similar exercise, Claytons were tested at 66A Polmadie, 66B Motherwell, 66C Hamilton and Larkfield Carriage Shed during early 1972, although whether any of them were used 'in anger' is unknown. The NCTS magazine (April-June 1972) reported that D8552 and D8574 were at Hamilton and Motherwell respectively on 27 February 1972, with D8529/62 apparently 'working' at Polmadie on the same date.

22.2 Re-Railing Exercises.

D8539 and D8616 were used on re-railing exercises at Eastfield and Dundee respectively during 1975. *The Railway Magazine* (May 1975) reported: 'it (*D8616*) has been moved to the depot to enable the shed staff to gain practical experience in the use of some new jacks that have recently been delivered to the depot.'

22.3 Great Northern (G.N.) Electrification Project.

The February 1974 edition of the *Railway Observer* carried the following paragraph:

'It has been reported that twelve of the withdrawn Class 17's now stored at the yard alongside Polmadie m.p.d. are to be reinstated and transferred to the Kings Cross division to assist with the G.N. electrification project. Certain of the other members of the class now stored at St Rollox will be sent back to Polmadie to provide spares for the twelve reprieved locomotives.'

At the end of 1973 the Claytons stored at 66A Polmadie were D8504/39/42/6/8/50/1/7/63/73, D8607/8/12/3/6, fifteen in total to choose from.

However, nothing came of this initiative, with the *Railway Observer* reporting as follows:

'…With the G.N. electrification project now at an advanced stage, it would seem unlikely that any member of this class will now be reinstated and transferred to the Eastern Region as was once planned.' (June 1975)

22.4 Deployment Abroad?

The Railway Magazine (March 1974) made an alternative suggestion that the Claytons stored at St Rollox Works were to be used on the G.N. suburban electrification work, whereas 'a batch of 'withdrawn '17s', at present in store at Polmadie Depot, is expected to be sold for further use abroad.' Whether this was pure speculation on the part of a correspondent is unknown, but it certainly never happened.

22.5 Research Department.
22.5.1 D8512

After withdrawal D8512/21 were sent from Carlisle to Derby on 23 July 1969 for use by the Research Centre (RCD) and in the ensuing two and a half years, D8512 was used extensively on (a) cab-signalling and remote-controlled

D8512, 9A Longsight, undated. Blue livery, but note the 'non-standard' application of the yellow front ends (not extended onto the bonnet roof), the very small double-arrow emblem and old-style numbers minus the D-prefix. (Colour-Rail)

multiple-working trials with Test Coach 'Hermes' on the Styal-Wilmslow line, based at Longsight depot and (b) providing traction for the 'POP' train which was used to test train-tilting on articulated bogies on the Old Dalby Test Track in advance of the Advanced Passenger Train prototype (undertaken during October/November 1971).

According to Steve Allsop (*Derby and the R.T.C.*, Traction No.11):

'After an initially trouble-free beginning D8512 began to suffer from engine problems, including blowing cylinder heads and torn engine mountings. It was continually patched up at Etches Park but eventually, in 1971, a seized turbo-charger saw its demise.'

Peter Hall (RCTS *Diesel Dilemmas*) indicates that withdrawal took place in October 1971; five sightings in October position D8512 in the Derby area. However, during December 1971 and January 1972 D8512 was to be found back at its old haunt of Longsight MPD which would suggest that repairs to the turbo-charger(s) had been undertaken to secure a return to traffic on RCD work. In fact Mark Alden (CD&E3) records that: 'between 7th and 9th December (*1971*), it was seen hauling an AM4 EMU and test coach 'Hermes' over the 'Styal Line' carrying out speed related trials.'

Between 23 February and 10 March 1972, D8512 moved to Stafford for emergency generator duties with D8545/67 (see Section 22.1.1). On return to Derby, Mark Alden states that: 'D8512 had been damaged too badly to be economically repaired.' The exact nature of the damage sustained is unknown, but it was clearly sufficient to trigger the requirement for a replacement locomotive.

The chosen replacement for D8512, Beyer Peacock-built D8598 (withdrawn on 31 December 1971), was still in Scotland in February and even possibly early-March 1972. Conflicting sighting information place D8598 at both Derby RCD and Longsight on 12 March 1972. During late-March 1972 D8598 was repainted at Derby RCD, and in mid-April it turned up at Longsight for operational duties in lieu of D8512.

After the GEC generator duties, D8512 was stored at Stoke Cockshute and Longsight (spares recovery for D8598?), before movement to St Rollox Works during July and August 1972 for final disposal.

At some point during its RCD career D8512 was repainted from GSY to BFY livery.

22.5.2 D8521 / S18521

D8521 arrived at Derby Research Centre in July 1969 and, unlike D8512, led a very sedentary life at Derby and the nearby test facilities. D8521 spent two <u>extended</u> periods in Derby Works i.e. November 1970 to June 1971, and, circa February 1974 to March 1975. After the first visit the Clayton was released in all-over blue livery, renumbered S18521, and with the following details painted on the cab sides immediately below the driver's windows:

Mobile Power Station
Electrical Research Div.
Derby
Nº. S18521

D8521, 16C Derby, 15 November 1970. Still in 'as-withdrawn' condition. (Colour-Rail)

S18521, Derby Works (Klondyke Sidings), early-March 1974. All-over blue livery. Converted to a 'mobile power station'; rendered non-self-propelling it avoided the necessity for yellow ends. Note the large 'terminal board' mounted on the No.1 bonnet nose end and the heavy-duty cabling on the bonnet roof connecting the generator and terminal board. Note also the additional small grille in the tablet-catcher recess. (Rail-Online)

In addition to re-painting, considerable modifications were undertaken as described by Steve Allsop (*Derby & the R.T.C.*, Traction No.11):

D8521 . . . 'was converted in Derby Locomotive Works into a 'mobile power station', its traction motors were disconnected, the switchgear mounted in the centre of the driving cab and a large terminal board mounted on the nose at one end. The original driving controls were retained. It was probably intended for use with the 'plasma torch' adhesion experiments but, instead, languished unused in various places over the years, returning occasionally to Etches Park for anti-freeze checks, etc, and to have its engines run up. The possibility of using it during the 1974 miners' strike as an emergency generator was investigated, but its DC output was incompatible with the standard industrial 3-phase AC supplies. It visited Etches Park to be 'fettled-up' for this purpose and this was probably the last occasion on which its engines ran.'

The photograph above illustrates the substantial re-work undertaken on the No.1 bonnet roof and end. With its traction motors isolated, the Clayton was unable to move under its own power.

On 17 August 1972, S18521 was transferred from Derby RCD to the Mickleover test track by D8598; whether this was for storage or for 'power station' usage is unknown.

February 1974 saw S18521 back at Derby. A photograph taken on 24 February, not long after its arrival at Derby Works for its second protracted visit, shows the locomotive in a somewhat sorry state. The fact that D8521's visit to Derby Works (sightings 24 February 1974-16 March 1975) substantially overlap the dates for D8598 (18 March 1973-20 October 1974) might suggest that D8521 was cannibalised to assist with the return to traffic of D8598.

After ejection from Derby Works S18521 lingered around the Derby and Mickleover area for several years before its final movement to St Rollox Works in October 1978 (with D8598), with disposal undertaken in April the following year.

22.5.3 D8598

D8598 arrived at the Derby Research Centre in March 1972, and following repainting into blue livery (with non-standard red buffer beams), moved on to Longsight to carry on where D8512 left off on the Styal line test trains, duties which it operated throughout the remainder of 1972 and well into 1973.

Occasional excursions broke up the monotony of the Styal test trains including:

- the movement of E26048 to Crewe Works before carrying on to the Mickleover test track on 5 May 1972.

D8598, Derby Works (Erecting Shop), 10 March 1974. No.2 end nearest to the camera. The blue livery was applied whilst under RCD jurisdiction in April 1972 and apart from the small double-arrow emblem, original style numbers (with no D-prefix) and red buffer beams, it very closely matched the St Rollox Works repaints. *(Colour-Rail)*

- a visit to Swindon Works on 16 August 1972 to collect Test Coach 'Hermes' for return to Derby after overhaul.
- an outing to Mickleover test track from Derby RCD on 17 August 1972 hauling S18521 and D5901.
- a visit to Slochd on the Perth-Inverness main line on 13 February 1973 with *Hermes* to test the effectiveness of APT transponders in snowy conditions.

According to Steve Allsop (*Derby & the RTC* (Traction No.11)):

'D8598 spent a considerable amount of time running on the Edwalton test track hauling the "POP" train [prior to receiving modifications at Derby Works - see below]. This consisted of two skeletal vehicles articulated together and mounted on three APT-E trailer car bogies. A converted Mk.1 coach ran as a 'match' vehicle at each end.'

D8598 spent a somewhat protracted period between late February/early March 1973 and late October/early November 1974 in Derby Works. It was during this time that D8598 received significant modifications as described once again by Steve Allsop:

'D8598 ran in its original condition for a while until the . . .[Derby RCD] . . .faction asserted itself. It could already operate in push and pull mode with 'Hermes' and 'Mercury' with the vacuum brake in operation, as could any other "Blue Star" locomotive. To enable it to operate with the two test cars under air-braked conditions, it was the subject of an extremely "Heath Robinson" . . .dual brake and multiple unit working conversion, becoming a rather complicated toy in the process. Whilst retaining its conventional Westinghouse vacuum/straight air brake and "Blue Star" multiple working equipment, it was also fitted with Southern Region type waist height flexible air brake and main reservoir pipes and a

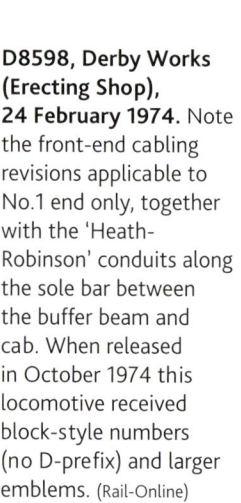

D8598, Derby Works (Erecting Shop), 24 February 1974. Note the front-end cabling revisions applicable to No.1 end only, together with the 'Heath-Robinson' conduits along the sole bar between the buffer beam and cab. When released in October 1974 this locomotive received block-style numbers (no D-prefix) and larger emblems. (Rail-Online)

27-way EPB multiple working jumper cable at its No.1 end only. A Westinghouse triple valve, a Westinghouse EP brake control unit and a 'Westcode' relay valve for pneumatic engine speed control were fitted into an equipment frame installed in the centre of the (fortunately) large driver's cab. A Westinghouse driver's automatic air brake control valve of the traditional non-self lapping type was installed at the No.2 driving position only.

'Both automatic brake systems applied the locomotive's own air brakes, but did not interact with each other. The EP brake would only apply the locomotive's brakes when operated from a remote driving position, but not from either of the locomotive's own driving positions . . .Hermes is equipped with a Southern Region-style driving compartment at its No.1 end. The driver's controls are similar to those of a post-1963 Southern Region EPB but an air/vacuum accelerator valve is fitted to enable vacuum or auto air-braked 'Blue Star' locos to be controlled. The alterations to D8598 meant that it could now be controlled from 'Hermes' either under the "Blue Star" or the EPB systems when trailing with its No.1 end coupled to the train, but could only be controlled by the "Blue Star" system when coupled by its No.2 end.'

During 1975 and 1976 D8598 was intermittently reported working test trains in the Styal/Wilmslow area. According to Steve Allsop the ultimate demise of D8598 resulted from the fact that:

'One of its engines was not loading up and a misbehaving load regulator was suspected. No technical information was available as usual and attention on a load bank was recommended . . . Toton's load bank was out of action and so it [D8598] followed all but one of its sisters to the scrap yard.'

D8598 (with D8521) was booked to be transferred from Derby to St Rollox during October 1978 in two stages (9Z10 11.15 Derby RCD-Carlisle on 7 October, then Carlisle-St Rollox Works on 9 October) for scrapping. However, sightings of D8521/98 at Chesterfield on the 7th followed by Rotherham on the 9th indicate that this movement was delayed. Both Claytons were noted at St Rollox Works on 14 October and on a visit to the Scrapping Area at St Rollox Works by myself on 21 April 1979 only the cut-up remains of D8598 were visible. D8598 thus became the last of the Beyer Peacock machines to be destroyed.

D8512 (with Test Coach "Hermes"), Location unknown, slides catalogued under 1971. (Author's Collection)

Chapter 23
INDUSTRIAL SERVICE

In 1972, D8568 was purchased from BR by the Hemel Hempstead Lightweight Concrete Co of Cupid Green in Hertfordshire. After some fettling work at Polmadie, D8568 arrived at its new home on 11 September 1972 with the transit from Scotland being undertaken under its own power. With the 'Hemlite' plant being located at the end of a six mile long branch from Harpenden Junction on the Midland Main Line, the company was looking for a more powerful locomotive with a significantly greater hauling capacity than the existing shunters, thereby reducing the number of journeys required along the branch.

Over the ensuing few years, the increasing costs of rail transportation resulted in a greater reliance on road haulage, and D8568 was put up for sale in 1977. D8568 was subsequently bought by the Ribblesdale Cement Co. (Ribble Cement, later Castle Cement) for use at their Horrocksford facility, near Clitheroe, Lancashire, leaving Cupid Green on 15 June 1977 and arriving at Horrocksford on 24 June. The

D8568, Hemel Hempstead Lightweight Concrete Ltd, Cupids Green, 19 June 1976.
(RCTS Archive)

Clayton was used to shunt the heavy cement trains and worked on the one-mile branch which connected the cement works to the BR Blackburn-Hellifield line at Pimlico.

Whilst at Clitheroe, D8568 was re-painted into the Ribble Cement livery of light grey (body work) and dark green (base frame and underframe equipment) with 'RIBBLE CEMENT' lettering applied to the body side doors and the Castle Cement logo to the bonnet ends and cab sides.

D8568 was retired from service in 1982 and put up for sale once more. This time it was acquired for preservation and was moved to the North Yorkshire Moors Railway (NYMR); D8568 left Clitheroe on 9 February 1983 and arrived at Pickering on the NYMR on 11 February 1983.

D8568, Ribble Cement, Clitheroe, 1978.
(Rail-Photoprints)

Chapter 24
PRESERVATION

Following industrial service, D8568 was sold to the Diesel Traction Group in 1983. It was taken to Pickering for a new life in preservation on the North Yorkshire Moors Railway arriving there on 11 February 1983. Its first run in preservation was on 1 April 1983.

Today D8568 is based on the Chinnor & Princes Risborough Railway, having arrived there on 25 April 1992, and has subsequently made numerous visits to Open Days and Diesel Galas across the country.

D8568, NYMR, Grosmont, 23 April 1988. Still in Ribble Cement grey and dark green livery after five years in preservation. Numbered 290-068-6, mimicking the Deutsche Bahn (DB) Class 290 centre-cab diesel-hydraulic locomotives. Complete with 45-gallon tank on the bonnet roof to prevent fuel starvation. (Anthony Sayer)

D8568, Old Oak Common, 5 August 2000. Open Day exhibit. (Anthony Sayer)

Chapter 25
CONCLUDING REMARKS

Clayton Equipment's design interpretation of the BR outline stipulations for the 'Standard' Type 1 locomotive was excellent, fully meeting the requirement to dramatically improve driver's all-round visibility, a criterion which was essential in the context of trip-freight and local yard shunting operations. The Clayton design ensured that their product was shortlisted, with final selection based on price. In practice too, the design was well liked by the driving fraternity and, as one observer has pointed out, it was truly a driver's locomotive.

After reading a considerable amount of published literature on the class, the over-riding impression is one of failure largely resulting from issues with the Paxman engine. However, following further reading of archive material it has become abundantly clear that this need not necessarily have been the case:

(a) prototype testing, even over six months, would have highlighted the major issues with the flat Paxman engine (crankshaft breakages, cylinder head and crankcase fractures, fuel supply issues, oil seepage from the exhaust bellows and consequent contamination of the generator) before volume production.

(b) if BR had listened to Paxman in the first place and deployed engines with cast iron crankcases and cylinder heads most of the engine problems would have been short-circuited.

Money, conflicting organisational objectives (particularly the policies surrounding spare part ordering and stock management) and the intransigence associated with the trials programmes (the cylinder head joints debacle was a classic!) all conspired against the fleet once in traffic.

It should also be remembered that it was not just the Paxman engine which caused all the problems. The Napier turbo-charger was a major problem for the class. Pre-trialling again would have highlighted the issue and the Holset alternative may well have avoided the problems in this area.

Ultimately, though, it was market conditions and an over-ordering of Type 1 locomotives by BR which ultimately consigned the Claytons to an early demise. That BR ordered 100 new Type 1's, of the English Electric (EE) variety, for delivery in 1966/67, and then for the 1967 Traction Plan to indicate that over 250 of the national fleet of Class 17 and Class 20 Type 1's would be superfluous by 1974 seems beyond incredible even with hindsight. The decline in freight traffic generally, and the impact of the Beeching decisions in particular, inevitably had a disproportionately large impact on trip-freight and local shunting traffic for which the Type 1's were associated; and this was known in 1964 when the 100 new Type 1s were ordered!

By the early 1970s and despite the best efforts of the P&SP Committee to improve Clayton availability to nearly 80%, the EE Type 1's still enjoyed a ten percentage point advantage over the Claytons and with the numerical superiority of the English Electric fleet (228 v. 117) it was inevitable that the axe would fall on the Claytons.

Had the Clayton's availability been improved through the efforts of the P&SP group to a level exceeding, say, 85%, would they have lasted longer than the average of six or seven years actually achieved? Well, probably not, again because of the numerical superiority of the EE Type 1s. The centre-cab visibility advantage of the Claytons was progressively lost with the demise of trip-freight work, and with the trend towards heavier freight trains the visibility disadvantage of the EE locomotives was mitigated by deploying them in pairs with the cabs at the outer ends. And in such situations there were now clear economic and operational advantages in favour of the EE product i.e. only two engines to maintain instead of four, and 2000hp available instead of 1800hp.

It is certainly easy to conclude that the Claytons were a very expensive mistake. However, looking at matters in a different way and, yes, with the benefit of significant amounts of hindsight, things could have been so different. With prototype testing, the major issues could have been designed-out of the large production orders with any outstanding issues resolved through effective modification programmes, enabling more than 85 per cent availability. Recognising that there really was no need to order the 100 Type 1s in 1965 for 1966/67 delivery, then the Claytons could have lasted into the early 1980s and allowing a more realistic return on investment. Spending half of the £6m laid out on the 100 new EE Type 1s in 1966/67 to finally resolve the Clayton issues would have been financially more prudent (even if this involved squadron re-engining with Rolls-Royce power units) and the fitting of Holset turbo-chargers.

Finally a quote from Colonel H.C.B. Rogers O.B.E. (*Transition from Steam*, 1980):

'It looked so good and first tests were so promising that its ultimate failure was something of a tragedy.'

As with previous books in this serious, I have compiled a list of 'wants' to help address outstanding gaps in the history of the Claytons. Any help will be greatly appreciated and should be sent to the editorial address, please. Thank you.

1. DLRCs for D8500-2/4/44/54/71/2/8/89/90.
2. Information re. painting of D8503 into BFY livery (1967/68 St Rollox Works sightings, ex-Works sightings, etc.)
3. Information to confirm whether D8594-6/9 carried GFY livery.
4. Details surrounding whether D8586 or D8587 received Paxman engines in later life?
5. Transfer dates from Carlisle Kingmoor (Steam) depot to Carlisle Yard (circa February 1969).
6. Transfer date of D8511/7/23/4 from Carlisle to Polmadie (1969).
7. Transfer date of D8532-4 from Carlisle to Glasgow (1970).
8. Transfer dates from Polmadie to Ardrossan (28 locomotives) (1971/72).
9. Transfer dates from Millerhill Yard to St Rollox Works (15 locomotives) (1972).
10. Observations/photographs of D8547 at Bird's scrapyard, Cardiff (1970)?
11. Photographs of D8589 dumped at 52A Gateshead following accident damage (1970/71).
12. Observations/photographs of D8513 and D8530 at St Rollox Works after withdrawal (1971).
13. Additional dated photographs of Claytons in the scrapping area at St Rollox Works.
14. Confirmation of the identities of the two Claytons which lasted in the St Rollox Works Scrapping Area for over a year in part-cut condition (1974/75).
15. Photographs (ideally dated) of Claytons at J. Cashmore, Great Bridge (1975).
16. Additional photographs (ideally dated) at J. MacWilliam, Shettleston (1969-71, 1975).

D8559 (with D8505), St Rollox Works (Scrapping Area), 19 May 1974. (Photographer unknown)

Regarding my "Wants List" No.14, a photograph dated 19 May 1974 has belatedly come to light illustrating the two part-cut Claytons in the Scrapping Area at St Rollox Works, with the number of D8559 visible. Fortuitously, everything above the frame of D8559 had been removed with the exception of the cab side panels.

REFERENCES & SOURCES

Books:
Bond, R.C., *A Lifetime with Locomotives*, Goose & Son, 1975.

Booth, A., *Ex-BR Diesels in industry*, 8th Edition, Industrial Railway Society, 2019.

Butlin, A., *Diesel & Electric Locomotives for Scrap*. Oxford Publishing, 2015.

Clough, D.N., *British Rail Standard Diesels of the 1960s*, Ian Allan, 2009.

Derrick, K., *Diesels & Electrics to the Scrapyards 1959-1989*, Strathwood, 2018.

Haresnape, B., *British Rail Fleet Survey: 4. Production Diesel-Electrics Type 1-3*, Ian Allan, 1983.

Harris, R., *The Allocation History of B.R. Diesels & Electrics: Parts 1, 4, 5, 6A, 6B and Final Update (Third & Final Edition)*, 2002/04/05/06/06/16 respectively. (AHBRDE1/4/5/6A/6B/FU respectively)

Hooper, J., *An Illustrated Historical Review of the Clayton Type 1 Bo-Bo Diesel-Electric Locomotives - Class 17*, Book Law, 2016.

Marsden, C.J., *The Complete UK Modern Traction Locomotive Directory*, TheRailwayCentre.com, 2011.

Rogers, Colonel H.C.B., *Transition from Steam*, Ian Allan, 1980.

Scotney, I. & Egan, B., *Draper's Scrapyard, Hull*, Steam Memories: 1950's-1960's No.74, Book Law, 2015.

Strickland, D.C., *Locomotive Directory - Every Single One There Has Ever Been*, D&EG, 1983, plus Supplements No.1-7 (1983-87). (LocDir)

Tufnell, R.M., *The Diesel Impact on British Rail*, Mechanical Engineering Publications, 1979.

Magazines:
Diesel Railway Traction, specifically:

November 1962, pp411-415, *Small General-Purpose Locomotives for BR*

Modern Railways (MR), specifically:

December 1962, pp380-382, *The Standard Diesel-Electric Type 1 Locomotive.*

Railway World (RW)

The Railway Magazine (RM), specifically:

January 1963, pp55-57, *Type 1 Diesel-Electric Locomotives for British Railways.*

December 1998, pp67-71, *A Costly Failure*.

The Railway Gazette, specifically:

17 December 1965, pp980-981, *900hp Locomotives with Rolls-Royce Engines*.

Modern Locomotives Illustrated, No.186, *The Class 14, Class 15, Class 16 and Class 17s*, December 2010-January 2011, The RailwayCentre.com. (MLI186)

British Railways Illustrated, specifically:

April 2007, pp298-299, *Diesel Decline*.

June 2007, p413, *A Reader Writes: The Ardrossan Type 1 Fleet*, Ian D. Osborne.

Railway Bylines, specifically:

June 1998, pp174-176, *A Clayton at Clitheroe*, A.Booth and T.Heavyside.

Classic Diesels & Electrics, specifically:

Issue 2, October / November 1997, pp10-17, 'Claytons Part 1: Classic Clayton', M.Alden.

Issue 3, December 1997 / January 1998, pp40-43, 'Claytons Part Two: Clayton Countdown', M.Alden.

Traction, specifically:

October/November 1994, pp6-9, *A BR Driver's Guide to Main-Line Traction*, H.Friend.

June 1995, pp42-45, *St Rollox Apprentice*, M.McManus.

September 1995, pp4-8, *Derby and the RTC*, S.Allsop.

June 1998, p42, *Clayton Place*, N.Ross.

Model Rail, specifically:

October 2005, pp18-26, *Just Seventeen*, P.Dunn.

Railway Observer (Railway Correspondence & Travel Society). (RO)

Journal (Stephenson Locomotive Society).

Bulletin (Locomotive Club of Great Britain).

Railway Locomotives (British Locomotive Society). (RL)

Buckley Wells Transport Enthusiasts (BWTE), subsequently *Northern Counties Transport Society* (NCTS).

Archive Sources:

Engine History Cards (EHC) and *Diesel Locomotive Record Cards* (DLRC).

BTC and BRB: Various Committee Meeting Minutes and Supporting Papers.

Main-Line Diesel Locomotives - Limitation of Variety, BTC Joint Report from R.C.Bond (Chief Mechanical Engineer) and S.Warder (Chief Electrical Engineer), 26 July 1957.

Main-Line Diesel Locomotives: The Approach to Standardisation, R.C.Bond (BTC Technical Advisor), 10 November 1958.

Standardisation of Main-Line Diesel Locomotives, BR General Staff, 8 June 1959.

Standardisation of Main-Line Diesel-Electric Locomotives - Types 1 and 4, Memorandum to the Technical Committee, 15 September 1960.

Fires on Diesel Train Locomotives Reports 1961-68, Locomotive Performance & Efficiency Development Unit, Derby. (FDTL)

National Traction Plans (February 1965, November 1967, December 1968). (NTP)

BRB/Davey Paxman Liaison Meeting minutes, 1966/67. (BRB/DP)

Performance and Service Problems of Clayton Type 1 Locomotives Meeting minutes, 1967-71. (P&SP)

Tender documents for D8588-D8616 batch.

Paxman Archive Trust.

Official BR Documents:-

Main-Line Diesel Locomotive Layout Diagrams (various editions).

Main-Line Diesel Locomotive Diagram Book, June 1969.

Websites:

'Clayton Equipment Ltd' (claytonequipment.co.uk)

'Railscot' (railscot.co.uk).
'RMWeb' Forum (rmweb.co.uk), specifically *On This Day in History*,

Modernisation Plan Diesels, and *Diesel Classes that didn't Transition*.

'Richard Carr's Paxman History Pages' (paxmanhistory.org.uk).

'The Rail Photo Archive' (railphotoarchive.org)
'WNXX Forum' (wnxxforum.co.uk).

Images.
See Acknowledgements.

Sightings.
Jim Alexander, Geoff Arnold, Garry Brookes, Richard Boyd, Chris Capewell, Chris Coates, Dave Coddington, Dennis Dey, Patrick Evans (Shed Master Archives), Ken Fairey, John Frisby, Vic Forster, John Glover, Sean Greenslade, Peter Hall, Bill Hamilton, Keith Long, Steve Marshall, Laurie Mulrine, D. Nicholson, Steve Perkins, Kevin Redwood, Anthony Sayer, Tony Skinner, Richard Strange, Brian Thomas, Peter Ventham.

D8516, between D8559 and Class 29s 6100/30/21/07/03, St Rollox Works (Old Paint Shop Yard), 29 August 1972. During the second half of 1972 the breaking-up of the Claytons was put on hold to facilitate the elimination of the remaining NBL Type 2 Class 29s. The life and times of these locomotives was covered by another one of my books in the Pen & Sword "Locomotive Portfolios" series entitled "The North British Type 2 Bo-Bo Diesel-Electric Classes 21 & 29". (Stewart Blencowe)